T0292852

Mathematical Biology

This text serves as an exploration of the beautiful topic of mathematical biology through the lens of discrete and differential equations. Intended for students who have completed differential and integral calculus, *Mathematical Biology: Discrete and Differential Equations* allows students to explore topics such as bifurcation diagrams, nullclines, discrete dynamics, and SIR models for disease spread, which are often reserved for more advanced undergraduate or graduate courses. These exciting topics are sprinkled throughout the book alongside the more typical first- and second-order linear differential equations and systems of linear differential equations.

This class-tested text is written in a conversational, welcoming voice, which should help invite students along as they discover the magic of mathematical biology and both discrete and differential equations. A focus is placed on examples with solutions written out step by step, including computational steps, with the goal of being as easy as possible for students to independently follow along.

Rich in applications, this book can be used for a semester-long course in either differential equations or mathematical biology. Alternatively, it can serve as a companion text for a two-semester sequence beginning with discrete-time systems, extending through a wide array of topics in differential equations, and culminating in systems, SIR models, and other applications.

Christina Alvey is an associate professor of mathematics at Mount Saint Mary College in Newburgh, NY. She earned a PhD in mathematics from Purdue University. Her current research investigates mathematical models in biology and epidemiology, as well as current trends and developments in the field of mathematics education.

Daniel Alvey is a data scientist at Accenture Federal Services. Prior to this, he was an assistant professor of mathematics at Manhattan College in the Bronx, NY, and a visiting assistant professor of mathematics at Trinity College in Hartford, CT. He earned his PhD in mathematics from Wesleyan University, where his research focused on homogeneous dynamics and metric number theory.

Textbooks in Mathematics
Series editors:
Al Boggess, Kenneth H. Rosen

Computational Optimization: Success in Practice
Vladislav Bukshtynov

Computational Linear Algebra: with Applications and MATLAB® Computations
Robert E. White

Linear Algebra With Machine Learning and Data
Crista Arangala

Discrete Mathematics with Coding
Hugo D. Junghenn

Applied Mathematics for Scientists and Engineers
Youssef N. Raffoul

Graphs and Digraphs, Seventh Edition
Gary Chartrand, Heather Jordon, Vincent Vatter and Ping Zhang

An Introduction to Optimization with Applications in Data Analytics and Machine Learning
Jeffrey Paul Wheeler

Encounters with Chaos and Fractals, Third Edition
Denny Gulick and Jeff Ford

Differential Calculus in Several Variables
A Learning-by-Doing Approach
Marius Ghergu

Taking the "Oof!" out of Proofs
A Primer on Mathematical Proofs
Alexandr Draganov

Vector Calculus
Steven G. Krantz and Harold Parks

Intuitive Axiomatic Set Theory
José Luis García

Fundamentals of Abstract Algebra
Mark J. DeBonis

A Bridge to Higher Mathematics
James R. Kirkwood and Raina S. Robeva

Advanced Linear Algebra, Second Edition
Nicholas Loehr

Mathematical Biology: Discrete and Differential Equations
Christina Alvey and Daniel Alvey

Numerical Methods and Analysis with Mathematical Modelling
William P. Fox and Richard D. West

https://www.routledge.com/Textbooks-in-Mathematics/book-series/CANDHTEXBOOMTH

Mathematical Biology
Discrete and Differential Equations

Christina Alvey and Daniel Alvey

CRC Press
Taylor & Francis Group
Boca Raton   London   New York

CRC Press is an imprint of the
Taylor & Francis Group, an **informa** business
A CHAPMAN & HALL BOOK

First edition published 2025
by CRC Press
2385 Executive Center Drive, Suite 320, Boca Raton, FL 33431

and by CRC Press
4 Park Square, Milton Park, Abingdon, Oxon, OX14 4RN

CRC Press is an imprint of Taylor & Francis Group, LLC

Library of Congress Cataloging-in-Publication Data
Names: Alvey, Christina, author. | Alvey, Daniel, author.
Title: Mathematical biology : discrete and differential
equations / Christina Alvey and Daniel Alvey.
Description: First edition. | Boca Raton, FL : CRC Press, 2025. | Includes
bibliographical references and index.
Identifiers: LCCN 2024006530 | ISBN 9781032288253 (hbk) | ISBN
9781032288246 (pbk) | ISBN 9781003298663 (ebk)
Subjects: LCSH: Biomathematics.
Classification: LCC QH323.5 .A445 2025 | DDC 570.1/51--dc23/eng/20240529
LC record available at https://lccn.loc.gov/2024006530

ISBN: 978-1-032-28825-3 (hbk)
ISBN: 978-1-032-28824-6 (pbk)
ISBN: 978-1-003-29866-3 (ebk)

DOI: 10.1201/9781003298663

Typeset in CMR10 font
by KnowledgeWorks Global Ltd.

Dedicated to our girls,
Josephine and Abigail.

Contents

Preface

Differential equations is a large and beautiful subject with applications across mathematics and the sciences. Mathematical biology similarly is a vast tapestry of topics. Especially in the recent past, population level disease dynamics have entered the cultural consciousness in a way they never have before. However, many texts in differential equations have only a small number of applications in biology included. Additionally, many math biology textbooks are fantastic resources filled with detailed examples, but they can be technical, geared towards graduate students, and tend to be inaccessible to early undergraduates.

This text is designed specifically for an elective course in differential equations or mathematical biology. Such a course may be an elective for mathematics majors or an elective open to students in the life sciences. We assume only a knowledge of derivatives and integrals from a standard differential and integral calculus sequence. No knowledge of multi-variable calculus is assumed, nor any knowledge of linear algebra topics such as determinants, eigenvalues, and eigenvectors which are required to treat the material. Rather, these topics are introduced and explained in the text as new material when required by the topic at hand. Class tested in both a differential equations and a mathematical biology course at Mount Saint Mary College and in a differential equations course at Manhattan College, this text is written in a conversational, welcoming voice which should help invite students along as they discover the magic of mathematical biology and differential equations.

A focus is placed on examples with solutions written out step by step, including computational steps, with the goal to be as easy as possible for students to independently follow along. Exercises mirror the examples, so the instructor both has a rich source of examples to pull from with detailed steps available, and can feel confident assigning work without students feeling blindsided by surprising challenges. Additionally, all chapters include a project, which is typically an application of the material, aside from Chapters 2 and 6. Chapter 6 intentionally has fewer exercises than other chapters, and in lieu of a project, the authors encourage instructors to incorporate reading from the literature into the coverage of Chapter 6. By the end of the chapter, students should possess sufficient knowledge to read some of the available literature.

When class tested at Mount Saint Mary College, the first author included presentations of recent articles as a final project. Some articles will use a basic Susceptible-Infected-Recovered (SIR) or Susceptible-Exposed-Infected-Recovered (SEIR) model as covered in this text, yet will omit the explicit computation of the basic reproduction number, R_0. Other articles will use more complex models not covered in this text, but which should be understandable to the students. Instructors may assign articles of these types to students and have them either compute R_0 explicitly to confirm the findings in the article, or examine the structure of a more complex model as such a project.

A textbook complete with every possible topic in either mathematical biology or differential equations, even at only a mild depth, would be an intimidating volume, let alone a text discussing both. Focus is placed in this book on the basic building blocks: first- and second-order differential equations and systems of differential equations. This allows the student to be exposed to these topics without having to devote a large portion of the course to their treatment. Other topics, such as numerical methods, higher order linear equations, Laplace transforms, series solutions, chaos, and partial differential equations, have been omitted.

This discretion with standard topics has allowed us to include less common topics. Among these are discrete models, bifurcation diagrams, nullclines, and SIR models. These are topics which may be more common in dedicated mathematical biology courses, but their inclusion here allows an instructor of a differential equations course the ability to forego more mainstream additional topics while showcasing qualitative analysis of systems and SIR models if they so desire.

Applications of discrete and differential equations are numerous and varied. Mathematical biology is, itself, a wide field, and there is no hope that every topic could be covered in a one semester course, even a dedicated mathematical biology course. As such, the mathematical biology topics included in this text are focused on population dynamics, such as exponential and logistic growth, discrete fishery models, interactions between species, and SIR models. Population dynamics is embedded into Chapters 1 and 3, and species interactions into Chapters 1, 2, and 5. SIR models have been given their own chapter, Chapter 6, which scratches the surface of the exceptionally deep study of disease modeling. The modeling of these scenarios is relatively easy to understand and there is no assumed knowledge of, for example, the biology of the cell or specific traits of viruses. Additionally, the topic of disease modeling via SIR type models, and understanding the proliferation of disease using the basic reproduction number has, unfortunately, become much more pertinent to day-to-day life over the past few years with the recent coronavirus pandemic. Allowing students the ability to further their understanding of disease modeling will hopefully lay bare some pressing motivating factors behind the study and analysis of differential equations.

Content Summary

Chapter 1 introduces the student to difference equations and discrete models. Cobweb analysis is introduced, as is the concept of fixed points and stability. We examine discrete population models such as exponential, linear, and logistic growth models as well as the Beverton-Holt and Ricker models. We then explore methods for finding the solutions of linear first- and second-order difference equations. Additionally, systems of difference equations are introduced along with a few methods for determining stability of difference equations. These tools are used to determine the stability of discrete interacting species models.

Chapter 2 serves as the introductory chapter to the study of ordinary differential equations. This chapter introduces and explains vector fields, as well as discussing some standard classification definitions and existence and uniqueness theorems.

Chapter 3 explores first-order differential equations and various methods for finding their solutions. Rare for an introduction to differential equations text is the inclusion of a section devoted to bifurcation diagrams. Additionally, a number of standard applications of first-order differential equations are presented, including problems in population dynamics, Newton's Law of Cooling, and mixing problems.

Chapter 4 introduces second-order differential equations. For simplicity, while higher order differential equations are noted, this text only focuses on first- and second-order differential equations. The Wronskian and the characteristic equation are introduced, followed by solutions of linear second-order differential equations and reduction of order. Applications to spring mass systems and circuits are introduced which utilize the methods developed for finding solutions of second-order differential equations. Though spring mass systems and electrical circuits are not directly biological systems, spring mass systems provide students with a concrete, physical model through which they can then view other types of oscillation. Similarly, electrical circuit diagrams are a useful tool which can be helpful in understanding models of neurons which are beyond the scope of this text but modeled in a similar fashion.

Chapter 5 covers systems of both linear and nonlinear differential equations. As in Chapter 4, while higher dimensional systems are noted, focus is restricted to systems of two differential equations. Sections 5.1–5.4 focus on the stability and solutions of linear systems, which include applications to models of multiple tank mixing problems and parallel circuit diagrams. Meanwhile, Sections 5.5–5.6 focus on nonlinear systems. A focus is placed here on qualitative analysis of the phase plane without the use of technology. As such, nullclines are introduced and used in conjunction with the stability criteria to draw rough sketches of solutions without generating full vector fields or using numerical methods. Finally, nonlinear systems are used to model various interactions between species. Knowledge of Sections 5.2–5.4 is not assumed for the rest of the chapter. As such, Chapter 5 can be covered excluding these sections if the course is designed to focus solely on qualitative analysis.

Chapter 6 presents the standard SIR model, both with and without natural birth and death. The basic reproduction number R_0, and a method for calculating it, is introduced. Subsequently, more complex disease models are presented, including models with exposed and vaccinated classes, and with different types of interactions and progressions between the classes. Finally, sensitivity analysis is presented which helps determine the effect of the various parameters on R_0.

Recommended Chapter Lists

For a differential equations course, Chapter 1 can be omitted. Chapters 2–4 serve as the main body of the standard course load for such a class covering first- and second-order differential equations. The instructor may then choose as additional topics Sections 5.1–5.4 which covers solutions of linear systems and their applications, Sections 5.5 and 5.6 which covers qualitative analysis of nonlinear systems, or Chapter 6 which covers SIR-type models. It is recommended that Chapter 6 follow Sections 5.5–5.6, but Sections 5.1–5.4 are not assumed for Chapter 6.

When used as the text for a mathematical biology course focusing on qualitative analysis of differential equations as applied to mathematical biology, focus can be shifted entirely from finding solutions of differential equations to understanding their long term behavior. For a course of this type, Chapters 1 and 2, along with Sections 3.1–3.3, followed by Sections 5.1, 5.5, and 5.6, and culminating in Chapter 6 can be used independent of the rest of the text.

A Note to Students and Instructors

It is the authors' sincere hope that students will enjoy reading from, and instructors will enjoy teaching from, this text as much as the authors have enjoyed writing it. These topics have held a warm spot in both of their hearts for a decade and a half as one of the first topics they both fell in love with as nascent students of mathematics. In their first exposure to the topic the authors fell in love with the topic and each other. A joint project devoted to increasing students' understanding and passion for this very special topic has been a long term dream which they are thrilled to have seen come true. They hope that you will enjoy this journey through the wonderful world of mathematical biology and differential equations.

Chapter 1

Modeling with Discrete Equations

Our first glimpse at using mathematics to model biological systems will be in the framework of discrete models and difference equations. Discrete models are a tool for modeling phenomena which change discretely over time, as opposed to continuously. Discrete models measure quantities at specific instances in time (e.g., daily, weekly, monthly, etc.), and say nothing about quantities in between time intervals. Therefore, rather than solving for a function $y = f(x)$ which defines the behavior of a biological system, here we are searching for a sequence

$$x_0, x_1, x_2, x_3, x_4 \ldots$$

which discretely defines the behavior at specific time intervals.

For example, the number of rabbits in a habitat may be measured every day, so modeling their population discretely would make sense. This is in contrast with measuring the volume of a draining reservoir, whose volume is changing continuously over time as it drains.

The remaining chapters of this text dive deeply into the use of continuous models. However, discrete models are a rich topic which we will explore in this chapter. Here, we will first introduce difference equations and how to calculate time steps iteratively. We will examine several common discrete models, such as the exponential, linear, and logistic models. We will also examine equations developed to model fishery populations such as the Beverton-Holt model and the Ricker model.

Throughout this chapter, we will explore these many interesting examples, while also focusing on solving these types of difference equations. We'll begin by exploring difference equations and their notation. Then we will see how to find fixed points and classify their stability. When we see how to find explicit solutions of linear first- and second-order difference equations, we will be able to find a solution of not only the exponential and linear population models, but also the Beverton-Holt model which happens to be nonlinear. Then, when we examine fixed points and stability for systems of difference equations, we will use those tools to find the stable equilibrium for interacting species models which share a common feature with the Ricker model. Finally, we will explore age-structured Leslie matrix models which will allow us to consider the growth of a population in which the population can be segmented into different age cohorts which may reproduce, die off, and age into the next cohort at different rates.

DOI: 10.1201/9781003298663-1

1.1   Introduction to Difference Equations

To begin, it will be helpful to precisely define discrete model and difference equations.

> **Definition:** Let $x_0, x_1, x_2, \ldots, x_n$ be measurements of the quantity of the variable x at the nth time step. A **discrete** model uses **difference equations** that are rules (or functions) describing how the quantities change. Difference equations take the form
>
> $$x_n = f(x_0, x_1, x_2, \ldots, x_{n-1}).$$
>
> In other words, a difference equation is a rule for finding the quantity of the variable x at the nth time step given information about one or more of the preceding time steps.

Typically, difference equations are accompanied by an **initial condition**, often x_0, which gives information about the quantity of x at a given initial time.

Example: Consider the difference equation $x_n = 3x_{n-1} - 4$, where $x_0 = 3$. Here the initial condition is $x_0 = 3$, which allows us to find the first few quantities of x. Then we can find:

$$
\begin{aligned}
x_1 &= 3x_0 - 4 &&= 3(3) - 4 &&= 5 \\
x_2 &= 3x_1 - 4 &&= 3(5) - 4 &&= 11 \\
x_3 &= 3x_2 - 4 &&= 3(11) - 4 &&= 29 \\
x_4 &= 3x_3 - 4 &&= 3(29) - 4 &&= 83.
\end{aligned}
$$

Notice that the calculation for each subsequent value depends on the previous answer. This is because our function is defined iteratively.

○○

Example: Let's compute the next four values x_1, x_2, x_3, and x_4 determined by the difference equation:

$$x_n = -2x_{n-1} + 5, \quad x_0 = 4.$$

Again, taking the initial condition of $x_0 = 4$ and substituting it into the difference equation when $n = 1$ results in the value for x_1. Continuing this process with $n = 2, 3, 4$, we can obtain values for x_2, x_3, and x_4. We obtain:

$$x_1 \;=\; -2x_0 + 5 \;=\; -2(4) + 5 \;=\; -3$$

$$x_2 \;=\; -2x_1 + 5 \;=\; -2(-3) + 5 \;=\; 11$$

$$x_3 \;=\; -2x_2 + 5 \;=\; -2(11) + 5 \;=\; -17$$

$$x_4 \;=\; -2x_3 + 5 \;=\; -2(-17) + 5 \;=\; 39.$$

Thus, the difference equation $x_n = -2x_{n-1} + 5$ with initial condition $x_0 = 4$ produces the sequence of numbers $4, -3, 11, -17, 39, \ldots$.

○○○

Example: Some difference equations rely on more than one previous value. For example, consider the difference equation $x_n = 5x_{n-1} + 2x_{n-2}$. Now, in order to calculate the first few terms for this model, we would need two initial values. For example, let's assume $x_0 = 3$ and $x_1 = 10$. Then we can calculate x_2 and x_3:

$$x_2 \;=\; 5x_1 + 2x_0 \;=\; 5(10) + 2(3) \;=\; 56$$

$$x_3 \;=\; 5x_2 + 2x_1 \;=\; 5(56) + 2(10) \;=\; 300.$$

Not only does the calculation for each value depend on the previous answer, but it depends on the answer before that one as well. Difference equations, like this one, which rely on two previous time steps are called *second-order* difference equations, as opposed to *first-order* difference equations that only need one previous value.

○○

Example: Write out the next four terms of the difference equation:

$$x_n = 3x_{n-1} - 2x_{n-2}, \qquad x_0 = 1, \quad x_1 = 5.$$

Again, notice that we'll need to substitute the two previous values in each iteration of this equation. In this case, we have:

$$x_2 \;=\; 3x_1 - 2x_0 \;=\; 3(5) - 2(1) \;=\; 13$$

$$x_3 \;=\; 3x_2 - 2x_1 \;=\; 3(13) - 2(5) \;=\; 29$$

$$x_4 \;=\; 3x_3 - 2x_2 \;=\; 3(29) - 2(13) \;=\; 61$$

$$x_5 \;=\; 3x_4 - 2x_3 \;=\; 3(61) - 2(29) \;=\; 125.$$

Thus the difference equation with given initial values produces the sequence of values $1, 5, 13, 29, 61, 125, \ldots$.

○○

Example: One famous difference equation is

$$x_n = x_{n-1} + x_{n-2}, \qquad x_0 = 0, \quad x_1 = 1.$$

This difference equation defines what is called the **Fibonacci sequence**, named after an Italian mathematician known as Fibonacci. He used it in the early 1200s to model the growth of a rabbit population. As before, we'll need to substitute the two previous values in each iteration of this equation. In this case, we have:

$$
\begin{aligned}
x_2 &= x_1 + x_0 &= 1 + 0 &= 1 \\[2mm]
x_3 &= x_2 + x_1 &= 1 + 1 &= 2 \\[2mm]
x_4 &= x_3 + x_2 &= 2 + 1 &= 3 \\[2mm]
x_5 &= x_4 + x_3 &= 3 + 2 &= 5 \\[2mm]
x_6 &= x_5 + x_4 &= 5 + 3 &= 8.
\end{aligned}
$$

This sequence continues, and is: $0, 1, 2, 3, 5, 8, 13, 21, 34, 55, 89, 144, \ldots$.

1.1.1 Exponential, Linear Difference, and Logistic Models

Some discrete models are common enough to warrant their separate treatment. The exponential model, for example, is used when modeling a population which grows or decays at a rate proportional to the existing population. Consider the following example:

Example: Let's write out the first few terms of the difference equation

$$x_n = 2x_{n-1}, \qquad x_0 = 1.$$

We find that:

$$
\begin{aligned}
x_1 &= 2x_0 &= 2(1) &= 2 \\[2mm]
x_2 &= 2x_1 &= 2(2) &= 4 \\[2mm]
x_3 &= 2x_2 &= 2(4) &= 8 \\[2mm]
x_4 &= 2x_3 &= 2(8) &= 16.
\end{aligned}
$$

We can see that at each time step, the quantity of the variable x *doubles*. We can also graph the results in a solution plot:

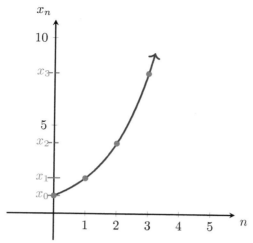

Notice that the graph looks like an exponential function. This is no coincidence!

The difference equation $x_n = rx_{n-1}$ represents the exponential model and

- if $0 < r < 1$, then x decays exponentially,
- if $r > 1$, then x grows exponentially,
- if $r = 1$, then x will stay constant.

There are a few common types of difference equations which lead to distinct discrete models. We just discussed the exponential model and will derive the solution of the difference equation given an initial condition in a later section. The exponential model is, in fact, a special case of the more general **linear** model.

The linear difference model has equation $x_n = rx_{n-1} + k$. We often use this model to illustrate growth of a population, and in this case, $r > 0$ is the growth rate of the population and k accounts for constant migration.

Example: A pepper plant is infested with spider mites. The plant initially hosts 500 spider mites. The gardener decides to sacrifice this plant, and observe the growth of the spider mite population. The gardener removes 50 spider mites each day. If the growth rate of the spider mites is $r = 1.5$, how many spider mites will be on the plant after 3 days?

As the gardener is removing 50 mites every day, we have $k = -50$, and $r = 1.5$ has been given. We have the difference equation $p_n = 1.5p_{n-1} - 50$. Because there are initially 500 spider mites, we also know that $p_0 = 500$, so we can find the number of mites for the first three days using the difference equation:

$$p_1 = 1.5p_0 - 50 = 1.5(500) - 50 = 700$$

$$p_2 = 1.5p_1 - 50 = 1.5(700) - 50 = 1000$$

$$p_3 = 1.5p_2 - 50 = 1.5(1000) - 50 = 1450.$$

Therefore, we can see that after three days, there will be 1450 spider mites on the pepper plant.

If we sketch the solution plot for this linear difference model, we obtain the following result:

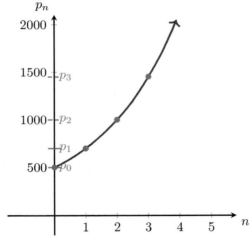

This graph also looks like an exponential function, and is, in fact, just a transformation of one.

○○○

Example: Not all linear difference models behave like the model from the previous example. Let's consider the linear difference model $x_n = 0.5x_{n-1} + 50$ with initial condition $x_0 = 20$. Finding the first three terms for this model, we obtain:

$$x_1 = 0.5x_0 + 50 = 0.5(20) + 50 = 60$$

$$x_2 = 0.5x_1 + 50 = 0.5(60) + 50 = 80$$

$$x_3 = 0.5x_2 + 50 = 0.5(80) + 50 = 90.$$

It's hard to tell what will happen over time by just calculating the first three terms, but you may have noticed that these numbers aren't increasing quite as fast as in the previous example. We can sketch the solution plot for this linear difference model to really understand the behavior.

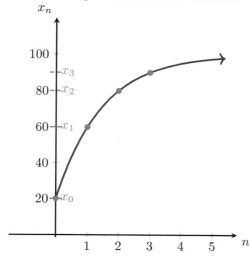

Notice that, in this case, the population size grows quickly at first, but then levels off and approaches a limit. This type of model describes populations that do not change much over time in the long run because of this.

○○○

Another type of model is the **logistic growth model**. This model takes into account the fact that some populations cannot grow without bound, and often live in an environment which has a natural carrying capacity for that population.

The difference equation for the logistic growth model is

$$p_n = p_{n-1} + r(M - p_{n-1})p_{n-1},$$

where $r > 0$ is the growth constant and M is the carrying capacity for the population in the given environment.

The $r(M - p_{n-1})p_{n-1}$ term of the logistic growth difference equation is what gives this model its distinctive shape. As p_{n-1} approaches M, this term approaches zero, at which point $p_n = p_{n-1}$, so the value of p_n would not change from the previous iteration. Therefore, the logistic growth model levels off (reaches equilibrium) at $p = M$.

○○

Example: Consider the discrete model

$$p_n = p_{n-1} + 0.0006(150 - p_{n-1})p_{n-1}, \quad p_0 = 10.$$

If we plot the solution curve, we obtain the following graph:

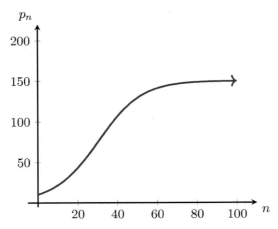

In this case, we can see that the population starts at size 10. It quickly increases, and then gradually levels off around the carrying capacity of 150.

It is worth thinking about what happens when the population begins above the carrying capacity in this model. For example, consider the logistic model:

$$p_n = p_{n-1} + 0.0006(150 - p_{n-1})p_{n-1}, \quad p_0 = 250.$$

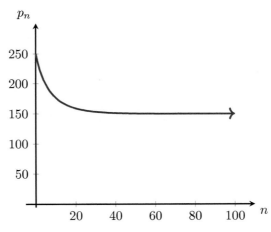

As you can see, the population decreases until reaching the carrying capacity in this case. This is desirable; for example, it would be a poor model if somehow starting above the carrying capacity allowed the population to grow without bound.

1.1.2 Fishery Models

There are two specific models worth examining which can both be used to model the population dynamics of fisheries. These are the Beverton-Holt model[1] and the Ricker model.[2] Both of these models are interesting to us for different reasons. The Beverton-Holt model is a clearly nonlinear difference equation which we will nevertheless be able to find a solution. Meanwhile, the Ricker model's fixed points display interesting behavior as one of the parameters is varied.

The Beverton-Holt model is given by:

$$p_n = \frac{k p_{n-1}}{1 + \frac{k-1}{M} p_{n-1}},$$

where p_n is the population, k is an inherent growth rate, and M is a carrying capacity.

Notice that this could be rewritten as:

$$p_n = \frac{k}{1 + \frac{k-1}{M} p_{n-1}} p_{n-1}$$

in which case we can think of it as an exponential model but with a variable growth rate which depends on p_{n-1},

$$g(p_{n-1}) = \frac{k}{1 + \frac{k-1}{M} p_{n-1}}$$

such that we have $p_n = g(p_{n-1}) p_{n-1}$. Notice that as p_{n-1} approaches the carrying capacity, M, $g(p_{n-1})$ approaches one. In fact, evaluating $g(p_{n-1})$ at M, we have:

$$
\begin{aligned}
g(M) &= \frac{k}{1 + \frac{k-1}{M} M} \\
&= \frac{k}{1 + k - 1} \\
&= \frac{k}{k} \\
&= 1.
\end{aligned}
$$

Thus, when $p_{n-1} = M$, the model reduces to $p_n = p_{n-1}$. Also note that if the carrying capacity is very large, $g(p_{n-1}) \approx k$, the model reduces to $p_n \approx k p_{n-1}$.

○○○

---

[1]Beverton, R.J.H., & Holt, S.J. (1957). *On the dynamics of exploited fish populations.* (Ser. Fishery investigations, series 2, 19). Her Majesty's Stationary Office.
[2]Ricker, W.E. (1954). Stock and recruitment. *Journal of the Fisheries Research Board of Canada*, 11(5): 559–623.

Example: Consider the Beverton-Holt model with $k = 2$ and $M = 100$ with $p(0) = 50$. Then we have:

$$p_{n-1} = \frac{2p_{n-1}}{1 + 0.01p_{n-1}}, \quad p(0) = 50.$$

We can compute the first few terms for this model, obtaining:

$$p_1 = \frac{2p_0}{1 + 0.01p_0} = \frac{2(50)}{1 + 0.01(50)} \approx 66.67$$

$$p_2 = \frac{2p_1}{1 + 0.01p_1} = \frac{2(66.67)}{1 + 0.01(66.67)} = 80$$

$$p_3 = \frac{2p_2}{1 + 0.01p_2} = \frac{2(80)}{1 + 0.01(80)} \approx 88.89$$

$$p_4 = \frac{2p_3}{1 + 0.01p_3} = \frac{2(88.89)}{1 + 0.01(88.89)} \approx 94.12$$

$$p_5 = \frac{2p_4}{1 + 0.01p_4} = \frac{2(94.12)}{1 + 0.01(94.12)} \approx 96.97$$

$$p_6 = \frac{2p_5}{1 + 0.01p_5} = \frac{2(96.97)}{1 + 0.01(96.97)} \approx 98.46.$$

Note that even after just these six terms, we already see that p_n appears to be approaching and leveling off at $M = 100$. We can see this behavior in the below graph.

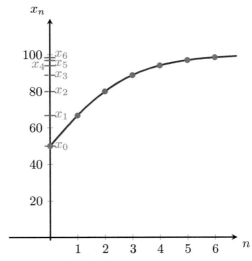

We will see in Section 1.3 how to find an explicit solution of this model which will allow us to predict the population at any time step without manually computing them all iteratively.

○○

The second model we will consider is the Ricker Model.

> The Ricker Model is given by:
>
> $$p_n = e^{r(1 - p_{n-1}/M)} p_{n-1},$$
>
> where r is a growth rate and M is the carrying capacity.

It is worth noticing that this model contains an exponential function in the difference equation, which is unlike the other models we have seen so far. We can rewrite the model as:

$$p_n = e^r e^{-r p_{n-1}/M} p_{n-1}$$

or

$$p_n = R e^{-r p_{n-1}/M} p_{n-1}$$

in which case we can think of the e^r term as a reproduction factor and the $e^{-r p_{n-1}/M}$ term as a mortality factor which depends on the density of the population. The Ricker model is interesting especially because the carrying capacity is not a hard and fast limit. In fact, we can see some interesting behavior depending on the value of r.

○○

Example: Consider the Ricker model with $r = 2.25$ and $M = 100$ with $p(0) = 50$. Then we have:

$$p_n = e^{2.25(1 - p_{n-1}/100)} p_{n-1}, \quad p(0) = 50.$$

We can compute the first few terms for this model, obtaining:

$$p_1 = e^{2.25(1 - p_0/100)} p_0 = e^{2.25(1 - 50/100)} 50 \approx 154.01$$

$$p_2 = e^{2.25(1 - p_1/100)} p_1 = e^{2.25(1 - 154.01/100)} 154.01 \approx 45.69$$

$$p_3 = e^{2.25(1 - p_2/100)} p_2 = e^{2.25(1 - 45.69/100)} 45.69 \approx 155.07$$

$$p_4 = e^{2.25(1 - p_3/100)} p_3 = e^{2.25(1 - 155.07/100)} 155.07 \approx 44.92$$

$$p_5 = e^{2.25(1 - p_4/100)} p_4 = e^{2.25(1 - 44.92/100)} 44.92 \approx 155.12$$

$$p_6 = e^{2.25(1 - p_5/100)} p_5 = e^{2.25(1 - 155.12/100)} 155.12 \approx 44.88$$

$$p_7 = e^{2.25(1 - p_6/100)} p_6 = e^{2.25(1 - 44.88/100)} 44.88 \approx 155.12.$$

As you may suspect, the population never stabilizes, but will oscillate between a value of 44.88 and 155.12 for all time. This model does not always have end behavior like this. For some values of r, the population stabilizes at the carrying capacity while for others there is a stable cycle as in this example. For yet other values of r we can observe chaotic cycle behavior which never stabilizes into a noticeable pattern. In Section 1.2, we will explicitly find the fixed points of this model and use the stability theorem and cobweb analysis to determine the stability and behavior around the fixed points.

1.1 Exercises

Compute the next four terms for the following difference equations:

1. $x_n = 5x_{n-1} + 2, \qquad x_0 = -3$

2. $x_n = -3x_{n-1} + 7, \qquad x_0 = 4$

3. $x_n = -4x_{n-1} - 1, \qquad x_0 = -2$

4. $x_n = 2x_{n-1} - 6, \qquad x_0 = 3$

5. $p_n = p_{n-1} + 0.02(500 - p_{n-1})p_{n-1}, \qquad p_0 = 5$

6. $p_n = 1.5p_{n-1} - 5, \qquad p_0 = 10$

7. $x_n = 3x_{n-1} - x_{n-2}, \qquad x_0 = 5, \quad x_1 = 2$

8. $x_n = -2x_{n-1} + 4x_{n-2}, \qquad x_0 = 0, \quad x_1 = -1$

9. $x_n = x_{n-1} - 3x_{n-2}, \qquad x_0 = 3, \quad x_1 = -2$

10. $x_n = 2x_{n-1} + 3x_{n-2}, \qquad x_0 = -4, \quad x_1 = 1$

11. $x_n = 2x_{n-1} + 3x_{n-2}, \qquad x_0 = -4, \quad x_1 = 1$

12. For the following difference equation, $x_n = 4x_{n-1} - 3$, write out the next five terms of the sequence, with the given initial conditions. Plot the results on nx_n-plane. Describe the behavior of each sequence. What do you notice about your answers?

    (a) $x_0 = 0$
    (b) $x_0 = 1$
    (c) $x_0 = 2$

13. For the linear difference model,

$$p_n = 1.5p_{n-1} - 10, \quad p_0 = 20$$

   (a) What does the value 1.5 represent?
   (b) What does the value 10 represent?
   (c) What does the value 20 represent?
   (d) write out the next five terms of the sequence, and plot the results on np_n-plane.

14. For the exponential model, $p_n = 1.2p_{n-1}$, write out the next five terms of the sequence, with the given initial conditions. Plot the results on np_n-plane. Describe the behavior of each sequence.

   (a) $p_0 = 10$
   (b) $p_0 = 2$
   (c) $p_0 = 25$
   (d) $p_0 = 100$

15. For the following logistic model, $p_n = p_{n-1} + 1.2(100 - p_{n-1})p_{n-1}$, write out the next five terms of the sequence with the given initial conditions. Plot the results on np_n-plane. Describe the behavior of each sequence. What do you notice about your answers?

   (a) $p_0 = 25$
   (b) $p_0 = 75$
   (c) $p_0 = 125$
   (d) $p_0 = 200$

16. For the following difference equation, $x_n = 4 - x_{n-1}$, write out the next five terms of the sequence, with the given initial conditions. Plot the results on nx_n-plane. Describe the behavior of each sequence. What do you notice about your answers?

   (a) $x_0 = -5$
   (b) $x_0 = 1$
   (c) $x_0 = 2$
   (d) $x_0 = 3$
   (e) $x_0 = 10$

17. For the exponential model, $p_n = 0.9p_{n-1}$, write out the next five terms of the sequence, with the given initial conditions. Plot the results on np_n-plane. Describe the behavior of each sequence.

   (a) $p_0 = 10$

   (b) $p_0 = 2$

   (c) $p_0 = 25$

   (d) $p_0 = 100$

18. For the Beverton-Holt model,

$$p_n = \frac{kp_{n-1}}{1 + \frac{k-1}{M}p_{n-1}}$$

write out the next five terms of the sequence with the given initial conditions and parameter values. Describe the behavior of each sequence. Make a hypothesis about the long-term behavior of the sequence. (For example, if $p_0 > M$ and $k > 1$, the long-term behavior of the sequence is...)

   (a) $p_0 = 10$, $k = 1.2$, $M = 100$

   (b) $p_0 = 150$, $k = 2.5$, $M = 100$

   (c) $p_0 = 50$, $k = 0.75$, $M = 100$

19. For the Ricker model,

$$p_n = e^{r(1-p_{n-1}/M)}p_{n-1}$$

write out the next six terms of the sequence with the given initial conditions and parameter values. Describe the behavior of each sequence. Make a hypothesis about the long-term behavior of the sequence. (For example, if $p_0 > M$ and $r < 2$ the long-term behavior of the sequence is...)

   (a) $p_0 = 200$, $r = 0.5$, $M = 100$

   (b) $p_0 = 50$, $r = 0.5$, $M = 100$

   (c) $p_0 = 40$, $r = 1.5$, $M = 100$

   (d) $p_0 = 200$, $r = 1.5$, $M = 100$

   (e) $p_0 = 75$, $r = 2.25$, $M = 100$

   (f) $p_0 = 250$, $r = 2.25$, $M = 100$

1.2 First-Order Difference Equations and Fixed Points

As you may suspect, the long-term behavior of a discrete model may vary wildly depending on the difference equation for the model and the initial condition. In this section we will explore a few methods for describing this long term behavior without finding an actual solution of the discrete model. One such method, cobweb analysis, allows us to graphically determine some of the values for x at different time steps and observe the long-term behavior of x given a certain initial condition. We will also find **fixed points** algebraically and determine their stability using calculus.

1.2.1 Cobweb Analysis

A cobweb diagram is a tool which can be used to help describe the behavior of a discrete model, and requires both a difference equation $x_n = f(x_{n-1})$ and an initial condition x_0.

In order to construct a cobweb diagram for the discrete model $x_n = f(x_{n-1})$ with initial condition x_0, we graph the function $x_n = f(x_{n-1})$ and the function $x_n = x_{n-1}$ on the $x_{n-1}x_n$-plane. We generally graph this with x_n on the "y"-axis and x_{n-1} on the "x"-axis. The "cobweb" part of the cobweb diagram is now ready to be constructed.

Starting at $(x_0, 0)$, draw a vertical line up/down to $(x_0, f(x_0))$. Then, draw a horizontal line left/right from this point to the point $(f(x_0), f(x_0))$ on the line $x_n = x_{n-1}$. Make sure to identify this point, which we will call x_1, on your graph, as these points on the $x_n = x_{n-1}$ curve will be crucial in determining the behavior of the given discrete model.

From this point, we now draw another vertical line to the point $(x_1, f(x_1))$ on the graph of $x_n = f(x_{n-1})$, and then draw another horizontal line from this point to the point $(f(x_1), f(x_1))$ on the line $x_n = x_{n-1}$. We'll label this point x_2. This process continues indefinitely.

Notice that, in effect, we have "calculated" the first few time steps of this discrete model simply by sketching the cobweb diagram. To determine the behavior of the model, we can now use our sequence of points on the $x_n = x_{n-1}$ line and observe their limit graphically.

Example: Let's assume that our population has an initial size $x_0 = 1$ and triples at each step. Then, the difference equation representing the population size at time n is:

$$x_n = 3x_{n-1}, \quad x_0 = 1.$$

To sketch the cobweb diagram, we will plot both this difference equation and $x_n = x_{n-1}$. Then, beginning at $x_0 = 1$, we first go up to the curve $x_n = 3x_{n-1}$ and then over to $x_n = x_{n-1}$. We complete our cobweb diagram by repeating this process, labeling the points on the $x_n = x_{n-1}$ curve along the way.

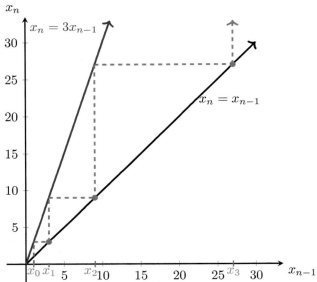

Now we can examine our cobweb diagram and the sequence of values $x_0, x_1, x_2, x_3, \ldots$ to describe the behavior of our model. Because our cobweb consistently grows up and to the right and will do so forever, we can describe the solution of our discrete model as being unbounded. We know this is true simply because our population triples at each time step and because we know that $x_n = 3x_{n-1}$ represents exponential growth, so we are confirming this result with the cobweb in this example.

○○

Example: Consider the discrete model:

$$p_n = p_{n-1} + 0.02(100 - p_{n-1})p_{n-1}, \quad p_0 = 120.$$

Recall that this is a logistic growth model with a carrying capacity $M = 100$. We would expect that the cobweb diagram will show the population approaching this value of $p = 100$. To sketch this, we will plot the difference equation and $p_n = p_{n-1}$, and then begin the process of cobwebbing.

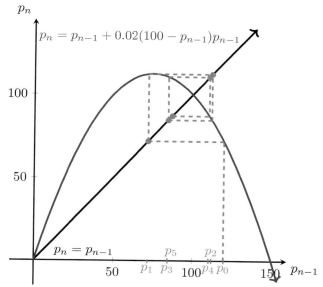

$$p_n = p_{n-1} + 0.02(100 - p_{n-1})p_{n-1}$$

We can see from our cobweb diagram that the values seem to be circling around the intersection point between $p_n = p_{n-1} + 0.02(100 - p_{n-1})p_{n-1}$ and $p_n = p_{n-1}$, which is at the carrying capacity of $p = 100$. If we did our cobweb analysis for a few more iterations, we would see that we would continue to approach this carrying capacity. Further, looking only at the sequence of values $p_0, p_1, p_2, p_3, p_4, p_5, \ldots$ we observe that their limit approaches $p = 100$ as well.

○○

Example: Consider the Ricker model with $r = 2.1$ and $M = 100$:

$$p_n = e^{2.1(1-0.01p_{n-1})}p_{n-1}, \quad p(0) = 25.$$

Recall that we observed cyclic behavior from this model in the previous section with $r = 2.25$. Should we expect the same behavior with $r = 2.1$? To sketch this, we will plot the difference equation and $p_n = p_{n-1}$, and then begin the process of cobwebbing.

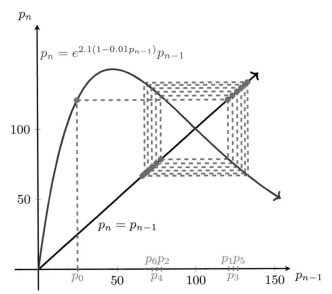

We can see from our cobweb diagram that the values seem to be circling out from the intersection point between $p_n = e^{2.1(1-0.01p_{n-1})}p_{n-1}$ and $p_n = p_{n-1}$, which is at the carrying capacity of $p = 100$. If we did our cobweb analysis for a few more iterations, we would see that we would continue to cycle out by less and less each iteration. This sequence eventually cycles between $p = 62.93$ and $p = 137.07$.

○○

Example: Now, Consider the Ricker model with $r = 1.9$ and $M = 100$:

$$p_n = e^{1.9(1-0.01p_{n-1})}p_{n-1}, \quad p(0) = 25.$$

We observed cyclic behavior from this model in the previous example with $r = 2.1$ and in the previous section with $r = 2.25$. Should we expect similar behavior with $r = 1.9$? To sketch this, we will plot the difference equation and $p_n = p_{n-1}$, and then begin the process of cobwebbing.

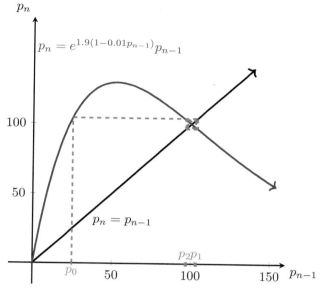

$$p_n = e^{1.9(1-0.01p_{n-1})}p_{n-1}$$

$$p_n = p_{n-1}$$

We can see from our cobweb diagram that the values seem to be circling inward toward the intersection point between $p_n = e^{2.1(1-0.01p_{n-1})}p_{n-1}$ and $p_n = p_{n-1}$, which is at the carrying capacity of $p = 100$. If we did our cobweb analysis for a few more iterations, we would see that we would continue to cycle in each iteration, eventually converging on $p = 100$.

Note that in the previous example, when $r = 2.1$, we cycled out from $p = 100$ and settled on a cycle between two values. Conversely, here with $r = 1.9$, we cycled in and converged on a **fixed point** of $p = 100$. How could we have predicted this different behavior on either side of $r = 2$? In the following subsection, we will explore how to identify and classify fixed points. We will see that we can find an explicit condition required for the point $p = 100$ to be stable which would have predicted this bifurcated behavior dependent on the value of r.

○○

Example: Consider the discrete model:

$$x_n = 2\sin(x_{n-1}), \quad x_0 = \frac{\pi}{4}.$$

We can compute the first time step by hand to be $x_1 = 2\sin(\frac{\pi}{4}) = \sqrt{2}$, but trying to find the second time step, $x_2 = 2\sin(\sqrt{2})$ would prove to be a challenge without a calculator. However, with a cobweb diagram, we can estimate future time steps of x and visually see that they eventually oscillate around the intersection point in the diagram.

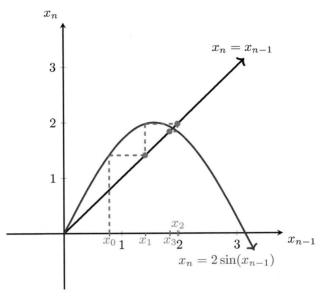

$$x_n = 2\sin(x_{n-1})$$

In this case, we see that the values of x tend toward the intersection of $x_n = 2\sin(x_{n-1})$ with $x_n = x_{n-1}$, which occurs at the x value of approximately 1.89549. This intersection point is referred to as a fixed point and because our sequence $x_0, x_1, x_2, x_3, \ldots$ tends toward this value, we will call this fixed point stable, similar to the previous example. Note that if we were to plot the solution of our discrete model with $x_0 = \frac{\pi}{4}$, we would notice that the solution is bounded and approaches this same x value over time.

1.2.2 Fixed Points and Stability

As we just saw, one way to determine stability and behavior of our model is through cobweb analysis, with the ultimate goal of determining the stability of a fixed point. You may have noticed that in the event we had a fixed point, it occurred where the graphs of $x_n = f(x_{n-1})$ and $x_n = x_{n-1}$ intersect. Thus, you may suspect that an approach to algebraically determine the fixed points for a difference equation would be to solve the equation $f(x) = x$. This is, in fact, the correct approach.

> **Definition: A fixed point** (or **equilibrium** or **steady state**) of a first-order discrete model is a point where $f(x) = x$.

Example: Consider the difference equation $x_n = 3x_{n-1} - 4$. To find the fixed point, we set $f(x) = 3x - 4$ equal to x and solve:

$$
\begin{aligned}
3x - 4 &= x \\
2x - 4 &= 0 \\
2x &= 4 \\
x &= 2.
\end{aligned}
$$

Therefore, we can see that the fixed point is at $x^* = 2$.

Now, we are confronted with the challenge of classifying this fixed point. What is meant by this? Roughly, we want to know if the long-term behavior of x tends towards x^* or away from x^* for various initial conditions. We can classify fixed points in the following ways.

> **Definition:** A fixed point x^* is **stable** if every solution curve near x^* tends towards x^* over time. In contrast, a fixed point is **unstable** if every solution curve near x^* tends away from x^* over time.

Once we have found a fixed point, we can then use the following stability theorem to determine if it is stable or unstable.

> **Stability Theorem for First-Order Discrete Models:** Let x^* be a fixed point of the difference equation $x_n = f(x_{n-1})$.
>
> - x^* is stable if $|f'(x^*)| < 1$.
>
> - x^* is unstable if $|f'(x^*)| > 1$.
>
> - If $|f'(x^*)| = 1$, then the test is inconclusive.

Example: Returning to our previous example, we can see that $f(x) = 3x - 4$ gives $f'(x) = 3$. For our fixed point at $x^* = 2$, we have:

$$|f'(x^*)| = |f'(2)| = |3| > 1.$$

Thus, the stability theorem says that $x^* = 2$ is an unstable fixed point.

While simply finding and classifying our fixed points can be very helpful, at times, it can be illuminating to illustrate our stability results using a solutions plot. Knowing that this fixed point is unstable allows us to determine that solutions with $x_0 > 2$ will increase away from the fixed point. Similarly, solutions with $x_0 < 2$ will decrease away from the fixed point.

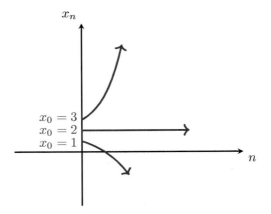

Hence, it makes sense to refer to this fixed point as unstable as all the solutions aside from the one where $x_0 = 2$ diverge away.

○○

Example: Consider the difference equation:

$$p_n = \frac{p_{n-1}^2 + 2}{3}.$$

We can find the fixed points for this difference equation by setting $f(p) = p$ and solving. We can see that:

$$\frac{p^2 + 2}{3} = p$$

$$p^2 + 2 = 3p$$

$$p^2 - 3p + 2 = 0$$

$$(p - 1)(p - 2) = 0,$$

so $p^* = 1, 2$ are fixed points for this difference equation. To determine their stability we find $f'(p^*)$ for each fixed point and use the stability theorem to classify them. We can see that:

$$f(p) = \frac{p^2 + 2}{3}$$

$$f'(p) = \frac{2p}{3},$$

so $f'(1) = \frac{2}{3}$ and $f'(2) = \frac{4}{3}$. As $|f'(1)| < 1$, $p^* = 1$ is a stable fixed point, whereas $|f'(2)| > 1$, so $p^* = 2$ is an unstable fixed point.

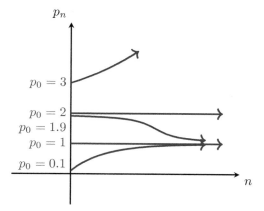

We can see from the solutions plot that if $p_0 > 2$, the solution grows without bound. However, if $1 < p_0 < 2$ or if $p_0 < 1$, we can see that the solution tends towards the stable fixed point $p_0 = 1$.

Note that the above drawing of the solutions plot is not always a helpful tool. For example, we saw that with the Ricker model, the sequence p_n can cycle about a fixed point and either converge on it or move away. The blue lines of our solutions plot here are continuous curves we are fitting to the discrete sequences p_n generated by the difference equations and it is worth noting that the sequence p_n will not take every value on the solutions plot.

○○

Example: Consider the logistic model:

$$p_n = p_{n-1} + 0.005(100 - p_{n-1})p_{n-1},$$

where the carrying capacity $M = 100$, and the growth constant $r = 0.005$. We can let $f(p) = p + 0.005(100 - p)p$ and see that:

$$p = f(p)$$

$$p = p + 0.005(100 - p)p$$

$$0 = 0.005(100 - p)p,$$

so we can see that $p^* = 0, 100$ are the fixed points of the model. To determine stability, we compute:

$$f'(p) = 1.5 - 0.01p$$

and see that

$$|f'(0)| = |1.5 - 0.01(0)| = 1.5 > 1,$$

and
$$|f'(100)| = |1.5 - 0.01(100)| = 0.5 < 1.$$

Thus, the stability theorem tells us that $p^* = 0$ is unstable, while $p^* = 100$ is a stable fixed point.

○○

Example: Consider the general Ricker model,
$$p_n = e^{r(1-p_{n-1}/M)}p_{n-1}.$$

We can let $f(p) = pe^{r(1-p/M)}$ and see that:

$$p = f(p)$$
$$p = pe^{r(1-p/M)}$$
$$0 = pe^{r(1-p/M)} - p$$
$$0 = p\left(e^{r(1-p/M)} - 1\right).$$

Thus we have $p^* = 0$ or:

$$0 = e^{r(1-p/M)} - 1$$
$$1 = e^{r(1-p/M)}$$
$$0 = r(1 - p/M)$$
$$0 = 1 - p/M$$
$$p/M = 1$$
$$p = M,$$

so we have that $p^* = M$ is another fixed point of the model. To determine the stability, we can compute:
$$f'(p) = \left(1 - \frac{rp}{M}\right)e^{r(1-p/M)}$$

and find that
$$|f'(0)| = \left|\left(1 - \frac{r(0)}{M}\right)e^{r(1-0/M)}\right| = e^r$$

and
$$|f'(M)| = \left|\left(1 - \frac{r(M)}{M}\right)e^{r(1-M/M)}\right| = |1 - r|.$$

Therefore, the fixed point at $p^* = 0$ is stable when $e^r < 1$, or in other words when $r < 0$, and unstable when $r > 0$. Furthermore, $p^* = M$ is stable when $|1 - r| < 1$, or in other words when $0 < r < 2$, and unstable when $r \leq 0$ or $r \geq 2$.

This explains the behavior we saw with our cobweb diagrams in the previous subsection. When we chose $r = 2.1$ we saw an unstable sequence alternating between a value above the fixed point and a value below it. However, when we chose $r = 1.9$ we saw the sequence alternate and converge towards the value $p = 100$, which was the carrying capacity for that problem. This value for the parameter r at which the fixed point at M transitions from a stable fixed point to an unstable one is called a **bifurcation value**. We will not explore the concept of a bifurcation diagram in the discrete setting, but will address this concept for first-order differential equations in Section 3.3.

1.2 Exercises

For the following cobweb diagram problems you should not actually calculate any time steps, but rather use the diagram to graphically find the points. It is acceptable to have a small deviation, it shouldn't affect your solutions for these problems.

1. Use a cobweb diagram to graph the first 4 time steps of the difference equation
$$x_n = 1.5x_{n-1} - 5$$
with $x_0 = 15$. What do you expect the long-term behavior of this sequence will be?

2. Use a cobweb diagram to graph the first 4 time steps of the difference equation
$$x_n = 1.5x_{n-1} - 5$$
with $x_0 = 25$. What do you expect the long-term behavior of this sequence will be?

3. Use a cobweb diagram to graph the first 4 time steps of the difference equation
$$x_n = -0.1(x_{n-1} - 20)^2 + 25$$
with $x_0 = 35$. What do you expect the long-term behavior of this sequence will be?

4. Use a cobweb diagram to graph the first 4 time steps of the difference equation
$$x_n = 15 - 0.5x_{n-1}$$
    with $x_0 = 20$. What do you expect the long-term behavior of this sequence will be?

5. Use a cobweb diagram to graph the first 4 time steps of the difference equation
$$x_n = e^{-x_{n-1}^2}$$
    with $x_0 = 1$. What do you expect the long-term behavior of this sequence will be?

For the following problems, find the fixed point(s) of the given difference equation and use the stability theorem to classify them.

6. $x_n = 2x_{n-1} - 50$

7. $x_n = x_{n-1}^2 - 5x_{n-1} + 6$

8. $p_n = -0.2p_{n-1} + 50$

9. $x_n = 0.5x_{n-1}^2$

10. $x_n = 5e^{-x_{n-1}}$

11. $p_n = p_{n-1} + 0.001(500 - p_{n-1})p_{n-1}$

12. $p_n = e^{1.5(1-0.05p_{n-1})}p_{n-1}$

1.2 Project: Fishery Model Comparison

In this project we aim to compare the two fishery models we have seen in this chapter: the Beverton-Holt model and the Ricker model. In this project your goal is to compare these two models and their parameters.

1. Find the fixed points and determine their stability for the Ricker model with $r = 1$ and $M = 300$, and for the Beverton-Holt model with $k = 2.5$ and $M = 300$. Are these the same? Do they have the same stability?

2. Find the fixed points and determine their stability for the Ricker model with $r = 1.5$ and $M = 300$, and for the Beverton-Holt model with $k = 1.5$ and $M = 300$. Are these the same? Do they have the same stability?

3. Find the fixed points and determine their stability for the Ricker model with $r = 2.1$ and $M = 300$, and for the Beverton-Holt model with $k = 2.5$ and $M = 300$. Are these the same? Do they have the same stability?

4. Using a computer, calculate the first 30 values for both models with $p_0 = 25$. For the Ricker model, use $r = 1.5$ and $M = 300$, and for the Beverton-Holt model, use $k = 1.5$ and $M = 300$.

5. Compare the behavior of these two models. Do they reach the same fixed point? Which model reaches the fixed point fastest?

6. What do you notice about the relative magnitude of the parameters k and r?

Recall that the Ricker model can also be written as:

$$p_n = e^r e^{-rp_{n-1}/M} p_{n-1},$$

where the e^r term is a reproduction factor and the $e^{-rp_{n-1}/M}$ term is a mortality factor. As such, when $r = 1.5$, it is scaling the model by a factor of $e^{1.5} \approx 4.48$.

7. Using this insight, let $k = 1.5$, let $r = \ln 1.5 \approx 0.405$, and let $M = 300$. Graph the first 30 terms of the sequences for both models. What do you notice?

8. Graph the models for a few other values of r and M, keeping the same relationship between r and k. What do you notice? Does this seem like a good relationship?

1.3 Solutions of Linear First-Order Difference Equations

In Section 1.1 we introduced both exponential and linear difference equations. Both of these are **first-order** difference equations because x_n depends only on x_{n-1}, or in other words, we can write the difference equation as:

$$x_n = f(x_{n-1}).$$

Furthermore, both of these types of difference equations are *linear*, meaning that the function above is a linear function of x_{n-1}. For some difference equations, we can solve the equation to find an expression for x_n given only the initial condition and this is what we aim to do in this section.

These first few examples are special cases that we can solve by simply deriving the formula. In other words, if we write out the first few terms of these models, we should be able to see a pattern that will result in the final statement of the solution of the given difference equation. Let's see how this works with the following exponential function.

○○

Example: In Section 1.1, we discussed the exponential function $x_n = 2x_{n-1}$ with $x_0 = 1$. To see why this truly represents an exponential function, we can solve to get x_n as a function of n and which only depends on x_0. Essentially, we are eliminating the dependence on x_{n-1} and will no longer need to calculate each term one after the next. We'll obtain a formula that will allow us to calculate any term, just by knowing the initial condition. For example, we know that $x_1 = 2x_0$, simply by using the given difference equation. Similarly, we know that $x_2 = 2x_1$, and if we substitute our expression for x_1 into this equation, we arrive at $x_2 = 2x_1 = 2 \cdot 2x_0 = 2^2 x_0$. Continuing in this manner and working forward from x_0, we can see that:

$$
\begin{aligned}
x_1 &= 2x_0 \\
x_2 &= 2x_1 &= 2 \cdot 2x_0 &= 2^2 x_0 \\
x_3 &= 2x_2 &= 2 \cdot 2^2 x_0 &= 2^3 x_0 \\
&\vdots &\vdots &\vdots \\
x_n &= 2x_{n-1} &= 2 \cdot 2^{n-1} x_0 &= 2^n x_0.
\end{aligned}
$$

Therefore, we have a solution for the difference equation, $x_n = 2^n x_0$ in terms of the initial condition x_0. For this difference equation, the initial condition is sufficient to completely describe the future behavior of x. Furthermore, we were given the initial condition $x_0 = 1$, so we can simplify our final solution to get $x_n = 2^n$, an exponential function!

○○

We can use a similar process to find solutions of any difference equation that has the form:

$$ x_n = Ax_{n-1} \qquad \text{or} \qquad x_n = x_{n-1} + B, $$

where A and B are constants. Let's look at this second case in another example.

Example: Consider the difference equation $x_n = x_{n-1} + 5$ with initial condition $x_0 = 2$. We can follow the same idea of calculating the terms one by one and substituting our previous answer in at each step.

$$
\begin{aligned}
x_1 &= x_0 + 5 \\
x_2 &= x_1 + 5 &= x_0 + 5 + 5 &= x_0 + 5(2) \\
x_3 &= x_2 + 5 &= x_0 + 5(2) + 5 &= x_0 + 5(3) \\
&\vdots &\vdots &\vdots \\
x_n &= x_{n-1} + 5 &= x_0 + 5(n-1) + 5 &= x_0 + 5n.
\end{aligned}
$$

By recognizing that we are simply just adding 5 in each step, we can see that a pattern emerges and our solution is $x_n = x_0 + 5n$. Of course, we were also given the initial condition $x_0 = 2$ and so the solution of the given difference equation is $x_n = 2 + 5n$. We can now use this equation to calculate any and all terms for our model, simply by knowing n.

○○○

So far we have found the solution of the exponential difference equation and of one very special case of the linear difference equation, but we will see that we can find the solution of linear difference equations more generally.

A linear first-order difference equation with constant coefficients can be written as

$$x_n = Ax_{n-1} + B,$$

where A and B are constants. A difference equation of this form with $A \neq 1$ will have general solution

$$x_n = c_1 A^n + c_2,$$

where c_1 and c_2 are constants.

In the event that $A = 1$ we will have the general solution

$$x_n = x_0 + Bn.$$

These constants can be determined by using the initial condition for the difference equation, as we will see in the following example.

○○○

Example: Note that the difference equation for the exponential model is, in fact, just a special case of this linear model. For example, consider the difference equation $x_n = 2x_{n-1}$ with initial condition $x_0 = 25$. This may model, for example, the growth of an invasive species. We found earlier in this section that $x_n = 2^n x_0 = 25(2)^n$, so we should expect that we will obtain the same solution using this procedure.

As this is a linear difference equation with $A = 2$, it will have a general solution:

$$x_n = c_1 2^n + c_2.$$

We need to determine the two constants, c_1, c_2. We should expect, given what we know about the exponential model, that c_1 will be equal to x_0 and c_2 will be zero, but let's see this in action. Finding $x_1 = 2x_0 = 2(25) = 50$, we have that at $n = 0$,

$$x_0 = c_1 2^0 + c_2$$

$$25 = c_1 + c_2$$

and at $n = 1$, we have that

$$x_1 = c_1 2^1 + c_2$$

$$50 = 2c_1 + c_2.$$

Therefore, we have a system of equations:

$$\begin{cases} 25 = c_1 + c_2 \\ 50 = 2c_1 + c_2, \end{cases}$$

from which we can see that $c_1 = 25$ and $c_2 = 0$, as expected; and thus the solution of this difference equation is $x_n = 25(2)^n$.

Therefore, we can see that the solutions we obtain from this method and the rule for exponential models, in particular, do, in fact, align. While it may take fewer calculations if we recognize an exponential difference equation as one, rather than treating it as a more general linear difference equation, we can rest assured that in any event, we have the tools to obtain the solution.

○○○○○○○○○○○○○○○○○○○○○○○○○○○○○○○○○○○○○○○

Example: Consider the difference equation:

$$x_n = 4x_{n-1} + 7, \quad x_0 = 1.$$

This difference equation will have general solution of the form $x_n = c_1 4^n + c_2$ as $A = 4$. Now we need to solve for the two constants. In order to do this, we first compute the first time step, $x_1 = 4(1) + 7 = 11$. We will then use the general solution and the first two time steps to generate a system of equations which can be used to solve for c_1 and c_2.

At $n = 0$, we have that

$$x_0 = c_1 4^0 + c_2$$

$$1 = c_1 + c_2$$

and at $n = 1$, we have that

$$x_1 = c_1 4^1 + c_2$$

$$11 = 4c_1 + c_2.$$

Therefore, we have a system of equations:

$$\begin{cases} 1 = c_1 + c_2 \\ 11 = 4c_1 + c_2. \end{cases}$$

We can solve this system of equations and see that $c_1 = \frac{10}{3}$, and $c_2 = -\frac{7}{3}$, so the solution of this difference equation with initial condition $x_0 = 1$ is $x_n = \frac{10}{3}(4)^n - \frac{7}{3}$.

○○

Example: Let's find the solution of the linear difference equation:

$$x_n = 3x_{n-1} - 5, \quad x_0 = 10.$$

As $A = 3$, we know the solution will have the form $x_n = c_1 3^n + c_2$, where c_1 and c_2 are constants. Using this general solution and the fact that $x_0 = 10$, we have for $n = 0$:

$$x_0 = c_1 3^0 + c_2$$

$$10 = c_1 + c_2.$$

Similarly, we know that $x_1 = 3x_0 - 5 = 3(10) - 5 = 25$ and can use this and $n = 1$ in the general solution to obtain:

$$x_1 = c_1 3^1 + c_2$$

$$25 = 3c_1 + c_2.$$

Finally, we need to solve the system of equations:

$$\begin{cases} 10 = c_1 + c_2 \\ 25 = 3c_1 + c_2, \end{cases}$$

and we arrive at $c_1 = \frac{15}{2}$ and $c_2 = \frac{5}{2}$. Thus, the solution of the difference equation is $x_n = \frac{15}{2}(3)^n + \frac{5}{2}$.

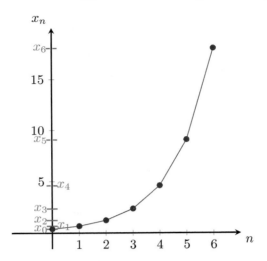

As we can see in the above figure, this sequence grows exponentially and monotonically; that is, it is only increasing and never decreases.

○○○○○○○○○○○○○○○○○○○○○○○○○○○○○○○○○○○○○

Example: Consider the difference equation:

$$x_n = -5x_{n-1} + 2, \quad x_0 = 3.$$

Then, because $A = -5$, we have a general solution:

$$x_n = c_1(-5)^n + c_2.$$

In order to find c_1 and c_2, we can substitute $n = 0$ into this equation to obtain:

$$x_0 \;=\; c_1(-5)^0 + c_2$$

$$3 \;=\; c_1 + c_2.$$

We want to do a similar thing with $n = 1$, but we'll need to first calculate the value of x_1. We have $x_1 = -5x_0 + 2 = -5(3) + 2 = -13$. Then, substituting this into the general solution, we have:

$$x_1 \;=\; c_1(-5)^1 + c_2$$

$$-13 \;=\; -5c_1 + c_2.$$

Putting this together, we need to solve the following system of equations:

$$\begin{cases} 3 \;=\; c_1 \;+\; c_2 \\ -13 \;=\; -5c_1 \;+\; c_2 \end{cases}$$

to get $c_1 = \frac{8}{3}$ and $c_2 = \frac{1}{3}$. Therefore, the solution of the difference equation is
$$x_n = \frac{8}{3}(-5)^n + \frac{1}{3}.$$

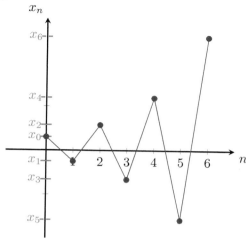

In contrast to the previous example, this sequence is not monotonic and rather, alternates from positive to negative while growing exponentially in magnitude. This is because our difference equation, $x_n = Ax_{n-1} + B$, has $A = -5$ for this example. Negative values of A with $|A| > 1$ result in this type of oscillatory behavior.

○○○

Example: Consider the general Beverton-Holt model:
$$p_n = \frac{kp_{n-1}}{1 + \frac{k-1}{M}p_{n-1}}.$$

Note that this model is certainly not a linear difference equation. However, there is a clever trick which will allow us to use the tools in this section to find the solution of this difference equation. Note that if we take the reciprocal of both sides, we have:
$$\frac{1}{p_n} = \frac{1 + \frac{k-1}{M}p_{n-1}}{kp_{n-1}}$$
$$= \frac{1}{k}\frac{1}{p_{n-1}} + \frac{k-1}{kM}$$

and thus, if we allow $q_n = 1/p_n$, we have the linear difference equation
$$q_n = \frac{1}{k}q_{n-1} + \frac{k-1}{kM}.$$

This has the general solution:
$$q_n = c_1 \left(\frac{1}{k}\right)^n + c_2.$$

Given the initial condition $q_0 = 1/p_0$, we can compute $q_1 = \frac{1}{kp_0} + \frac{k-1}{kM}$. Then we have the system of equations:

$$
\begin{cases}
\dfrac{1}{p_0} = c_1 + c_2 \\[2mm]
\dfrac{1}{kp_0} + \dfrac{k-1}{kM} = \dfrac{1}{k}c_1 + c_2.
\end{cases}
$$

We can solve this system for c_1 and c_2 and we see that:

$$
c_1 = \frac{1}{p_0} - \frac{1}{M}, \qquad c_2 = \frac{1}{M}.
$$

Thus, we have the solution:

$$
q_n = \left(\frac{1}{p_0} - \frac{1}{M} \right) \left(\frac{1}{k} \right)^n + \frac{1}{M}.
$$

Finally, taking the reciprocal, we have:

$$
\begin{aligned}
p_n &= \frac{1}{\left(\dfrac{1}{p_0} - \dfrac{1}{M} \right) \left(\dfrac{1}{k} \right)^n + \dfrac{1}{M}} \\[4mm]
&= \frac{M}{\left(\dfrac{M - p_0}{p_0} \right) \left(\dfrac{1}{k} \right)^n + 1} \\[4mm]
&= \frac{M p_0}{(M - p_0) \left(\dfrac{1}{k} \right)^n + p_0}
\end{aligned}
$$

as a solution of the population model.

○○○

Example: Find a solution of the Beverton-Holt model given the parameters $k = 1.25$, $M = 100$, and an initial condition of $p_0 = 10$.

This gives us the difference equation:

$$
p_n = \frac{1.25 p_{n-1}}{1 + 0.0025 p_{n-1}}, \qquad p_0 = 10.
$$

Following the steps in the preceding example, we see that the solution is:

$$
\begin{aligned}
p_n &= \frac{M p_0}{(M - p_0) \left(\dfrac{1}{k} \right)^n + p_0} \\[4mm]
p_n &= \frac{100(10)}{(100 - 10) \left(\dfrac{1}{1.25} \right)^n + 10} \\[4mm]
p_n &= \frac{1000}{90(0.8)^n + 10}.
\end{aligned}
$$

1.3 Exercises

For the following difference equations, write $x_1, x_2, x_3, \ldots, x_n$ in terms of only x_0, and use these to find the solution of the difference equation.

1. $x_n = 3x_{n-1}, \quad x_0 = -4$

2. $x_n = -5x_{n-1}, \quad x_0 = 2$

3. $x_n = x_{n-1} - 12, \quad x_0 = 5$

4. $x_n = x_{n-1} + 7, \quad x_0 = -3$

Find the solution of the following difference equations.

5. $x_n = 2x_{n-1} + 10, \quad x_0 = -1$

6. $x_n = -3x_{n-1} - 1, \quad x_0 = 2$

7. $x_n = -8x_{n-1} + 2, \quad x_0 = 1$

8. $x_n = 6x_{n-1} - 3, \quad x_0 = 0$

9. $x_n = -x_{n-1} + 7, \quad x_0 = -2$

10. $x_n = 5x_{n-1} - 4, \quad x_0 = -3$

11. $x_n = 4 - x_{n-1}, \quad x_0 = 5$

12. $x_n = 15 - 2x_{n-1}, \quad x_0 = 2$

13. Consider the Beverton-Holt model with $k = 1.75$, $M = 500$, and initial condition of $p_0 = 25$.

    (a) Write the corresponding difference equation with initial conditions.

    (b) Find the solution of the difference equation you found.

    (c) Use your solution to compute the exact value for p_5, p_{10}, and p_{20}.

14. Consider the Beverton-Holt model with $k = 1.75$, $M = 500$, and initial condition of $p_0 = 750$.

    (a) Write the corresponding difference equation with initial conditions.

    (b) Find the solution of the difference equation you found.

    (c) Use your solution to compute the exact value for p_5, p_{10}, and p_{20}.

15. Consider the Beverton-Holt model with $k = 1.25$, $M = 500$, and initial condition of $p_0 = 10$.

    (a) Write the corresponding difference equation with initial conditions.

    (b) Find the solution of the difference equation you found.

    (c) Use your solution to compute the exact value for p_5, p_{10}, and p_{20}.

    (d) Use your solution to determine the value of n for which the inequality $|p_n - M| < 1$ holds.

1.4   Solutions of Linear Homogeneous Second-Order Difference Equations

So far, we have only been able to find the solutions of *first-order* linear difference equations. It is natural to wonder about our ability to find the solution of higher order linear difference equations. These are equations of the form:

$$x_n = f(x_0, x_1, \ldots, x_{n-1}).$$

In the Section 1.3, we found the solutions of first-order linear difference equations, in which f was only a function of x_{n-1}. In this section, we will find the solutions of *second-order* linear difference equations, i.e., equations of the form

$$x_n = f(x_{n-2}, x_{n-1}).$$

As we are asking for this to be a linear equation, we can rewrite this as:

$$x_n = Ax_{n-1} + Bx_{n-2} + C$$

or, in standard form,

$$x_n - Ax_{n-1} - Bx_{n-2} = C.$$

Furthermore, we are going to insist that the difference equations in question be *homogeneous*. By this we mean that $C = 0$, so all the difference equations which we will be concerned with in this section will be of the form:

$$x_n - Ax_{n-1} - Bx_{n-2} = 0.$$

Rest assured, there is a well-developed theory for solutions of nonhomogeneous difference equations, and difference equations of order higher than two. However, these cases are beyond the scope of this book and we will not discuss them here.

The approach here is to assume that one solution of this difference equation takes the form $x_n = r^n$ for some constant r. As we have seen, the solution of a linear first-order difference equation takes the form $x_n = c_1 A^n + c_2$, so it is reasonable to expect that solutions of the second-order equations will take a similar form.

If we assume, then, that $x_n = r^n$, we also have that $x_{n-1} = r^{n-1}$ and $x_{n-2} = r^{n-2}$. If we substitute these values into the original difference equation, we have:

$$
\begin{aligned}
x_n - Ax_{n-1} - Bx_{n-2} &= 0 \\
r^n - Ar^{n-1} - Br^{n-2} &= 0 \\
r^{n-2}\left(r^2 - Ar - B\right) &= 0.
\end{aligned}
$$

If $r = 0$, our whole process will have been for nothing as $x_n = r^n = 0$ and every single value in our list of numbers would also be 0. So, assuming $r \neq 0$, we will consider just the equation:

$$r^2 - Ar - B = 0.$$

This equation is called the **characteristic equation** for the difference equation:

$$x_n - Ax_{n-1} - Bx_{n-2} = 0.$$

We know how to find the solutions of this equation, either through factoring or utilizing the quadratic formula. As you might remember, however, a quadratic equation may have two distinct real roots, one repeated real root, or two complex roots which are conjugate. These three cases lead to different general solutions of the difference equation in question, so we will treat them separately in the following subsections.

1.4.1 Distinct Roots

In the case when we have two distinct real roots, we can name the solutions of the characteristic equation r_1, r_2, and so two possible solutions of the difference equation will be $x_n = r_1^n$ and $x_n = r_2^n$. As these two are possible solutions, a linear combination of them will also be a solution, so we take:

$$x_n = c_1 r_1^n + c_2 r_2^n$$

to be the general solution of the difference equation, and all solutions of this difference equation will have this same form.

The linear second-order difference equation

$$x_n - Ax_{n-1} - Bx_{n-2} = 0$$

has characteristic equation $r^2 - Ar - B = 0$, where A and B are constants. If r_1 and r_2 are distinct real roots of the characteristic equation, then

$$x_n = c_1 r_1^n + c_2 r_2^n$$

is the general solution of the difference equation, where c_1 and c_2 are constants.

Example: Let's find the general solution of the difference equation:

$$x_n = 3x_{n-1} - 2x_{n-2}.$$

We first rewrite this in standard form as:

$$x_n - 3x_{n-1} + 2x_{n-2} = 0,$$

which has characteristic equation:

$$r^2 - 3r + 2 = 0.$$

This factors as $(r - 2)(r - 1) = 0$, so we have solutions $r = 1, 2$ of the characteristic equation. Therefore, this difference equation has a general solution $x_n = c_1(2)^n + c_2(1)^n$, and as 1^n is equal to one for all n, we can simplify this solution to $x_n = c_1(2)^n + c_2$.

○○○

Example: Find the general solution of the difference equation:

$$x_n - 18x_{n-1} + 77x_{n-2} = 0.$$

First, notice that this equation is already in standard form. We can write and then factor the characteristic equation as:

$$r^2 - 18r + 77 \quad = \quad 0$$

$$(r - 7)(r - 11) \quad = \quad 0.$$

Thus, $r = 7$ and $r = 11$ are the roots of the characteristic equation. Hence, the general solution of the difference equation $x_n - 18x_{n-1} + 77x_{n-2} = 0$ is $x_n = c_1(7)^n + c_2(11)^n$.

○○○

We can also use this methodology to find solutions of difference equations with initial conditions. While for the exponential, linear, and logarithmic models, having one initial condition was sufficient, we will need two initial conditions to solve what will become a system of equations for our two constants, c_1 and c_2. It should make sense that we would need two initial conditions in any event, as we need to know the last *two* time steps to determine x_n rather than just the last one.

Example: Consider the difference equation:

$$x_n - 2x_{n-1} - 8x_{n-2} = 0$$

with initial conditions $x_0 = 1$ and $x_1 = 3$. The characteristic equation for this difference equation is:

$$r^2 - 2r - 8 = 0,$$

which factors as $(r - 4)(r + 2) = 0$ and has roots $r = 4, -2$. The general solution of this difference equation, therefore, is:

$$x_n = c_1(4)^n + c_2(-2)^n.$$

In order to find the solution for the given initial conditions, we substitute the two initial conditions into the general solution, and will obtain a system of equations which we can then solve to find our two constants, c_1 and c_2. Substituting $x_0 = 1$, we obtain:

$$x_0 = c_1(4)^0 + c_2(-2)^0$$

$$1 = c_1 + c_2,$$

while substituting $x_1 = 3$ yields:

$$x_1 = c_1(4)^1 + c_2(-2)^1$$

$$3 = 4c_1 - 2c_2.$$

Thus, we have obtained the system of equations:

$$\begin{cases} 1 = c_1 + c_2 \\ 3 = 4c_1 - 2c_2, \end{cases}$$

which we can solve to find that $c_1 = \frac{5}{6}$ and $c_2 = \frac{1}{6}$. Therefore, the solution of the difference equation $x_n - 2x_{n-1} - 8x_{n-2} = 0$ with initial conditions $x_0 = 1$ and $x_1 = 3$ is $x_n = \frac{5}{6}(4)^n + \frac{1}{6}(-2)^n$.

○○

Example: Let's find the solution of the difference equation:

$$x_n = 5x_{n-1} - 6x_{n-2}, \qquad x_0 = 1, \quad x_1 = -2.$$

First, notice that in standard form, our equation is:

$$x_n - 5x_{n-1} + 6x_{n-2} = 0.$$

In this case, we can write and factor our characteristic equation as:

$$r^2 - 5r + 6 = 0$$

$$(r - 2)(r - 3) = 0.$$

Then we have $r = 2, 3$. Hence, the general solution of this difference equation is:

$$x_n = c_1(2)^n + c_2(3)^n.$$

In order to find c_1 and c_2, we can use the given initial conditions. Substituting $x_0 = 1$ into the general solution equation, we see that:

$$x_0 \quad = \quad c_1(2)^0 + c_2(3)^0$$

$$1 \quad = \quad c_1 + c_2,$$

and similarly, by substituting $x_1 = -2$ into the general solution equation, we have:

$$x_1 \quad = \quad c_1(2)^1 + c_2(3)^1$$

$$-2 \quad = \quad 2c_1 + 3c_2.$$

This gives the following system of equations:

$$\begin{cases} 1 \quad = \quad c_1 \quad + \quad c_2 \\ -2 \quad = \quad 2c_1 \quad + \quad 3c_2 \end{cases},$$

which results in $c_1 = 5$ and $c_2 = -4$. Thus, the solution of the difference equation $x_n = 5x_{n-1} - 6x_{n-2}$ with initial conditions $x_0 = 1$ and $x_1 = -2$ is $x_n = 5(2)^n - 4(3)^n$.

○○○

Example: Find the solution of the difference equation:

$$x_n - 3x_{n-1} - 4x_{n-2} = 0, \qquad x_0 = 5, \quad x_1 = 10.$$

As the equation is already in standard form, we can write and solve the characteristic equation:

$$r^2 - 3r - 4 \quad = \quad 0$$

$$(r - 4)(r + 1) \quad = \quad 0,$$

to obtain $r = 4, -1$. Then the general solution of the difference equation is:

$$x_n = c_1(4)^n + c_2(-1)^n.$$

Next we substitute $x_0 = 5$ into this equation to obtain

$$x_0 \quad = \quad c_1(4)^0 + c_2(-1)^0$$

$$5 \quad = \quad c_1 + c_2,$$

and substitute $x_1 = 10$ into the general solution equation to get:

$$x_1 = c_1(4)^1 + c_2(-1)^1$$

$$10 = 4c_1 - c_2.$$

Putting these together, we can now solve the system of equations:

$$\begin{cases} 5 = c_1 + c_2 \\ 10 = 4c_1 - c_2 \end{cases}$$

to get $c_1 = 3$ and $c_2 = 2$. Hence, the solution of the difference equation with the given initial values is $x_n = 3(4)^n + 2(-1)^n$.

1.4.2 Repeated Roots

If the characteristic equation has a repeated root, we have to be a bit more careful. In this case, it would appear that we could simply use the general solution from the previous section. However, the solutions would be equal to each other, so we need to do something a bit more.

Notice that in the event that we have a repeated root, a, our characteristic equation must be able to be factored as $(r-a)(r-a) = 0$, so it must take the form $r^2 - 2ar + a^2 = 0$. This corresponds to an original difference equation $x_n - 2ax_{n-1} + a^2 x_{n-2} = 0$. The solution $x_n = a^n$ will satisfy this equation, but there is another possible solution! The solution $x_n = na^n$ will also satisfy this difference equation. Note that we have:

$$x_n = na^n$$

$$x_{n-1} = (n-1)a^{n-1}$$

$$x_{n-2} = (n-2)a^{n-2},$$

which when substituted into the difference equation gives us:

$$\begin{aligned} x_n - 2ax_{n-1} + a^2 x_{n-2} &= na^n - 2a(n-1)a^{n-1} + a^2(n-2)a^{n-2} \\ &= na^n - 2na^n + 2a^n + na^n - 2a^n \\ &= 0. \end{aligned}$$

Therefore, when the difference equation has a repeated root in the characteristic equation, r, the general solution is $x_n = c_1 r^n + c_2 n r^n$, the linear combination of the two possible roots $x_n = r^n$ and $x_n = nr^n$.

We can include this case in the following way.

The linear second-order difference equation:

$$x_n - Ax_{n-1} - Bx_{n-2} = 0$$

has characteristic equation $r^2 - Ar - B = 0$, where A and B are constants.

If r_1 and r_2 are distinct real roots of the characteristic equation, then:

$$x_n = c_1 r_1^n + c_2 r_2^n$$

is the general solution of the difference equation, where c_1 and c_2 are constants.

If r is a repeated root of the characteristic equation, then:

$$x_n = c_1 r^n + c_2 n r^n$$

is the general solution of the difference equation, where c_1 and c_2 are constants.

Example: Consider the difference equation $x_n = 4x_{n-1} - 4x_{n-2}$. We can write this in standard form as:

$$x_n - 4x_{n-1} + 4x_{n-2} = 0.$$

This has characteristic equation:

$$r^2 - 4r + 4 = 0$$

which factors as $(r-2)(r-2) = 0$. As we can see, this characteristic equation has one repeated root, $r = 2$, so the general solution of this difference equation is $x_n = c_1(2)^n + c_2 n(2)^n$.

○○

Example: As before, let's consider another difference equation which has initial conditions. Consider the difference equation:

$$x_n + 6x_{n-1} + 9x_{n-2} = 0$$

with initial conditions $x_0 = 4$ and $x_1 = 3$. This difference equation, happily, is already in standard form, so we can quickly see that the characteristic equation is:

$$r^2 + 6r + 9 = 0.$$

This factors as $(r+3)(r+3) = 0$, so has the repeated root $r = -3$. Therefore, the general solution is:

$$x_n = c_1(-3)^n + c_2 n(-3)^n.$$

Now, we can use the initial conditions to solve for c_1 and c_2. Substituting $x_0 = 4$, we obtain:

$$x_0 = c_1(-3)^0 + c_2(0)(-3)^0$$

$$4 = c_1$$

while substituting $x_1 = 3$ yields:

$$x_1 = c_1(-3)^1 + c_2(1)(-3)^1$$

$$3 = -3c_1 - 3c_2.$$

Thus, we have obtained the system of equations:

$$\begin{cases} 4 = c_1 \\ 3 = -3c_1 - 3c_2. \end{cases}$$

We already know that $c_1 = 4$ and can solve to find $c_2 = -5$. Therefore, the difference equation $x_n + 6x_{n-1} + 9x_{n-2} = 0$ with the initial conditions $x_0 = 4$ and $x_1 = 3$ has the solution given by $x_n = 4(-3)^n - 5n(-3)^n$.

○○○○○○○○○○○○○○○○○○○○○○○○○○○○○○○○○○○○○○○

Example: Consider the difference equation:

$$x_n = 8x_{n-1} - 16x_{n-2}, \qquad x_0 = 3, \quad x_1 = -8.$$

We first need to put this into standard form:

$$x_n - 8x_{n-1} + 16x_{n-2} = 0.$$

Then we can write and factor the characteristic equation:

$$r^2 - 8r + 16 = 0$$

$$(r - 4)^2 = 0,$$

which has repeated root $r = 4$. Then the general solution of the difference equation is:

$$x_n = c_1(4)^n + c_2 n(4)^n.$$

To solve for c_1 and c_2, we once again substitute the initial conditions into this general solution. Substituting $x_0 = 3$, we have:

$$x_0 = c_1(4)^0 + c_2(0)(4)^0$$

$$3 = c_1.$$

Substituting this value of $c_1 = 3$ and the other initial condition of $x_1 = -8$ into the general solution equation, we have:

$$x_1 = c_1(4)^1 + c_2(1)(4)^1$$

$$-8 = 3(4) + 4c_2$$

$$-8 = 12 + 4c_2$$

$$-20 = 4c_2$$

$$-5 = c_2.$$

Thus $c_2 = -5$, and the solution of the difference equation $x_n = 8x_{n-1} - 16x_{n-2}$ with initial conditions $x_0 = 3$ and $x_1 = -8$ is $x_n = 3(4)^n - 5n(4)^n$.

○○

Example: Consider the difference equation:

$$x_n + 2x_{n-1} + x_{n-2} = 0, \qquad x_0 = -6, \quad x_1 = 2.$$

Note that this is already in standard form. Then the characteristic equation is:

$$r^2 + 2r + 1 = 0$$

$$(r+1)^2 = 0,$$

which has repeated root $r = -1$. Thus the general solution of the difference equation is:

$$x_n = c_1(-1)^n + c_2 n(-1)^n.$$

Substituting the value $x_0 = -6$ into this equation, we obtain:

$$x_0 = c_1(-1)^0 + c_2(0)(-1)^0$$

$$-6 = c_1.$$

Then we can substitute both $c_1 = -6$ and $x_1 = 2$ into the general solution equation to obtain:

$$
\begin{aligned}
x_1 &= c_1(-1)^1 + c_2(1)(-1)^1 \\
2 &= -6(-1) - c_2 \\
2 &= 6 - c_2 \\
-4 &= -c_2 \\
4 &= c_2.
\end{aligned}
$$

Hence, the solution of the difference equation with the given initial conditions is $x_n = -6(-1)^n + 4n(-1)^n$.

○○

Now we are only left with the case where the roots of the characteristic equation are complex.

1.4.3 Complex Roots

If the characteristic equation has a complex root, we have a bit more work to do. In this case, the difference equation $x_n - Ax_{n-1} - Bx_{n-2} = 0$ has characteristic equation $r^2 - Ar - B = 0$, as before, but now the roots of the characteristic equation take the form $x = a \pm bi$. Naively, we may think that we have two distinct solutions, so the general solution should be:
$$x_n = c_1(a + bi)^n + c_2(a - bi)^n.$$
This is the right idea, but unfortunately we now have a complex-valued function as our general solution; that won't do!

The way to realize this general solution in terms of real-valued functions is to rewrite the complex values $a + bi$ and $a - bi$ as $Re^{i\theta}$ and $Re^{-i\theta}$ where $R = \sqrt{a^2 + b^2}$ and $\theta = \tan^{-1}\left(\frac{b}{a}\right)$. These values, R and θ, correspond to the distance from the origin and the angle off the real axis of the complex number $a + bi$. Using these, we can rewrite our general solution as

$$
\begin{aligned}
x_n &= c_1(a + bi)^n + c_2(a - bi)^n \\
x_n &= c_1(Re^{i\theta})^n + c_2(Re^{-i\theta})^n \\
x_n &= c_1 R^n e^{i\theta n} + c_2 R^n e^{-i\theta n} \\
x_n &= c_1 R^n (\cos\theta n + i\sin\theta n) + c_2 R^n (\cos\theta n - i\sin\theta n) \\
x_n &= R^n [C_1 \cos\theta n + C_2 \sin\theta n].
\end{aligned}
$$

Note that in the last step, we combined the constants c_1 and c_2 to create new constants $C_1 = c_1 + c_2$ and $C_2 = ic_1 - ic_2$. While it may not be obvious, if we substitute $n = 0$ and $n = 1$ into our original equation for the general solution, we can solve for c_1 and c_2 in terms of x_0 and x_1, and it turns out that C_1 and C_2 are real-valued constants. Therefore, we have the following characterization of the general solution for a linear homogeneous second-order difference equation.

The linear second-order difference equation $x_n - Ax_{n-1} - Bx_{n-2} = 0$ has characteristic equation $r^2 - Ar - B = 0$, where A and B are constants.

- If r_1 and r_2 are distinct real roots of the characteristic equation, then the general solution is: $x_n = c_1 r_1^n + c_2 r_2^n$.

- If r is a repeated root of the characteristic equation, then the general solution is: $x_n = c_1 r^n + c_2 n r^n$.

- If $r = a \pm bi$ are complex roots of the characteristic equation, then the general solution is: $x_n = R^n[c_1 \cos \theta n + c_2 \sin \theta n]$, where $R = \sqrt{a^2 + b^2}$ and $\theta = \tan^{-1}\left(\frac{b}{a}\right)$.

Example: Consider the difference equation:

$$x_n + 4x_{n-2} = 0.$$

This has characteristic equation:

$$r^2 + 4 = 0.$$

Solving, we obtain $r^2 = -4$, so $r = \pm 2i$. When written in $a \pm bi$ form, these are $r = 0 \pm 2i$. To find the general solution, we compute $R = \sqrt{0^2 + 2^2} = 2$.

We now need to compute $\tan^{-1}\left(\frac{2}{0}\right)$, which seems to be hopeless. However, if we think about the location of $2i$ in the complex plane, in which the real axis is the x-axis and the imaginary axis is the y-axis, we realize that it corresponds to the angle $\theta = \frac{\pi}{2}$.

As we now have $R = 2$ and $\theta = \frac{\pi}{2}$, the general solution of the difference equation is $x_n = 2^n[c_1 \cos\left(\frac{\pi n}{2}\right) + c_2 \sin\left(\frac{\pi n}{2}\right)]$.

○○○○○○○○○○○○○○○○○○○○○○○○○○○○○○○○○○○○○○

Example: Let's consider the difference equation:

$$x_n - 2x_{n-1} + 4x_{n-2} = 0, \qquad x_0 = 2, \quad x_1 = 3.$$

This difference equation has characteristic polynomial:

$$r^2 - 2r + 4 = 0$$

which has roots $r = 1 \pm \sqrt{3}i$. To find the general solution, we compute:

$$R = \sqrt{1^2 + \left(\sqrt{3}\right)^2} = 2$$

and

$$\theta = \tan^{-1}\left(\frac{\sqrt{3}}{1}\right) = \frac{\pi}{3}.$$

Therefore, the general solution is:

$$x_n = 2^n \left[c_1 \cos\left(\frac{\pi n}{3}\right) + c_2 \sin\left(\frac{\pi n}{3}\right)\right].$$

Substituting our two initial conditions into this general solution, we obtain the system of equations:

$$\begin{cases} x_0 &= 2^0[c_1 \cos(0) + c_2 \sin(0)] \\ x_1 &= 2^1[c_1 \cos\left(\frac{\pi}{3}\right) + c_2 \sin\left(\frac{\pi}{3}\right)] \end{cases}$$

which simplifies to

$$\begin{cases} 2 &= c_1 \\ 3 &= c_1 + \sqrt{3}c_2. \end{cases}$$

This system of equations can be solved to obtain $c_1 = 2$ and $c_2 = \frac{1}{\sqrt{3}}$. Therefore, the solution of the difference equation $x_n - 2x_{n-1} + 4x_{n-2} = 0$ with initial conditions $x_0 = 2$ and $x_1 = 3$ is $x_n = 2^n \left[2 \cos\left(\frac{\pi n}{3}\right) + \frac{1}{\sqrt{3}} \sin\left(\frac{\pi n}{3}\right)\right]$.

○○

Example: Let's consider the difference equation:

$$x_n - 4x_{n-1} + 13x_{n-2} = 0, \qquad x_0 = 5, \quad x_1 = 4.$$

This difference equation has characteristic polynomial:

$$r^2 - 4r + 13 = 0$$

which has roots $r = 2 \pm 3i$. Next we need to find R and θ. In this case, we have:

$$R = \sqrt{2^2 + 3^2} = \sqrt{13}$$

and

$$\theta = \tan^{-1}\left(\frac{3}{2}\right) \approx 0.9828.$$

Unfortunately, our value for θ is approximately 0.9828 radians, which is not a common angle measurement. Calculating the sine and cosine of this angle may not be pretty. Fortunately, there is an easier way. Let's leave $\theta = \tan^{-1}\left(\frac{3}{2}\right)$ for now. Then the general solution of the difference equation is:

$$x_n = \left(\sqrt{13}\right)^n \left[c_1 \cos\left(\tan^{-1}\left(\frac{3}{2}\right)n\right) + c_2 \sin\left(\tan^{-1}\left(\frac{3}{2}\right)n\right)\right].$$

Notice that when $n = 0$, we have:

$$x_0 = \left(\sqrt{13}\right)^0 \left[c_1 \cos\left(\tan^{-1}\left(\frac{3}{2}\right)\cdot 0\right) + c_2 \sin\left(\tan^{-1}\left(\frac{3}{2}\right)\cdot 0\right)\right]$$

$$5 = c_1,$$

and similarly, when $n = 1$ we have:

$$x_1 = \left(\sqrt{13}\right)^1 \left[c_1 \cos\left(\tan^{-1}\left(\frac{3}{2}\right)\cdot 1\right) + c_2 \sin\left(\tan^{-1}\left(\frac{3}{2}\right)\cdot 1\right)\right]$$

$$4 = \sqrt{13}\left[c_1 \cos\left(\tan^{-1}\left(\frac{3}{2}\right)\right) + c_2 \sin\left(\tan^{-1}\left(\frac{3}{2}\right)\right)\right]$$

$$4 = \sqrt{13}\left[\frac{2}{\sqrt{13}}c_1 + \frac{3}{\sqrt{13}}c_2\right]$$

$$4 = 2c_1 + 3c_2.$$

Note that we can find $\cos\left(\tan^{-1}\left(\frac{3}{2}\right)\right) = \frac{2}{\sqrt{13}}$ and $\sin\left(\tan^{-1}\left(\frac{3}{2}\right)\right) = \frac{3}{\sqrt{13}}$ by drawing a triangle.

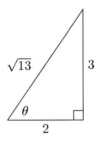

Also notice, that when given x_0 and x_1, as in this question and others, we will always obtain $x_0 = c_1$ and $x_1 = ac_1 + bc_2$. This is not a coincidence and is directly related to the nature of the solution and the trigonometry involved.

Putting this together, we need to solve the system of equations:

$$\begin{cases} 5 = c_1 \\ 4 = 2c_1 + 3c_2, \end{cases}$$

and this results in $c_1 = 5$ and $c_2 = -2$. Thus, the solution of the difference equation $x_n - 4x_{n-1} + 13x_{n-2} = 0$ with initial conditions $x_0 = 5$ and $x_1 = 4$ is $x_n = \left(\sqrt{13}\right)^n \left[5\cos\left(\tan^{-1}\left(\frac{3}{2}\right)n\right) - 2\sin\left(\tan^{-1}\left(\frac{3}{2}\right)n\right)\right]$.

○○○○○○○○○○○○○○○○○○○○○○○○○○○○○○○○○○○○○○○

Example: Consider the difference equation:

$$x_n = 8x_{n-1} - 17x_{n-2}, \qquad x_0 = 4, \quad x_1 = 10.$$

First, we write the difference equation in standard form:

$$x_n - 8x_{n-1} + 17x_{n-2} = 0.$$

Then the characteristic polynomial:

$$r^2 - 8r + 17 = 0$$

has roots $r = 4 \pm i$.

We calculate:

$$R = \sqrt{4^2 + 1^2} = \sqrt{17}$$

and

$$\theta = \tan^{-1}\left(\frac{1}{4}\right).$$

Thus, the general solution of the difference equation is:

$$x_n = \left(\sqrt{17}\right)^n \left[c_1 \cos\left(\tan^{-1}\left(\frac{1}{4}\right)n\right) + c_2 \sin\left(\tan^{-1}\left(\frac{1}{4}\right)n\right)\right].$$

From the previous example, we know that $x_0 = c_1$ and $x_1 = ac_1 + bc_2$, where $a \pm bi$ were the roots of the characteristic equation. In this case, we need to solve the system of equations:

$$\begin{cases} 4 = c_1 \\ 10 = 4c_1 + c_2, \end{cases}$$

and this results in $c_1 = 4$ and $c_2 = -6$. Thus, the solution of the difference equation $x_n = 8x_{n-1} - 17x_{n-2}$ with initial conditions $x_0 = 4$ and $x_1 = 10$ is $x_n = \left(\sqrt{17}\right)^n \left[4\cos\left(\tan^{-1}\left(\frac{1}{4}\right)n\right) - 6\sin\left(\tan^{-1}\left(\frac{1}{4}\right)n\right)\right]$.

1.4 Exercises

Find the general solution of the following difference equations.

1. $x_n - x_{n-1} - 20x_{n-2} = 0$
2. $x_n - 5x_{n-1} - 14x_{n-2} = 0$

3. $x_n = 10x_{n-1} - 25x_{n-2}$
4. $x_n - 4x_{n-1} + 8x_{n-2} = 0$

5. $x_n - 4x_{n-1} + 16x_{n-2} = 0$
6. $x_n = 16x_{n-1} - 64x_{n-2}$

7. $x_n - 120x_{n-1} + 2000x_{n-2} = 0$
8. $x_n - 2x_{n-1} + 5x_{n-2} = 0$

Find the solution of the following difference equations with initial conditions.

9. $x_n = 11x_{n-1} - 24x_{n-2}$, $x_0 = 17$, $x_1 = 6$

10. $x_n - 6x_{n-1} + 18x_{n-2} = 0$, $x_0 = -2$, $x_1 = 9$

11. $x_n = 6x_{n-1} - 5x_{n-2}$, $x_0 = -1$, $x_1 = -13$

12. $x_n + 3x_{n-1} - 28x_{n-2} = 0$, $x_0 = 5$, $x_1 = 9$

13. $x_n + 6x_{n-1} + 9x_{n-2} = 0$, $x_0 = 1$, $x_1 = 9$

14. $x_n = 2\sqrt{3}x_{n-1} - 4x_{n-2}$, $x_0 = 5$, $x_1 = 4\sqrt{3}$

15. $x_n + 25x_{n-2} = 0$, $x_0 = 4$, $x_1 = 1$

16. $x_n - 20x_{n-1} + 100x_{n-2} = 0$, $x_0 = -3$, $x_1 = 10$

17. $x_n - 14x_{n-1} + 40x_{n-2} = 0$, $x_0 = 7$, $x_1 = -2$

18. $x_n = 6x_{n-1} - 34x_{n-2}$, $x_0 = -4$, $x_1 = 23$

1.5 Systems of Difference Equations: Fixed Points and Stability

Thus far we have focused only on single difference equations which describe the value of a variable at different time steps. In this section, we will explore *systems* of difference equations, which describe the value of multiple variables at different time steps. These types of models can be extremely helpful, for example, for modeling how two species interact.

In general, we have two equations:

$$\begin{cases} x_n &= f(x_{n-1}, y_{n-1}) \\ y_n &= g(x_{n-1}, y_{n-1}) \end{cases}$$

Our goal is to find the equilibrium solutions (fixed points). We do this in a similar way as finding the fixed points of one difference equation. We simply solve $x = f(x, y)$ and $y = g(x, y)$ simultaneously.

ooooooooooooooooooooooooooooooooooooooo

Example: Let's find the fixed point(s) of the below system:

$$\begin{cases} x_n &=& y_{n-1}^2 \\ y_n &=& \dfrac{x_{n-1} - 3}{2}. \end{cases}$$

To find the fixed points, we drop the subscripts from the original equations and set $x = y^2$ and $y = \frac{x-3}{2}$. Then, substituting y^2 for x in the second equation, we have:

$$y = \frac{y^2 - 3}{2}$$

$$2y = y^2 - 3$$

$$0 = y^2 - 2y - 3$$

$$0 = (y - 3)(y + 1).$$

Therefore, the fixed points are at $y = -1, 3$. However, as we are working with a system of difference equations, our fixed points will be coordinates in the xy-plane, so we need to find the corresponding x-values. As $x = y^2$, we can see that the fixed points are $(1, -1)$ and $(9, 3)$.

ooooooooooooooooooooooooooooooooooooooo

Example: Consider the system:

$$\begin{cases} x_n &=& 2x_{n-1} - y_{n-1} \\ y_n &=& y_{n-1} - x_{n-1} + 2. \end{cases}$$

To find the fixed point(s), we drop the subscripts to get $x = 2x - y$ and $y = y - x + 2$. This second equation simplifies to $x = 2$. Substituting this into the first equation, we have $2 = 2(2) - y$, so that $y = 2$. Thus the fixed point is $(2, 2)$.

ooooooooooooooooooooooooooooooooooooooo

Example: Let's find the fixed point(s) of the system:

$$\begin{cases} x_n &=& -x_{n-1} + y_{n-1} \\ y_n &=& x_{n-1}^2 - 3. \end{cases}$$

Dropping the subscripts, we have $x = -x+y$ and $y = x^2 - 3$. The first equation simplifies to $y = 2x$. Substituting this into the second equation, we have:

$$
\begin{aligned}
y &= x^2 - 3 \\
2x &= x^2 - 3 \\
0 &= x^2 - 2x - 3 \\
0 &= (x-3)(x+1).
\end{aligned}
$$

Hence $x = 3, -1$. Remembering that $y = 2x$, we arrive at the fixed points of $(3, 6)$, and $(-1, -2)$.

○○○○○○○○○○○○○○○○○○○○○○○○○○○○○○○○○○○○○○

Now that we have found these fixed points, the natural question to ask is whether or not they are stable. We will need to take a slightly different approach to finding stability than when we only had one difference equation.

1.5.1 The Eigenvalue Approach

The first approach is the eigenvalue approach. In this method, we find the **Jacobian matrix** for the system of difference equations, then find the **eigenvalues** of this matrix.

Goal: Determine if the equilibrium of a system of difference equations is stable or unstable.

(1) Find the **Jacobian matrix**:

$$
J = \begin{pmatrix}
\dfrac{\partial f}{\partial x}(x,y) & \dfrac{\partial f}{\partial y}(x,y) \\[2ex]
\dfrac{\partial g}{\partial x}(x,y) & \dfrac{\partial g}{\partial y}(x,y)
\end{pmatrix}
$$

(2) Substitute the fixed point into the matrix J.

(3) Find the **eigenvalues** λ of the new matrix J. These are solutions of $\det(J - \lambda I) = 0$.

(4) Use the following theorem to determine stability. The equilibrium is:

- stable if all eigenvalues of J have magnitude < 1
- unstable if one or more eigenvalue has magnitude > 1

Note that the entries in J are found by computing partial derivatives, which is similar to performing implicit differentiation. Remember that to find $\frac{\partial f}{\partial x}(x,y)$, for example, we treat y as a constant and take the derivative of $f(x,y)$ with respect to x. Sometimes the partial derivative will still be a function of either x and/or y, depending on the original difference equations. Of course, once we substitute the fixed point into J, we should have a real-valued matrix.

Also, recall that the determinant of a 2×2 matrix $A = \begin{pmatrix} a & b \\ c & d \end{pmatrix}$ is

$$\det(A) = \begin{vmatrix} a & b \\ c & d \end{vmatrix} = ad - bc.$$

oooooooooooooooooooooooooooooooooooooooooo

Example: Let's determine the stability of the equilibrium of the system of difference equations:

$$\begin{cases} x_n &= x_{n-1} - y_{n-1} \\ y_n &= \frac{3}{4}x_{n-1} - y_{n-1}. \end{cases}$$

To find the equilibrium, we set $x = x - y$, which gives $y = 0$. Substituting this into $y = \frac{3}{4}x - y$, we get $x = 0$. Hence, we have the equilibrium $(0,0)$.

Next, we use $f(x,y) = x - y$ and $g(x,y) = \frac{3}{4}x - y$ to find the Jacobian:

$$J = \begin{pmatrix} 1 & -1 \\ \frac{3}{4} & -1 \end{pmatrix}.$$

At this point, we will typically substitute our equilibrium $(0,0)$ into the Jacobian matrix. However, our Jacobian simply consists of constants, not functions, so this is unnecessary for this example. Now, we can find the eigenvalues for this matrix by solving the equation $\det(J - \lambda I) = 0$. In this case, we have:

$$\begin{aligned} 0 &= \left| \begin{pmatrix} 1 & -1 \\ \frac{3}{4} & -1 \end{pmatrix} - \begin{pmatrix} \lambda & 0 \\ 0 & \lambda \end{pmatrix} \right| \\ &= \begin{vmatrix} 1-\lambda & -1 \\ \frac{3}{4} & -1-\lambda \end{vmatrix} \\ &= (1-\lambda)(-1-\lambda) + \frac{3}{4} \\ &= \lambda^2 - 1 + \frac{3}{4} \\ &= \lambda^2 - \frac{1}{4} \\ &= (\lambda - \frac{1}{2})(\lambda + \frac{1}{2}). \end{aligned}$$

Hence, $\lambda = \pm\frac{1}{2}$. Because both eigenvalues have $|\lambda| = \frac{1}{2} < 1$, our equilibrium $(0,0)$ is stable.

○○○○○○○○○○○○○○○○○○○○○○○○○○○○○○○○○○○○○○○

Example: Consider the system from an earlier example,

$$
\begin{cases}
x_n &= y_{n-1}^2 \\
y_n &= \dfrac{x_{n-1} - 3}{2}
\end{cases}
$$

In this case, $f(x,y) = y^2$ and $g(x,y) = \dfrac{x-3}{2}$. We find the Jacobian to be

$$
J = \begin{pmatrix} 0 & 2y \\ \frac{1}{2} & 0 \end{pmatrix}.
$$

To find the eigenvalues corresponding to the fixed point $(9,3)$, we first substitute this point into the Jacobian matrix:

$$
J(9,3) = \begin{pmatrix} 0 & 6 \\ \frac{1}{2} & 0 \end{pmatrix},
$$

Then we can find the eigenvalues for this matrix by solving the equation $\det(J - \lambda I) = 0$. Therefore, we set:

$$
\left| \begin{pmatrix} 0 & 6 \\ \frac{1}{2} & 0 \end{pmatrix} - \begin{pmatrix} \lambda & 0 \\ 0 & \lambda \end{pmatrix} \right| = \begin{vmatrix} -\lambda & 6 \\ \frac{1}{2} & -\lambda \end{vmatrix} = 0,
$$

which leads to the equation $\lambda^2 - 3 = 0$. Therefore, the eigenvalues of $J(9,3)$ can be found by solving $\lambda^2 - 3 = 0$, so $\lambda^2 = 3$, and $\lambda = \pm\sqrt{3}$. We can see that the eigenvalues of the Jacobian for $(9,3)$ have a magnitude greater than 1, so $(9,3)$ is an unstable fixed point.

Similarly, we can find the eigenvalues for the fixed point $(1,-1)$. In this case,

$$
J(1,-1) = \begin{pmatrix} 0 & -2 \\ \frac{1}{2} & 0 \end{pmatrix},
$$

so to find the eigenvalues we are solving the following equation:

$$
\left| \begin{pmatrix} 0 & -2 \\ \frac{1}{2} & 0 \end{pmatrix} - \begin{pmatrix} \lambda & 0 \\ 0 & \lambda \end{pmatrix} \right| = \begin{vmatrix} -\lambda & -2 \\ \frac{1}{2} & -\lambda \end{vmatrix} = 0.
$$

Hence, the eigenvalues will be the roots of the equation $\lambda^2 + 1 = 0$, so $\lambda^2 = -1$, and $\lambda = \pm i$. The magnitude of the eigenvalues for $J(1,-1)$ are equal to 1, and this test is inconclusive. To determine stability in this case, we would need to use another method such as plotting the solutions or sketching the phase plane.

○○

Now that we know all the tools required to determine the stability of a system of difference equations using eigenvalues, let's consider another example, and progress through the entire process. We will first find the fixed point(s), then we will find the corresponding Jacobian matrix for each one, and then we will find the eigenvalues. We will conclude by using the stability theorem to determine stability based off of the eigenvalues of the Jacobian.

○○○

Example: Consider the system of difference equations:

$$\begin{cases} x_n &= 6x_{n-1} + 2y_{n-1} - 5 \\ y_n &= 3x_{n-1} + y_{n-1} \end{cases}.$$

The first step is to find the fixed point(s) by solving the following system of equations:

$$\begin{cases} x &= 6x + 2y - 5 \\ y &= 3x + y \end{cases}.$$

The second equation implies that $x = 0$, so substituting that value into the first equation, we see that:

$$x = 6x + 2y - 5$$

$$0 = 6(0) + 2y - 5$$

$$0 = 2y - 5$$

$$5 = 2y$$

$$\frac{5}{2} = y.$$

Therefore, the fixed point for this system of difference equations is $(0, \frac{5}{2})$. The next step is to find the Jacobian for this system. We can see that:

$$J = \begin{pmatrix} 6 & 2 \\ 3 & 1 \end{pmatrix},$$

and therefore,

$$J\left(0, \frac{5}{2}\right) = \begin{pmatrix} 6 & 2 \\ 3 & 1 \end{pmatrix}$$

as well. Solving the equation $\det(J - \lambda I) = 0$, we have:

$$
\begin{aligned}
0 &= \begin{vmatrix} 6 - \lambda & 2 \\ 3 & 1 - \lambda \end{vmatrix} \\
&= (6 - \lambda)(1 - \lambda) - 6 \\
&= \lambda^2 - 7\lambda \\
&= \lambda(\lambda - 7).
\end{aligned}
$$

Therefore, the eigenvalues of $J(0, \frac{5}{2})$ are $\lambda = 0, 7$. By the stability theorem, as one of the eigenvalues has magnitude greater than 1, the fixed point $(0, \frac{5}{2})$ is unstable.

○○

Example: Let's consider the system of difference equations:

$$
\begin{cases} x_n &= y_{n-1} + 1 \\ y_n &= x_{n-1}^2 + y_{n-1} - 4 \end{cases}.
$$

To find the fixed point(s), we drop the subscripts and solve the following system of equations:

$$
\begin{cases} x &= y + 1 \\ y &= x^2 + y - 4 \end{cases}.
$$

Rearranging and solving the second equation, we have:

$$
\begin{aligned}
y &= x^2 + y - 4 \\
0 &= x^2 - 4 \\
0 &= (x - 2)(x + 2).
\end{aligned}
$$

Thus, $x = \pm 2$. Substituting these values into the other equation, $x = y + 1$, we obtain the fixed points $(2, 1)$ and $(-2, -3)$. We'll examine each of these fixed points separately in a moment.

Next, we need to find the Jacobian matrix. For our system of difference equations, we have:

$$
J = \begin{pmatrix} 0 & 1 \\ 2x & 1 \end{pmatrix}.
$$

For the fixed point $(2, 1)$, we have:

$$J(2, 1) = \begin{pmatrix} 0 & 1 \\ 4 & 1 \end{pmatrix}.$$

Then, solving the equation $\det(J - \lambda I) = 0$, we obtain:

$$\begin{aligned} 0 &= \begin{vmatrix} -\lambda & 1 \\ 4 & 1 - \lambda \end{vmatrix} \\ &= -\lambda(1 - \lambda) - 4 \\ &= \lambda^2 - \lambda - 4, \end{aligned}$$

which has roots $\lambda = \frac{1 \pm \sqrt{17}}{2}$. As both of these eigenvalues have magnitude greater than 1, the stability theorem says the fixed point $(2, 1)$ is unstable.

Similarly, we can check the fixed point $(-2, -3)$. We have:

$$J(-2, -3) = \begin{pmatrix} 0 & 1 \\ -4 & 1 \end{pmatrix}.$$

Then the equation $\det(J - \lambda I) = 0$ results in:

$$\begin{aligned} 0 &= \begin{vmatrix} -\lambda & 1 \\ -4 & 1 - \lambda \end{vmatrix} \\ &= -\lambda(1 - \lambda) + 4 \\ &= \lambda^2 - \lambda + 4, \end{aligned}$$

which has roots $\lambda = \frac{1 \pm \sqrt{15}i}{2}$. To determine the magnitude of these complex valued eigenvalues $\lambda = a \pm bi$, we find $|\lambda| = \sqrt{a^2 + b^2}$. In this case, we have magnitude $|\lambda| = \sqrt{(\frac{1}{2})^2 + (\frac{\sqrt{15}}{2})^2} = 2 > 1$. Thus, the fixed point $(-2, -3)$ is unstable by the stability theorem.

1.5.2 The Jury Condition

Rather than use the eigenvalues of the Jacobian to determine stability, an alternative method is to use the **Jury Condition**.[3] For this method, we will need to find the trace of the matrix J, which is the sum of the main diagonal entries. For a 2×2 matrix $A = \begin{pmatrix} a & b \\ c & d \end{pmatrix}$, we have $\text{tr} A = a + d$. Now we can give the statement of the Jury Condition.

---

[3] Jury, E.I. (1963). On the roots of a real polynomial inside the unit circle and a stability criterion for linear discrete systems. *IFAC Proceedings Volumes*, 1(2): 142–153.

The Jury Condition states that if

$$|\mathrm{tr}J| < 1 + \det J < 2,$$

then the fixed point is stable.

However, if $|\mathrm{tr}J| > 1 + \det J$ or if $1 + \det J > 2$, the fixed point is unstable. If any of these inequalities are equalities, the test is inconclusive.

Example: Let's use the Jury Condition to determine the stability of the fixed point for the system:

$$\begin{cases} x_n &= 6x_{n-1} + 2y_{n-1} - 5 \\ \\ y_n &= 3x_{n-1} + y_{n-1}. \end{cases}$$

In a previous example we found the fixed point to be $(0, \frac{5}{2})$, and determined it to be an unstable fixed point. We'll confirm this result using the Jury condition. The Jacobian is:

$$J\left(0, \frac{5}{2}\right) = \begin{pmatrix} 6 & 2 \\ 3 & 1 \end{pmatrix},$$

which has trace $\mathrm{tr}J = 6 + 1 = 7$ and determinant $\det J = 0$. Now, we check to see if the following inequalities are true:

$$|\mathrm{tr}J| \overset{?}{<} 1 + \det J \overset{?}{<} 2$$

$$|7| \overset{?}{<} 1 + 0 \overset{?}{<} 2$$

$$7 \overset{?}{<} 1 \overset{?}{<} 2.$$

As these inequalities are not true, and more specifically because:

$$7 = |\mathrm{tr}J| > 1 + \det J = 1,$$

we can conclude that $(0, \frac{5}{2})$ is an unstable fixed point.

○○

Example: Use the Jury Condition to determine the stability of the fixed point(s) for the following system of difference equations:

$$\begin{cases} x_n &= x_{n-1} - \frac{1}{2}y_{n-1} \\ \\ y_n &= x_{n-1} - y_{n-1}^2 - 1. \end{cases}$$

First, we find the fixed points by dropping the subscripts and solving the system of equations:

$$\begin{cases} x &= x - \frac{1}{2}y \\ y &= x - y^2 - 1. \end{cases}$$

The equation $x = x - \frac{1}{2}y$ gives $y = 0$. Substituting this into the second equation, we obtain $x = 1$. Thus the fixed point of the system of difference equations is $(1, 0)$.

Next, we find the Jacobian matrix:

$$J = \begin{pmatrix} 1 & -\frac{1}{2} \\ 1 & -2y \end{pmatrix}.$$

Substituting the fixed point $(1, 0)$ into this Jacobian, we obtain:

$$J(1, 0) = \begin{pmatrix} 1 & -\frac{1}{2} \\ 1 & 0 \end{pmatrix},$$

which has trace $\operatorname{tr}J = 1 + 0 = 1$ and determinant $\det J = \frac{1}{2}$. Now, we check to see if the following inequalities are true:

$$|\operatorname{tr}J| \overset{?}{<} 1 + \det J \overset{?}{<} 2$$

$$|1| \overset{?}{<} 1 + \frac{1}{2} \overset{?}{<} 2$$

$$1 \overset{?}{<} \frac{3}{2} \overset{?}{<} 2.$$

As both of these inequalities are true, the fixed point $(1, 0)$ is stable.

○○

Example: Consider the system of difference equations:

$$\begin{cases} x_n &= 3y_{n-1} \\ y_n &= x_{n-1}\left(y_{n-1} - \frac{1}{6}\right). \end{cases}$$

To find the fixed points, we must solve the following system of equations:

$$\begin{cases} x &= 3y \\ y &= x\left(y - \frac{1}{6}\right). \end{cases}$$

If we substitute $x = 3y$ into the other equation, we obtain:

$$y = 3y\left(y - \tfrac{1}{6}\right)$$

$$y = 3y^2 - \tfrac{1}{2}y$$

$$0 = 3y^2 - \tfrac{3}{2}y$$

$$0 = 3y\left(y - \tfrac{1}{2}\right).$$

Thus $y = 0$ or $y = \tfrac{1}{2}$. Substituting $y = 0$ into the original equation $x = 3y$, we find a fixed point at $(0,0)$. Similarly, substituting $y = \tfrac{1}{2}$ into $x = 3y$ gives another fixed point at $(\tfrac{3}{2}, \tfrac{1}{2})$.

We will determine the stability for each fixed point using the Jury Condition. First, we must find the Jacobian:

$$J = \begin{pmatrix} 0 & 3 \\ y - \tfrac{1}{6} & x \end{pmatrix}.$$

For the fixed point $(0,0)$, we have:

$$J(0,0) = \begin{pmatrix} 0 & 3 \\ -\tfrac{1}{6} & 0 \end{pmatrix},$$

which has $\det J = \tfrac{1}{2}$ and $\operatorname{tr}J = 0$. We need to see if the Jury Condition holds:

$$|\operatorname{tr}J| \ \overset{?}{<} \ 1 + \det J \ \overset{?}{<} \ 2$$

$$0 \ \overset{?}{<} \ 1 + \tfrac{1}{2} \ \overset{?}{<} \ 2$$

$$0 \ \overset{?}{<} \ \tfrac{3}{2} \ \overset{?}{<} \ 2.$$

As this statement is true, the fixed point $(0,0)$ is stable.

Similarly, we can determine the Jacobian for the fixed point $(\tfrac{3}{2}, \tfrac{1}{2})$:

$$J\left(\frac{3}{2}, \frac{1}{2}\right) = \begin{pmatrix} 0 & 3 \\ \tfrac{1}{3} & \tfrac{3}{2} \end{pmatrix},$$

which has $\det J = -1$ and $\operatorname{tr}J = \tfrac{3}{2}$. Now we'll check stability using the Jury Condition:

$$|\operatorname{tr}J| \ \overset{?}{<} \ 1 + \det J \ \overset{?}{<} \ 2$$

$$\tfrac{3}{2} \ \overset{?}{<} \ 1 - 1 \ \overset{?}{<} \ 2$$

$$\tfrac{3}{2} \ \overset{?}{<} \ 0 \ \overset{?}{<} \ 2.$$

As this statement is false, and more particularly, $|\operatorname{tr}J| = \tfrac{3}{2} \not< 0 = 1 + \det J$, the fixed point $(\tfrac{3}{2}, \tfrac{1}{2})$ is unstable.

1.5.3 Discrete Interacting Species Models

When modeling the interaction of species with a discrete system, one natural approach may be to start with a general system:

$$
\begin{cases}
x_n &= x_{n-1}f(x_{n-1}, y_{n-1}) \\[2mm]
y_n &= y_{n-1}g(x_{n-1}, y_{n-1})
\end{cases}
$$

where the individual equations for the population of species x, x_n, and the population of species y, y_n, can be thought of as a generalized exponential growth model with a variable parameter, f or g, which is dependent on the values of x_{n-1}, y_{n-1}. These are very general models for which the stability of solutions have been carefully analyzed in detail.[4]

In this subsection, we will focus on a slightly more restrictive class of interacting species models: those that take the form:

$$
\begin{cases}
x_n &= x_{n-1}e^{r-ax_{n-1}-by_{n-1}} \\[2mm]
y_n &= y_{n-1}e^{s-cy_{n-1}-dx_{n-1}}.
\end{cases}
$$

While this may not seem like the most obvious model to select, recall the Ricker model from Section 1.1.2:

$$
p_n = e^{r(1-p_{n-1}/M)}p_{n-1}
$$

which we were able to rewrite as

$$
p_n = Re^{-rp_{n-1}/M}p_{n-1},
$$

where r is a growth rate, M is the carrying capacity, and $R = e^r$. Now consider the first equation of this system model and notice that in the absence of species y, we have:

$$
\begin{aligned}
x_n &= x_{n-1}e^{r-ax_{n-1}-by_{n-1}} \\[2mm]
&= x_{n-1}e^{r-ax_{n-1}} \\[2mm]
&= x_{n-1}e^r e^{-ax_{n-1}} \\[2mm]
&= Rx_{n-1}e^{-ax_{n-1}},
\end{aligned}
$$

which is exactly the Ricker model where the r parameters are identically the

---

[4] Hassell, M.P., & Comins, H.N. (1976). Discrete time models for two-species competition. *Theoretical Population Biology*, 9(2), 202–221.

same and the parameter a corresponds with the value r/M in the Ricker model. Likewise, in the absence of species x, the second equation gives us:

$$
\begin{aligned}
y_n &= y_{n-1}e^{s-cy_{n-1}-dx_{n-1}} \\
&= y_{n-1}e^{s-cy_{n-1}} \\
&= y_{n-1}e^{s}e^{-cy_{n-1}} \\
&= Sy_{n-1}e^{-cy_{n-1}},
\end{aligned}
$$

which similarly corresponds with the Ricker model. Notice that we can in fact rewrite our system in the form:

$$
\begin{cases}
x_n &= Rx_{n-1}e^{-ax_{n-1}}e^{-by_{n-1}} \\
y_n &= Sy_{n-1}e^{-cy_{n-1}}e^{-dx_{n-1}}
\end{cases}
$$

where the parameters $R = e^r$ and $S = e^s$ are growth rates, the $e^{-ax_{n-1}}$ and $e^{-cy_{n-1}}$ terms are mortality factors similar to the Ricker model, and the $e^{-by_{n-1}}$ and $e^{-dx_{n-1}}$ terms are interaction terms. The parameters a, b, c, and d can be positive or negative depending on the nature of the relationship between the two species.

○○○○○○○○○○○○○○○○○○○○○○○○○○○○○○○○○○○○○○○

Example: Consider the interacting species model:

$$
\begin{cases}
x_n &= x_{n-1}e^{2.5-0.005x_{n-1}-0.01y_{n-1}}, & x_0 = 50 \\
y_n &= y_{n-1}e^{0.75-0.0075y_{n-1}+0.00375x_{n-1}}, & y_0 = 50.
\end{cases}
$$

This may represent a host-parasite relationship of interacting species. We can see this based off the fact that in the absence of y, we have:

$$
x_n = x_{n-1}e^{2.5-0.005x_{n-1}}
$$

which is a Ricker model for the population of species x with $r = 2.5$ and carrying capacity $M = 2.5/0.005 = 500$ in which the population will grow and stabilize in a cycle around the fixed point $x = 500$. The interaction term $e^{-0.01y_{n-1}}$ has a negative exponent, and thus will *lower* the value of x_n, thus species x suffers in the presence of y. In contrast, in the absence of x, we have:

$$
y_n = y_{n-1}e^{0.75-0.0075y_{n-1}}
$$

which is a Ricker model with $s = 0.75$ in which the population will grow and limit at the fixed point at $y = 100$. The interaction term $e^{0.00375x_{n-1}}$ has a positive exponent and thus will *increase* the value of y_n, thus species y benefits from the presence of x. The population of species y will not die out in the absence of x which is what distinguishes this model from a predator-prey model.

We proceed by finding the fixed points and set:

$$\begin{cases} x &= xe^{2.5-0.005x-0.01y} \\ y &= ye^{0.75-0.0075y+0.00375x}. \end{cases}$$

Considering the first equation, we see that:

$$x = xe^{2.5-0.005x-0.01y}$$

$$0 = xe^{2.5-0.005x-0.01y} - x$$

$$0 = x\left(e^{2.5-0.005x-0.01y} - 1\right).$$

In this case, either $x = 0$ or:

$$0 = e^{2.5-0.005x-0.01y} - 1$$

$$1 = e^{2.5-0.005x-0.01y}$$

$$0 = 2.5 - 0.005x - 0.01y$$

$$0.005x + 0.01y = 2.5$$

$$x + 2y = 500.$$

Similarly, considering the second equation, we have:

$$y = ye^{0.75-0.0075y+0.00375x}$$

$$0 = ye^{0.75-0.0075y+0.00375x} - y$$

$$0 = y\left(e^{0.75-0.0075y+0.00375x} - 1\right).$$

So either $y = 0$ or:

$$0 = e^{0.75 - 0.0075y + 0.00375x} - 1$$

$$1 = e^{0.75 - 0.0075y + 0.00375x}$$

$$0 = 0.75 - 0.0075y + 0.00375x$$

$$-0.00375x + 0.0075y = 0.75$$

$$-x + 2y = 200.$$

Pairing $x = 0$ with $y = 0$, we see that $(0,0)$ is a fixed point of the system.

If we pair $x = 0$ with $-x + 2y = 200$, we see that $(0, 100)$ is a fixed point. This corresponds with the situation where species x dies off and species y reaches its carrying capacity.

Similarly, if we pair $y = 0$ with $x + 2y = 500$, we have the fixed point $(500, 0)$. This corresponds with the situation where species x reaches its carrying capacity and species y dies off.

Finally, to find the last fixed point of the system, we must solve:

$$\begin{cases} x + 2y = 500 \\ -x + 2y = 200. \end{cases}$$

This has solution $x = 150$ and $y = 175$, so the final fixed point is $(150, 175)$.

In summary, we have the fixed points $(0,0)$, $(0, 100)$, $(500, 0)$, and $(150, 175)$. These correspond with mutual die off, species x population dying off while the other reaches carrying capacity, species y population dying off while the other reaches carrying capacity, and an equilibrium of coexistence.

To determine stability we find the Jacobian to be:

$$J = \begin{pmatrix} (1 - 0.005x)e^{2.5 - 0.005x - 0.01y} & -0.01xe^{2.5 - 0.005x - 0.01y} \\ 0.00375ye^{0.75 - 0.0075y + 0.00375x} & (1 - 0.0075y)e^{0.75 - 0.0075y + 0.00375x} \end{pmatrix}.$$

We see that at $(0,0)$, we have:

$$J(0,0) = \begin{pmatrix} e^{2.5} & 0 \\ 0 & e^{0.075} \end{pmatrix}$$

which has $\text{tr} J = e^{2.5} + e^{0.075} \approx 13.26$ and $\det J = e^{2.5}e^{0.075} \approx 13.13$. Thus, as $|\text{tr} J| < 1 + \det J$ but $1 + \det J > 2$, we fail the Jury condition for stability, and can conclude that the fixed point at $(0,0)$ is unstable.

For the fixed point at $(0, 100)$, we have:

$$J(0, 100) = \begin{pmatrix} e^{1.5} & 0 \\ 0.0375 & 0.25 \end{pmatrix}$$

which has $\mathrm{tr}J = e^{1.5} + 0.25 \approx 4.73$ and $\det J = 0.25e^{1.5} \approx 1.12$. Thus, as $|\mathrm{tr}J| > 1 + \det J$ and $1 + \det J > 2$, we fail the Jury condition for stability, and can conclude that the fixed point at $(0, 100)$ is unstable.

For the fixed point at $(500, 0)$, we have:

$$J(500, 0) = \begin{pmatrix} -1.5 & -5 \\ 0 & e^{2.625} \end{pmatrix}$$

which has $\mathrm{tr}J = -1.5 + e^{2.625} \approx 12.30$ and $\det J = -1.5e^{2.625} \approx -20.71$. Thus, as $1 + \det J < 2$ but $|\mathrm{tr}J| > 1 + \det J$, we fail the Jury condition for stability, and can conclude that the fixed point at $(500, 0)$ is unstable.

Finally, for the fixed point at $(150, 175)$ we have:

$$J(150, 175) = \begin{pmatrix} 0.25 & -1.5 \\ 0.65625 & -0.3125 \end{pmatrix}.$$

In this case, we calculate the values $\mathrm{tr}J = 0.25 - 0.3125 = -0.0625$ and $\det J = 0.25(-0.3125) + 1.5(0.65625) = 0.90625$. Thus, the Jury condition is satisfied with $|\mathrm{tr}J| < 1 + \det J < 2$, and we can conclude that $(150, 175)$ is a stable fixed point. In fact, if we examine the plot of the x and y values for this system we can see that the populations of species x and y oscillate towards $x_n = 150$ and $y_n = 175$.

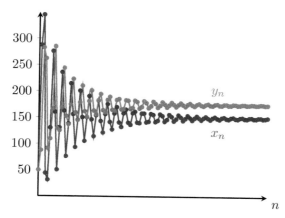

Sometimes it is helpful to visualize the system in the **phase plane**, where rather than graphing x_n and y_n against n, we graph the values for x_n and y_n in the x_n, y_n-plane.

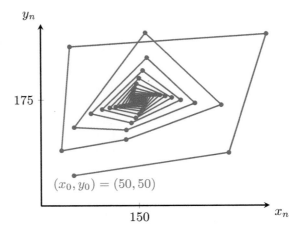

$$(x_0, y_0) = (50, 50)$$

Here we can see that in the phase plane, the populations appear to spiral inward towards the fixed point at $(150, 175)$. This corresponds with the species both reaching a population where they stabilize without cyclic variations long term.

○○○

Example: Consider the interacting species model:

$$\begin{cases} x_n = x_{n-1}e^{2.25-0.0045x_{n-1}-0.00675y_{n-1}}, & x_0 = 225 \\ \\ y_n = y_{n-1}e^{-0.75-0.0075y_{n-1}+0.005625x_{n-1}}, & y_0 = 75. \end{cases}$$

This may represent a predator-prey relationship of interacting species. We can see this based off the fact that in the absence of y, the population of species x will grow and stabilize at the carrying capacity of 500 while in the absence of x, the population of species y will die off. Also, note that the interaction term $e^{-0.00675y_{n-1}}$ has a negative exponent, so species x suffers in the presence of y, while the interaction term $e^{0.005625x_{n-1}}$ has a positive exponent, so species y benefits from the presence of x. In this case, x is the prey and y is the predator.

We proceed by finding the fixed points and set:

$$\begin{cases} x = xe^{2.25-0.0045x-0.00675y} \\ \\ y = ye^{-0.75-0.0075y+0.005625x}. \end{cases}$$

Notice that from the first equation, we have:

$$x = xe^{2.25-0.0045x-0.00675y}$$

$$0 = xe^{2.25-0.0045x-0.00675y} - x$$

$$0 = x\left(e^{2.25-0.0045x-0.00675y} - 1\right).$$

Hence, $x = 0$ or we have:

$$0 = e^{2.25-0.0045x-0.00675y} - 1$$

$$1 = e^{2.25-0.0045x-0.00675y}$$

$$0 = 2.25 - 0.0045x - 0.00675y$$

$$0.0045x + 0.00675y = 2.25$$

$$2x + 3y = 1000.$$

Similarly, notice that the second equation gives:

$$y = ye^{-0.75-0.0075y+0.005625x}$$

$$0 = ye^{-0.75-0.0075y+0.005625x} - y$$

$$0 = y\left(e^{-0.75-0.0075y+0.005625x} - 1\right).$$

So either $y = 0$ or we have:

$$0 = e^{-0.75-0.0075y+0.005625x} - 1$$

$$1 = e^{-0.75-0.0075y+0.005625x}$$

$$0 = -0.75 - 0.0075y + 0.005625x$$

$$-0.005625x + 0.0075y = -0.75$$

$$3x - 4y = 400.$$

Pairing $x = 0$ with $y = 0$, we see that $(0, 0)$ is a fixed point of the system.
    If we pair $x = 0$ with $3x - 4y = 400$, we see that $(0, -100)$ is a fixed point. As we can't have a negative values for a population size, we can discard this fixed point.

If we pair $y = 0$ with $2x + 3y = 1000$, we have the fixed point $(500, 0)$. This corresponds with the situation where species x reaches its carrying capacity and species y dies off.

Finally, to find the last fixed point of the system, we must solve the system of equations:

$$\begin{cases} 2x & + & 3y & = & 1000 \\ \\ 3x & - & 4y & = & 400 \end{cases}$$

which has the solution $(5200/17, 2200/17) \approx (305.88, 129.41)$. Thus we have fixed points $(0, 0)$, $(500, 0)$, and $(305.88, 129.41)$.

To determine stability we find the Jacobian to be

$$\begin{pmatrix} (1 - 0.0045x)e^{2.25-0.0045x-0.00675y} & -0.00675xe^{2.25-0.0045x-0.00675y} \\ 0.005625ye^{-0.75-0.0075y+0.005625x} & (1 - 0.0075y)e^{-0.75-0.0075y+0.005625x} \end{pmatrix}.$$

We see that at $(0, 0)$, we have:

$$J(0, 0) = \begin{pmatrix} e^{2.25} & 0 \\ 0 & e^{-0.75} \end{pmatrix}.$$

Using the eigenvalue approach, we see that solving $\det(J - \lambda I) = 0$, we have:

$$\begin{aligned} 0 &= \begin{pmatrix} e^{2.25} - \lambda & 0 \\ 0 & e^{-0.75} - \lambda \end{pmatrix} \\ &= (e^{2.25} - \lambda)(e^{-0.75} - \lambda) \\ &= \lambda^2 - (e^{2.25} + e^{-0.75})\lambda + e^{1.5} \end{aligned}$$

which has roots $\lambda \approx 9.81, 0.15$. As one of these eigenvalues has a magnitude greater than one, the stability theorem tells us that the fixed point at $(0, 0)$ is unstable.

For the fixed point at $(500, 0)$ we have:

$$J(500, 0) = \begin{pmatrix} -1.25 & -3.375 \\ 0 & e^{2.0625} \end{pmatrix}$$

which has eigenvalues $\lambda \approx 7.87, -1.25$. As both eigenvalues have a magnitude greater than one, the stability theorem tells us that the fixed point at $(500, 0)$ is unstable.

For the fixed point at $(305.88, 129.41)$, we'll use the exact values rather than the rounded values to ensure that we arrive at the correct stability results. We have:

$$J(305.88, 129.41) = \begin{pmatrix} -\frac{32}{85} & -\frac{351}{170} \\ \frac{99}{136} & \frac{1}{34} \end{pmatrix}$$

which has eigenvalues $\lambda \approx -0.17 \pm 1.21i$. To determine the magnitude of these complex valued eigenvalues, we find: $|\lambda| = \sqrt{(-0.17)^2 + (1.21)^2} \approx 1.22$. As this magnitude is greater than one, the stability theorem tells us that the fixed point at $(305.88, 129.41)$ is unstable. To support this, if we look at the graph of the x_n and y_n values against n we can see that they oscillate wildly about the values $x = 305.88$ and $y = 129.41$.

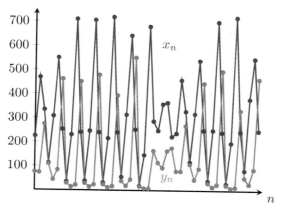

This behavior should feel reasonable if you recall that these models are similar to the Ricker model, and we saw that when $r > 2$ in the Ricker model, the fixed point is unstable and the value of x_n can alternate between two or more points but never grow without bound. Note that in this system, $r = 2.25$, so in the absence of the population y_n we would expect to see this type of behavior. If we examine the phase plane of this system we can see that the populations of species x and y cycle around the fixed point at $(305.88, 129.41)$, but never converge.

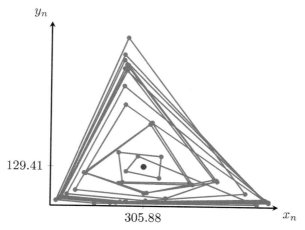

Hence, the stable state for these species is to have a cyclic population, with the predators eating too many prey, then starting to die off due to lack of food. In the absence of predation the prey population starts to grow again, prompting the predator population to increase due to the new bounty of food, only to over hunt, repeating the cycle.

1.5 Exercises

Find the fixed point(s) for the following systems of difference equations.

1.
$$\begin{cases} x_n = 4y_{n-1} - x_{n-1} \\ y_n = x_{n-1} - \frac{1}{2}y_{n-1} - 1 \end{cases}$$

2.
$$\begin{cases} x_n = x_{n-1} + 2y_{n-1} - 6 \\ y_n = x_{n-1}(y_{n-1} - 1) \end{cases}$$

3.
$$\begin{cases} x_n = 2x_{n-1} + 3y_{n-1} - 5 \\ y_n = 4x_{n-1} - y_{n-1} + 1 \end{cases}$$

4.
$$\begin{cases} x_n = y_{n-1} - 3 \\ y_n = x_{n-1}^2 + 1 \end{cases}$$

5.
$$\begin{cases} x_n = -x_{n-1} + y_{n-1} + 1 \\ y_n = y_{n-1} + x_{n-1}^2 - 4 \end{cases}$$

6.
$$\begin{cases} x_n = 5x_{n-1} - 2y_{n-1} \\ y_n = x_{n-1}^2 - y_{n-1} \end{cases}$$

Use the eigenvalue approach to determine the stability of all fixed point(s) for the following systems of difference equations.

7.
$$\begin{cases} x_n = \frac{1}{2}x_{n-1} + 2y_{n-1} \\ y_n = \frac{1}{4}x_{n-1} - y_{n-1} + 2 \end{cases}$$

8.
$$\begin{cases} x_n = \frac{1}{2}x_{n-1} + 1 \\ y_n = x_{n-1} + \frac{3}{4}y_{n-1} \end{cases}$$

9.
$$\begin{cases} x_n = -x_{n-1} + \frac{1}{2}y_{n-1} - 1 \\ y_n = y_{n-1} - x_{n-1} \end{cases}$$

10.
$$\begin{cases} x_n = y_{n-1}^2 \\ y_n = y_{n-1} + \frac{1}{2}x_{n-1} - 2 \end{cases}$$

11.
$$\begin{cases} x_n = x_{n-1} + y_{n-1} \\ y_n = x_{n-1}(y_{n-1} - \frac{1}{9}) \end{cases}$$

12.
$$\begin{cases} x_n = \frac{1}{2}y_{n-1}^2 \\ y_n = x_{n-1} - \frac{1}{2}y_{n-1} \end{cases}$$

Use the Jury Condition to determine the stability of all fixed point(s) for the following systems of difference equations.

13. $\begin{cases} x_n = x_{n-1} + y_{n-1} \\ y_n = y_{n-1}^2 - \frac{1}{2}x_{n-1} + 1 \end{cases}$

14. $\begin{cases} x_n = \frac{1}{3}y_{n-1} - x_{n-1} \\ y_n = y_{n-1} - x_{n-1}^2 + 4 \end{cases}$

15. $\begin{cases} x_n = \frac{1}{3}x_{n-1}y_{n-1} \\ y_n = x_{n-1} + y_{n-1} - 2 \end{cases}$

16. $\begin{cases} x_n = x_{n-1} - \frac{1}{2}y_{n-1} \\ y_n = \frac{1}{3}x_{n-1} - y_{n-1} \end{cases}$

17. $\begin{cases} x_n = \frac{1}{9}y_{n-1}^2 - 1 \\ y_n = y_{n-1} - x_{n-1} \end{cases}$

18. $\begin{cases} x_n = y_{n-1}(x_{n-1} - 2) \\ y_n = x_{n-1} - 2 \end{cases}$

19. Consider the interacting species model

$$\begin{cases} x_n = x_{n-1}e^{1.75 - 0.0035x_{n-1} - 0.007y_{n-1}} \\ y_n = y_{n-1}e^{-0.5 - 0.005y_{n-1} + 0.01x_{n-1}} \end{cases}$$

(a) Describe the nature of the relationship between the two species.

(b) Find the fixed points of the system.

(c) Determine the stability of the fixed points using either the eigenvalue approach or the Jury condition.

(d) Graph the first 50 time steps of this system with initial conditions $x_0 = 100$, $y_0 = 75$ with x_n and y_n versus n, and sketch the phase plane.

20. Consider the interacting species model

$$\begin{cases} x_n = x_{n-1}e^{1.75 - 0.007x_{n-1} - 0.014y_{n-1}} \\ y_n = y_{n-1}e^{1.5 - 0.006y_{n-1} - 0.0105x_{n-1}} \end{cases}$$

(a) Describe the nature of the relationship between the two species.

(b) Find the fixed points of the system.

(c) Determine the stability of the fixed points using either the eigenvalue approach or the Jury condition.

(d) Graph the first 50 time steps of this system with initial conditions $x_0 = 50$, $y_0 = 25$ with x_n and y_n versus n, and sketch the phase plane.

21. Consider the interacting species model

$$\begin{cases} x_n &=& x_{n-1}e^{-0.25-0.01x_{n-1}+0.025y_{n-1}} \\ y_n &=& y_{n-1}e^{0.1-0.0004y_{n-1}+0.001x_{n-1}} \end{cases}$$

(a) Describe the nature of the relationship between the two species.

(b) Find the fixed points of the system.

(c) Determine the stability of the fixed points using either the eigenvalue approach or the Jury condition.

(d) Graph the first 50 time steps of this system with initial conditions $x_0 = 175$, $y_0 = 150$ with x_n and y_n versus n, and sketch the phase plane.

22. Consider the interacting species model

$$\begin{cases} x_n &=& x_{n-1}e^{0.75-0.0075x_{n-1}-0.01125y_{n-1}} \\ y_n &=& y_{n-1}e^{1.5-0.015y_{n-1}-0.01125x_{n-1}} \end{cases}$$

(a) Describe the nature of the relationship between the two species.

(b) Find the fixed points of the system.

(c) Determine the stability of the fixed points using either the eigenvalue approach or the Jury condition.

(d) Graph the first 50 time steps of this system with initial conditions $x_0 = 50$, $y_0 = 12$ with x_n and y_n versus n, and sketch the phase plane.

1.6　Age-Structured Leslie Matrix Models

We have seen a few models of population dynamics thus far, but they have all had one thing in common. Each model assumes members within the population all behave the same way, regardless of various factors such as age or gender. However, there are often a lot of dynamics occurring within populations. Juveniles are born, they reach sexual maturity, adults reproduce, and members of the population from all age classes, inevitably, die. This dynamic is obscured by treating all members of the species as one class, and Leslie matrices allow us to uncover this behavior and understand it.

◦◦◦◦◦◦◦◦◦◦◦◦◦◦◦◦◦◦◦◦◦◦◦◦◦◦◦◦◦◦◦◦◦◦◦◦◦◦◦

Example: Consider a population of larks, which have a maximum lifespan of three years. We divide the lark population into three age cohorts for larks in their first year of life, second year of life, and third year of life. We know that 60% of larks in the first year of life survive to the second, and no larks in the first year of life reproduce. Larks in the second year of life have on average 2.3 offspring and 80% of larks survive the second year. Larks in the third year of life have on average 0.9 offspring and no larks survive the last age cohort. We can represent this scenario with the following system of difference equations:

$$\begin{cases} x_n &= 2.3y_{n-1} + 0.9z_{n-1} \\ y_n &= 0.6x_{n-1} \\ z_n &= 0.8y_{n-1}, \end{cases}$$

where x represents the population of larks in the first year of life, y represents the population of larks in the second year of life, z represents the population of larks in the final year of life, and n is measured in years. This system can be represented in matrix form as:

$$\begin{pmatrix} x_n \\ y_n \\ z_n \end{pmatrix} = \begin{pmatrix} 0 & 2.3 & 0.9 \\ 0.6 & 0 & 0 \\ 0 & 0.8 & 0 \end{pmatrix} \begin{pmatrix} x_{n-1} \\ y_{n-1} \\ z_{n-1} \end{pmatrix}.$$

Notice that we can easily see the effect of one age group on another by looking at the matrix. We can see all birth rates in the top row and survival rates along the subdiagonal.

◦◦◦◦◦◦◦◦◦◦◦◦◦◦◦◦◦◦◦◦◦◦◦◦◦◦◦◦◦◦◦◦◦◦◦◦◦◦◦

Matrices such as the one in the previous example are called *Leslie matrices*. We define a Leslie matrix to be a matrix of the form:

$$A = \begin{pmatrix} b_1 & b_2 & b_3 & \cdots & b_{n-1} & b_n \\ s_1 & 0 & 0 & \cdots & 0 & 0 \\ 0 & s_2 & 0 & \cdots & 0 & 0 \\ 0 & 0 & s_3 & \cdots & 0 & 0 \\ \vdots & \vdots & \vdots & \ddots & \vdots & \vdots \\ 0 & 0 & 0 & 0 & s_{n-1} & 0 \end{pmatrix}.$$

The s_i terms are *survival* terms. These denote the fraction of individuals which survive from one age cohort to the next. The b_i terms are *birth* terms, sometimes called *fecundity* terms. These denote the average number of offspring born to the first age cohort from parents in the ith age cohort.

Note that an assumption which is baked into the structure of this matrix is that no member of an age cohort can survive in the same age cohort into the next time step. This may at first seem like a large assumption to make. However, consider the species of humans. If we have time steps of one year, we can classify all one-year olds into age cohort one, all two-year olds into age cohort two, etc. Clearly, after a year, no one-year old is still a one-year old; they've all aged into the next cohort or did not survive the first year of infancy. Also note that for some Leslie matrices the fecundity terms will likely be pretty sparse at first. With humans, for example, children are typically not born to parents below a certain age nor above a certain age, and this should be reflected in our matrix.

○○○○○○○○○○○○○○○○○○○○○○○○○○○○○○○○○○○○○○○

Example: Consider a population of insects, which only survive for two days. Insects in the first day of life produce on average 1.2 offspring and 10% survive to the second day, while insects in the second day of life produce on average 5 offspring. This can be represented by the Leslie matrix:

$$\begin{pmatrix} 1.2 & 5 \\ 0.1 & 0 \end{pmatrix}.$$

If we know that there are initially 200 insects in their first day of life and none in the second day of life, we can use this matrix to calculate the number of insects in each day of life in the next time step. We do this by multiplying this matrix by the column vector $(200, 0)^t$, so we have:

$$\begin{pmatrix} 1.2 & 5 \\ 0.1 & 0 \end{pmatrix} \begin{pmatrix} 200 \\ 0 \end{pmatrix} = \begin{pmatrix} 240 \\ 20 \end{pmatrix}.$$

Therefore, after one day, there are 240 insects in their first day of life and 20 insects in their second day of life.

○○○○○○○○○○○○○○○○○○○○○○○○○○○○○○○○○○○○○○○

It is sometimes the case that we want to allow some members of an age cohort to remain in that age cohort into the next time step. For example, we might consider larger age groups so that children remain in a particular group for several years, while adults have the ability to survive into the following year without changing cohorts. In this case we have a matrix of the form:

$$A = \begin{pmatrix} \ell_1 & b_2 & b_3 & \cdots & b_{n-1} & b_n \\ s_1 & \ell_2 & 0 & \cdots & 0 & 0 \\ 0 & s_2 & \ell_3 & \cdots & 0 & 0 \\ 0 & 0 & s_3 & \cdots & 0 & 0 \\ \vdots & \vdots & \vdots & \ddots & \vdots & \vdots \\ 0 & 0 & 0 & 0 & s_{n-1} & \ell_n \end{pmatrix},$$

where the s_i terms still denote the survival rate of members of the ith age cohort into the next, b_i terms represent the average number of offspring born to the first age cohort from parents in the ith age cohort, and the ℓ_i terms denote the percentage of members of the ith age cohort which *linger* in that age cohort into the next time step. We will still refer to these as Leslie matrices, though it is sometimes the case that authors only use Leslie matrix to mean matrices of the first type.

○○○

Example: Consider a population of mice where we want to distinguish between juveniles and adults. In this case, let's say that every month, 20% of juveniles survive to adulthood, 40% remain juveniles, and 40% perish either from predation or other natural causes. Similarly, 60% of adults survive to the next month and 30% of the adults give birth every month. We may be tempted to assemble this into a system of difference equations, and we can! We can model this scenario, where x represents the number of juveniles and y represents the number of adults, as:

$$\begin{cases} x_n &= 0.4x_{n-1} + 0.3y_{n-1} \\ \\ y_n &= 0.2x_{n-1} + 0.6y_{n-1}. \end{cases}$$

We can write this particular system in matrix form as:

$$\begin{pmatrix} x_n \\ y_n \end{pmatrix} = \begin{pmatrix} 0.4 & 0.3 \\ 0.2 & 0.6 \end{pmatrix} \begin{pmatrix} x_{n-1} \\ y_{n-1} \end{pmatrix}.$$

Now we can use this matrix together with an initial condition to quickly compute the next time step. Suppose we have 50 juveniles and 10 adults; then we can compute:

$$\begin{pmatrix} 0.4 & 0.3 \\ 0.2 & 0.6 \end{pmatrix} \begin{pmatrix} 50 \\ 10 \end{pmatrix} = \begin{pmatrix} 23 \\ 16 \end{pmatrix}.$$

Thus, we can see that after one month there will be 23 juveniles and 16 adults.

○○○

Note that, in general, if we have a Leslie matrix A and an initial condition vector $\mathbf{x}_0$, then we can find the populations for the various age classes at time n by computing $A^n \mathbf{x}_0$.

Example: Consider an elephant population in a wildlife refuge. The elephants are divided into three classes: calves, yearlings, and adults. Among calves, 60% survive the first year and move on to be yearlings while the rest perish. Calves cannot reproduce. Among the yearlings, 85% of them survive to become adults while the rest perish. Yearlings similarly cannot reproduce. Among adults,

90% survive into the next year and on average every adult gives birth to 0.3 calves each year. This can be represented by the Leslie matrix:

$$\begin{pmatrix} 0 & 0 & 0.3 \\ 0.6 & 0 & 0 \\ 0 & 0.85 & 0.9 \end{pmatrix}.$$

Given that there are initially 100 calves, 200 yearlings, and 500 adults in the wildlife refuge, we can use this matrix to determine the number of elephants after one year by computing:

$$\begin{pmatrix} 0 & 0 & 0.3 \\ 0.6 & 0 & 0 \\ 0 & 0.85 & 0.9 \end{pmatrix} \begin{pmatrix} 100 \\ 200 \\ 500 \end{pmatrix} = \begin{pmatrix} 150 \\ 60 \\ 620 \end{pmatrix}.$$

Therefore, after one year there will be 150 calves, 60 yearlings, and 620 adult elephants in the wildlife refuge.

To calculate the number of elephants in the next time step, we could do this again by computing:

$$\begin{pmatrix} 0 & 0 & 0.3 \\ 0.6 & 0 & 0 \\ 0 & 0.85 & 0.9 \end{pmatrix} \begin{pmatrix} 150 \\ 60 \\ 620 \end{pmatrix} = \begin{pmatrix} 186 \\ 90 \\ 609 \end{pmatrix},$$

or we could simply calculate

$$\begin{pmatrix} 0 & 0 & 0.3 \\ 0.6 & 0 & 0 \\ 0 & 0.85 & 0.9 \end{pmatrix}^2 \begin{pmatrix} 100 \\ 200 \\ 500 \end{pmatrix} = \begin{pmatrix} 186 \\ 90 \\ 609 \end{pmatrix}.$$

Either way, we would find that there will be 186 calves, 90 yearlings, and 609 adult elephants in the refuge the following year.

○○

In some examples, we may be able to observe, just by calculating a few time steps, how slowly or quickly a population might be changing. How can we determine this behavior over the long run? The key to determining this is to find the eigenvalues of the Leslie matrix.

> Given an age structured population represented by the Leslie matrix A, the long term growth rate of the population is the dominant eigenvalue λ of the matrix A. For the long term stability of this population, this translates into the following rule:
>
> - If $\lambda > 1$, the population will grow without bound over time.
>
> - If $\lambda = 1$, the population will eventually reach an equilibrium point.
>
> - If $\lambda < 1$, the population will eventually die out.

Example: Consider an age-structured population described by the Leslie matrix:

$$A = \begin{pmatrix} 0.3 & 0.7 \\ 0.2 & 0.8 \end{pmatrix}.$$

To determine the long-term stability of this population, we compute the eigenvalues by solving the equation $\det(A - \lambda I) = 0$. In this case, we have:

$$\begin{aligned} 0 &= \begin{vmatrix} 0.3 - \lambda & 0.7 \\ 0.2 & 0.8 - \lambda \end{vmatrix} \\ &= (0.3 - \lambda)(0.8 - \lambda) - 0.14 \\ &= \lambda^2 - 1.1\lambda + 0.1. \end{aligned}$$

We can find the roots to this equation using the quadratic formula, and we see that the eigenvalues are $\lambda = 1, 0.1$. As the dominant eigenvalue is equal to one, this population will eventually stabilize.

○○

Example: Consider an age-structured population described by the Leslie matrix:

$$\begin{pmatrix} 0 & 0.2 \\ 0.8 & 0 \end{pmatrix}.$$

We can find the eigenvalues of this matrix by solving

$$\begin{aligned} 0 &= \begin{vmatrix} -\lambda & 0.2 \\ 0.8 & -\lambda \end{vmatrix} \\ &= \lambda^2 - 0.16. \end{aligned}$$

Therefore, we have that $\lambda = \pm 0.4$. As the dominant eigenvalue has a magnitude less than one, this population will die out over time.

○○

Example: Consider an age-structured population described by the Leslie matrix:

$$\begin{pmatrix} 0.5 & 2 \\ 0.6 & 0 \end{pmatrix}.$$

We can find the eigenvalues of this matrix by solving

$$\begin{aligned} 0 &= \begin{vmatrix} 0.5 - \lambda & 2 \\ 0.6 & -\lambda \end{vmatrix} \\ &= -\lambda(0.5 - \lambda) - (2)(0.6) \\ &= \lambda^2 - 0.5\lambda - 1.2. \end{aligned}$$

Therefore, we have eigenvalues $\lambda \approx -0.87, 1.37$. As $\lambda \approx 1.37$ is the dominant eigenvalue and its magnitude is greater than one, this population will grow without bound.

1.6 Exercises

Construct a Leslie matrix for the following situations.

1. A population of birds which begin as yearlings and only reproduce as adults after their second year. Exactly 70% of yearlings survive to adulthood, and the adult birds have on average 2.3 offspring. Additionally 80% of adults survive to the next year.

2. A population of Hippopotamus begins as calves, become adolescents, then finally adults. Every year 30% of calves become adolescents and 40% remain calves with the other 30% perishing. Additionally, 50% of adolescents become adults and 40% remain adolescents yearly with the other 10% perishing. Adults are the only hippos which reproduce, having on average 0.7 offspring each year. Due to conservation efforts 90% of adult hippos survive each year.

3. An invasive biennial plant only reproduces in the second year. Due to an eradication program, only 10% of the first-year plants survive to the second year. However, second-year plants on average produce 4.6 offspring plants and no second year plants survive to a third year.

4. A population of beetles has three life stages: larva, pupa, and adult. Of the larvae, 10% survive to the pupa, and of the pupae, 50% survive to adulthood. No adults survive, and on average adults produce 50 offspring.

For the following problems, determine the long-term behavior of the population modeled by the given Leslie matrix.

5. $A = \begin{pmatrix} 0.5 & 1.5 \\ 0.8 & 0.2 \end{pmatrix}$

6. $A = \begin{pmatrix} 0 & 5 \\ 0.2 & 0 \end{pmatrix}$

7. $A = \begin{pmatrix} 0.5 & 8 \\ 0.1 & 0 \end{pmatrix}$

8. $A = \begin{pmatrix} 0.5 & 10 \\ 0.1 & 0.5 \end{pmatrix}$

9. $A = \begin{pmatrix} 0.6 & 0.4 \\ 0.3 & 0.2 \end{pmatrix}$

10. $A = \begin{pmatrix} 1 & 2.5 \\ 0.5 & 0 \end{pmatrix}$

11. $A = \begin{pmatrix} 0.1 & 0.7 \\ 0.3 & 0.1 \end{pmatrix}$

12. $A = \begin{pmatrix} 0.4 & 0.8 \\ 0.5 & 0.25 \end{pmatrix}$

For the following problems you are given a Leslie matrix and partial information about the population. Use the data in the Leslie matrix to answer the given questions about the population

13. A population of turtles is divided into three age groups: yearlings, adults, and mature adults. The following Leslie matrix represents this population

$$\begin{pmatrix} 0.5 & 3 & 0.7 \\ 0.75 & 0.2 & 0 \\ 0 & 0.6 & 0.1 \end{pmatrix}$$

(a) How many yearlings survive to adulthood?

(b) Do adults or mature adults reproduce at a higher average rate?

(c) How many adults linger in the cohort every time step?

14. A population of whales is divided into four age groups: calves, yearlings, adults, and mature adults. The following Leslie matrix represents this population

$$\begin{pmatrix} 0 & 0 & 0.6 & 0.1 \\ 0.8 & 0 & 0 & 0 \\ 0 & 0.9 & 0 & 0 \\ 0 & 0 & 0.7 & 0 \end{pmatrix}$$

(a) How many yearlings survive to adulthood?

(b) Do adults or mature adults reproduce at a higher average rate?

(c) Do any whales linger in their cohort?

(d) Do yearlings reproduce? If so, at what rate?

(e) How many calves survive to become yearlings?

Chapter 2

Introduction to Ordinary Differential Equations

Differential equations arise from natural questions. What if the speed of a particle is proportional to its position? What if the growth of a population is proportional to its size? You have likely already found a solution of a differential equation without even realizing it. You've learned that if $f(x) = e^x$, then $f'(x) = e^x$. That's the beauty of the exponential function: it is its own derivative! That observation, that the function is equal to its own derivative, can be expressed as a differential equation! Let $y = e^x$. We noticed that $y' = e^x$, so $y' = y$. This is a differential equation.

We just found a function which we knew from our past experience is the solution of the differential equation $y' = y$, but what if we don't know the original function? What if we're given the differential equation $y' = 2y$, or $y'' - 9y = 0$. How can we find a solution of these? Is there only one?

Another natural question about $y' = y$: what *kind* of differential equation is it? The first classification question we should ask is whether this is an *ordinary* or a *partial* differential equation.

An **ordinary differential equation** (ODE) is an equation containing derivatives of a function in one independent variable, such as x or t, for time. A few examples of ODEs are:

$$y' = x + y,$$

$$(x + 2) \, dy = (y + 1) \, dx,$$

$$y'' + 2y' + 1 = \sin x.$$

A **partial differential equation** is an equation containing derivatives of a function in more than one independent variable. For example, the function $u(x, t)$ might describe the temperature of a rod which is being heated from one side. Both the variables x and t are required here because the temperature depends both on the position at which you measure (x), and at what time (t).

DOI: 10.1201/9781003298663-2

A few examples of partial differential equations are:

$$u_{xx} = u_t,$$

$$u_{tt} = 0,$$

$$u_x + u_t = u,$$

where u_{xx}, for example, is the second partial derivative of u with respect to x. In this text we will concern ourselves only with ODEs.

In general, we are interested in the solutions of differential equations, but a differential equation may have infinitely many solutions depending on the initial conditions. As such, we will concern ourselves with the solutions of **initial value problems** (IVPs), such as:

$$y' = xy, \quad (x_0, y_0) = (1, 1)$$

or

$$y'' + 2y' + 1 = 0, \qquad y(0) = 1, \quad y'(0) = 0.$$

An IVP is a combination of a differential equation, such as $y' = xy$, and an initial value or condition, such as $y(0) = 1$. We will see in the course of this chapter that ODEs have a *family* of solutions while IVPs with the requisite number of initial values have a *unique* solution.

You may also be confronted with a **boundary value problem**, which is most commonly seen while studying partial differential equations. However, these can also be seen with ODEs. For example, notice that:

$$y'' - 4y' + 5y = \cos x, \qquad y(0) = 0, \quad y'(0) = 1$$

is an IVP because two values are given: the initial value of y and the initial value of its derivative. Suppose that this differential equation is modeling a wave where y is the height of the wave and x is the horizontal position. Then what this IVP tells you is a rule governing the wave's motion, as well as an initial height and an initial "slope" of the wave.

Now consider the following:

$$y'' - 4y' + 5y = \cos x, \qquad y(0) = 0, \quad y(1) = 1.$$

This is the same differential equation but the given initial values are different. Here rather than the initial values of both y and its derivative, we are given the value of y at two different values of x. This is a boundary value problem. Supposing again that this is modeling a wave as above, this gives us a rule governing the wave's motion, as well as the height of the wave at two points (on the boundary of the wave). Unlike IVPs, boundary value problems may have infinitely many solutions. We will exclusively be considering IVPs in this text.

2.1 Classification of ODEs and the Verification of Their Solutions

You may notice something different between the following two ODEs:

$$y' = 2y \quad \text{and} \quad x^3 y'' + x^2 y' - 9xy = -16x^2.$$

This is because these two ODEs are of different **order**.

> **Definition:** An ordinary differential equation
>
> $$F(x, y, y', \ldots, y^{(n)}) = 0$$
>
> is said to have **order** n. In other words, the order of an ODE is the order of the highest derivative.

For the ODEs above, $y' = 2y$ is a *first-order* ODE while:

$$x^3 y'' + x^2 y' - 9xy = -16x^2$$

is a *second-order* ODE. There are other differences between these ODEs, which we will explore shortly, but for now let's focus on differentiating ODEs by order.

ooooooooooooooooooooooooooooooooooooooo

Example: Let's look at the order of a few ODEs.

$$
\begin{aligned}
y' - 5xy &= 17 & &\text{is a first-order ODE.} \\
y'' + y' + 2y &= 0 & &\text{is a second-order ODE.} \\
y^{(5)} + y^2 &= 0 & &\text{is a fifth-order ODE.} \\
y^{(2023)} &= \sin(y) & &\text{is a 2023rd order ODE.}
\end{aligned}
$$

We can tell the order of the ODE by the highest derivative that is a part of its equation.

You may have noticed that there is something different about the last two ODEs in the above example, aside from the fact that they're of higher order. This is because the first two ODEs are **linear**, while the last two are not.

> **Definition:** An ordinary differential equation
> $$F(x, y, y', \ldots, y^{(n)}) = 0$$
> of order n is said to be **linear** if it is a linear function of the variables $y, y', \ldots, y^{(n)}$. A function which is not linear in these variables is called **nonlinear**.

Example: Let's look at the linearity of a few ODEs.

$$y' - 5xy \;=\; 17 \qquad\qquad \text{is a linear ODE.}$$

$$y'' + y' + 2y \;=\; 0 \qquad\qquad \text{is a linear ODE.}$$

$$y^{(5)} + y^2 \;=\; 0 \qquad\qquad \text{is a nonlinear ODE, because of the } y^2.$$

$$y^{(2023)} \;=\; \sin(y) \qquad\qquad \text{is a nonlinear ODE, because of the } \sin(y).$$

$$y^{(82)} + 3y' \;=\; \cos(x) \qquad\qquad \text{is a linear ODE.}$$

As you can see, the *order* of an ODE does not matter when determining linearity, nor do terms in the ODE, which are solely functions of x. Note that the last two examples both have higher-order terms and trigonometric functions, so what is the difference? In the ODE $y^{(2023)} = \sin(y)$, the sine function is a function of y, thus it is nonlinear in y. However, in the ODE $y^{(82)} + 3y' = \cos(x)$, the cosine function is a function of x, so it is still linear in y.

○○

To classify an ODE is to determine both its order and whether or not it is linear. The classification can often help in deciding the appropriate method in which to solve an ODE. At other times, we may be able to make an educated guess as to what the solution might be. It can at times be the case that you have a function $y = f(x)$, which you may suspect to be a solution of a differential equation, $y' = F(x, y)$. Perhaps the shape of the function behaves in a manner in which you think the solution of the differential equation should behave.

For example, with the differential equation $y' = 2y$, it appears as if y should be constantly increasing as x goes to infinity. Perhaps you think that $y = x^2$ could be a good candidate. After all, y grows without bound as x goes to infinity, and the number 2 appears in the function; maybe that's the right choice! Alas, it is not, but how can we show this?

We can see that for $y = x^2$, $y' = 2x$. We can substitute $y = x^2$ into the differential equation $y' = 2y$ and see that this implies that $y' = 2y = 2x^2$, but we just showed that $y' = 2x$! Clearly, we didn't verify this as a solution of this ODE, so let's look at an example where we do verify that we have found the correct solution.

Suppose instead that you think that $y = e^{2x}$ is the right choice. After all, e^{2x} is increasing, and there's a two in the equation, so that's promising! Let's verify that the function $y = e^{2x}$ is a solution of the ODE $y' = 2y$. We simply take the derivative of $y = e^{2x}$ using the chain rule, so that $y' = 2e^{2x}$. Note that we can substitute y in for e^{2x}. We obtain $y' = 2e^{2x} = 2y$. Thus we have verified that $y = e^{2x}$ is a solution of the ODE $y' = 2y$.

Notice however, that $y = 3e^{2x}$ is also a solution of this differential equation, as $y' = 6e^{2x} = 2(3e^{2x}) = 2y$. In fact, the whole family of functions $y = Ce^{2x}$, where C is a real constant, are solutions of this differential equation. We will concern ourselves soon with how to find these families of solutions, but in this section we will instead focus on verifying whether a specific given function is a solution of an ODE or not.

○○

Example: Let's verify that $y = 5x^4 + 2$ is a solution of the ODE $xy' - 4y = -8$. We see that both y and y' are in the ODE, i.e., it is a first-order ODE. Therefore, in order to verify the solution we must first use the given function $y = 5x^4 + 2$ and find the derivative. We find that $y' = 20x^3$. Now we may substitute this y' and the original given y into the left-hand side (LHS) of the ODE and simplify:

$$
\begin{aligned}
LHS = xy' - 4y &= x(20x^3) - 4(5x^4 + 2) \\
&= 20x^4 - 20x^4 - 8 \\
&= -8 = RHS.
\end{aligned}
$$

As our final simplified expression is -8, which is also the original right-hand side (RHS) of the ODE, we have shown that $y = 5x^4 + 2$ does satisfy the ODE $y' - 4y = -8$. Hence, $y = 5x^4 + 2$ is a solution of the ODE $y' - 4y = -8$.

○○

Example: Consider the following ODE:

$$x^3 y''' + x^2 y'' - 4xy' = 0.$$

We would like to verify that $y = x^3$ is a solution of this third-order ODE. In the previous example we had a first order ODE and needed to take the first derivative of the candidate function. Here we will need to take the third derivative. You may suspect correctly that, in general, in order to verify that $y = f(x)$ is a solution of an nth order ODE, we need to take up to the nth derivative of $f(x)$.

Proceeding with our example, we see that we have:

$$y = x^3$$

$$y' = 3x^2$$

$$y'' = 6x$$

$$y''' = 6.$$

Substituting these into the left-hand side of the ODE, we see that:

$$LHS = x^3 y''' + x^2 y'' - 4xy' = x^3(6) + x^2(6x) - 4x(3x^2)$$

$$= 6x^3 + 6x^3 - 12x^3$$

$$= 0 = RHS.$$

Thus we can conclude that $y = x^3$ is, in fact, a solution of the ODE.

○○○○○○○○○○○○○○○○○○○○○○○○○○○○○○○○○○○○○○○

Example: We can also verify more general solutions. In this case, we can verify that $y = Cx^2 + 4$ is a solution of the ODE:

$$y'' x^2 - xy' = 0,$$

where C is an arbitrary constant. This will allow us to verify that the whole family of functions, for any constant C, is a solution of the given ODE. This ODE has both y' and y'' in the equation, so we must first derive these equations from the given equation for y:

$$y = Cx^2 + 4$$

$$y' = 2Cx$$

$$y'' = 2C.$$

We can then substitute y' and y'' into the original left-hand side of the ODE and simplify:

$$LHS = y'' x^2 - xy' = (2C)x^2 - x(2Cx)$$

$$= 2Cx^2 - 2Cx^2$$

$$= 0 = RHS.$$

As we arrived at the original right-hand side of the ODE, we have verified that $y = Cx^2 + 4$ is a solution of the ODE $y''x^2 - xy' = 0$ for any constant C.

○○○

Note that in this above case we did not show that just one function was a solution of the ODE. In fact, we showed that an entire *family* of functions were. If we want to find a function which is a *unique* solution we will need the requisite initial conditions. For now, let's see a few more examples of verifying that a family of functions are solutions of an ODE.

Example: We can verify that the family of functions $y = Cx^3 + 2x$ is a solution of the ODE:

$$x^3 y'' + x^2 y' - 9xy = -16x^2.$$

As both y' and y'' appear in the ODE, we must take the first and second derivatives of $y = Cx^3 + 2x$. Then:

$$y \;\;=\;\; Cx^3 + 2x$$

$$y' \;\;=\;\; 3Cx^2 + 2$$

$$y'' \;\;=\;\; 6Cx.$$

Now we substitute these expressions for y, y', and y'' into the left-hand side of the original ODE and simplify in the hopes of arriving at $-16x^2$, the original right-hand side of the equation. We see that:

$$
\begin{aligned}
LHS \;\;&=\;\; x^3 y'' + x^2 y' - 9xy \\
&=\;\; x^3(6Cx) + x^2(3Cx^2 + 2) - 9x(Cx^3 + 2x) \\
&=\;\; 6Cx^4 + 3Cx^4 + 2x^2 - 9Cx^4 - 18x^2 \\
&=\;\; 2x^2 - 18x^2 \\
&=\;\; -16x^2 \;\;=\;\; RHS.
\end{aligned}
$$

Hence $y = Cx^3 + 2x$ is a solution of the ODE $x^3 y'' + x^2 y' - 9xy = -16x^2$.

○○○

For some higher-order ODEs, there may be multiple constants in the solution. In these cases we typically use $c_1, c_2, \ldots, c_n$ to denote these constants. Note that these constants all may be different and are independent of each other.

Example: Verify that the family of functions $y = c_1 e^{-3t} + c_2 e^{-2t}$ is a solution of the ODE:

$$y'' + 5y' + 6y = 0.$$

As before we find the first two derivatives of the candidate solution, using the chain rule:

$$y = c_1 e^{-3t} + c_2 e^{-2t}$$

$$y' = -3c_1 e^{-3t} - 2c_2 e^{-2t}$$

$$y'' = 9c_1 e^{-3t} + 4c_2 e^{-2t}.$$

We proceed by substituting these derivative functions into the differential equation:

$$
\begin{aligned}
LHS &= y'' + 5y' + 6y \\
&= 9c_1 e^{-3t} + 4c_2 e^{-2t} + 5(-3c_1 e^{-3t} - 2c_2 e^{-2t}) + 6(c_1 e^{-3t} + c_2 e^{-2t}) \\
&= 9c_1 e^{-3t} + 4c_2 e^{-2t} - 15c_1 e^{-3t} - 10c_2 e^{-2t} + 6c_1 e^{-3t} + 6c_2 e^{-2t} \\
&= (9c_1 - 15c_1 + 6c_1)e^{-3t} + (4c_2 - 10c_2 + 6c_2)e^{-2t} \\
&= 0 = RHS.
\end{aligned}
$$

As the right and left hand sides of the equation are equal, we have verified that $y = c_1 e^{-3t} + c_2 e^{-2t}$ is a solution of the ODE for any constants c_1, c_2.

○○○

So far our examples have all been focused on the verification of solutions of ODEs. We can also verify that a function is a solution of an IVP. Let's explore that process with the below example.

Example: Consider the first-order ODE:

$$y' - 2y = 2e^{2x} \sin x.$$

Let's verify that $y = -2e^{2x} \cos x + 5e^{2x}$ is a solution of the ODE with the initial condition $y(0) = 3$. Using the product and chain rules, we begin by computing:

$$y = -2e^{2x} \cos x + 5e^{2x}$$

$$y' = -4e^{2x} \cos x + 2e^{2x} \sin x + 10e^{2x}$$

and substituting y and y' into the ODE:

$$
\begin{aligned}
LHS &= y' - 2y \\
&= -4e^{2x}\cos x + 2e^{2x}\sin x + 10e^{2x} - 2(-2e^{2x}\cos x + 5e^{2x}) \\
&= -4e^{2x}\cos x + 2e^{2x}\sin x + 10e^{2x} + 4e^{2x}\cos x - 10e^{2x} \\
&= 2e^{2x}\sin x \;=\; RHS.
\end{aligned}
$$

We also need to make sure that the initial condition is satisfied, so we end by verifying that, indeed, $y(0) = -2e^0\cos 0 + 5e^0 = 3$.

○○○

Now let's consider a second-order IVP. When we have two initial conditions, we must check them both, as well as checking that the function itself is a solution of the ODE.

Example: Let's verify that the function $y = \sin x$ is a solution of the IVP:

$$
y'' + y = 0, \qquad y(0) = 0, \quad y'(0) = 1.
$$

First, we will find the first two derivatives of the candidate function, as this is a second-order ODE. Note that:

$$
\begin{aligned}
y &= \sin x \\
y' &= \cos x \\
y'' &= -\sin x.
\end{aligned}
$$

Next, we will check that the function satisfies the ODE. Note that:

$$
\begin{aligned}
LHS = y'' + y &= -\sin x + \sin x \\
&= 0 = RHS.
\end{aligned}
$$

Finally, we will confirm that $y = \sin x$ satisfies the initial conditions. Note that $y(0) = \sin 0 = 0$ and $y'(0) = \cos 0 = 1$, thus the initial values are satisfied. We could have checked the initial conditions first and verify the ODE is satisfied after just as easily, and this can be the faster way to go in situations where we may be trying to disprove a solution rather than verify it. Nonetheless, for this example, as the candidate function satisfies both the ODE and the initial conditions, we can verify it is a solution of the IVP.

○○○

Example: Let's verify that the function $y = 4\sin 3x$ is a solution of the IVP

$$y'' + 9y = 0, \qquad y(\pi) = 0, \quad y'(0) = 12.$$

We will need to find the second derivative of $y = 4\sin 3x$ so that we can simplify the given expression. Note that:

$$y = 4\sin 3x$$

$$y' = 12\cos 3x$$

$$y'' = -36\sin 3x.$$

Then substituting y and y'' into the left-hand side of the original ODE, we have:

$$LHS = y'' + 9y = -36\sin 3x + 9(4\sin 3x)$$

$$= 0 = RHS.$$

We also must verify that this function satisfies the given initial conditions. In this case, we see that $y(\pi) = 4\sin(3\pi) = 0$ and $y'(0) = 12\cos(0) = 12$. Hence $y = 4\sin 3x$ is a solution of the given IVP.

○○○

Often we will find ourselves in the situation where we will, by some method, find a solution of an ODE. This solution will typically contain one or more free constants. Then we will be given one or more initial values. We will use those initial values to solve for the constants and find a specific function. The below examples will show us this process in detail.

Example: We can find the value of the constant C such that $y = Ce^{3x} - 2x + 1$ is a solution of the IVP

$$y' - 3y = 6x - 5, \qquad y(0) = 5.$$

First, we verify that the given function satisfies the differential equation. We find:

$$y = Ce^{3x} - 2x + 1$$

$$y' = 3Ce^{3x} - 2.$$

Substituting these functions into the left-hand side of the given ODE, we have:

$$LHS = y' - 3y = 3Ce^{3x} - 2 - 3(Ce^{3x} - 2x + 1)$$

$$= 3Ce^{3x} - 2 - 3Ce^{3x} + 6x - 3$$

$$= 6x - 5 = RHS.$$

Hence $y = Ce^{3x} - 2x + 1$ is a solution of the ODE $y' - 3y = 6x - 5$.

Furthermore, we need to find C such that the initial condition $y(0) = 5$ holds. Then $y(0) = Ce^{3(0)} - 2(0) + 1 = C + 1 = 5$, so $C = 4$. Thus, $y = 4e^{3x} - 2x + 1$ is the solution of the given IVP.

○○

We also may be confronted with a situation where there are two or more constants for which we must solve. For example, we verified in an earlier example that the family of functions $y = c_1 e^{-3t} + c_2 e^{-2t}$ was a solution of the ODE $y'' + 5y' + 6y = 0$. In the following example, we will be finding the specific values of c_1 and c_2 which will satisfy an IVP involving this ODE.

Example: Find values for c_1 and c_2 such that $y = c_1 e^{-3t} + c_2 e^{-2t}$ is a solution of the IVP:

$$y'' + 5y' + 6y = 0, \qquad y(0) = 2, \quad y'(0) = -1.$$

In the earlier example we already verified that this family of functions is a solution of the ODE. As such, we can proceed to find the requisite constant values by using the given initial conditions. We will substitute the $t = 0$ value into the solution and see that as $y(0) = 2$, we have:

$$2 = c_1 + c_2.$$

As you may know, one equation is not sufficient to solve for two unknowns. As such we need a second equation. What to do?

We should recall or compute that:

$$y' = -3c_1 e^{-3t} - 2c_2 e^{-2t}.$$

Using this equation for the derivative and the initial condition $y'(0) = -1$, we have that:

$$-1 = -3c_1 - 2c_2$$

or, rearranging,

$$3c_1 + 2c_2 = 1.$$

Now we have two equations and two unknowns, so we must solve the system of equations:

$$\begin{cases} c_1 + c_2 = 2 \\ 3c_1 + 2c_2 = 1. \end{cases}$$

Doing so, we see that we have solutions $c_1 = -3$ and $c_2 = 5$. Therefore, the function $y = -3e^{-3t} + 5e^{-2t}$ is a solution of the IVP.

2.1 Exercises

For the following exercises, determine the order of the given ODE.

1. $y'' + x^2 y = 1$ 2. $xy^2 + y''' - y = y'$

3. $y' = \dfrac{x^2 \sin xy}{y^3 e^{x^2} y^6}$ 4. $x'' - x't = 0$

5. $Q'' - tQ' + t^2 Q = \sin t^2$ 6. $u^{(4)} - uu'' = x.$

For the following exercises, determine if the given ODE is linear. If it is not linear, explain why not.

7. $y'' + 4y' = 0$ 8. $xy''' + x^2 y^2 = 0$

9. $y' = \dfrac{xy - \sin x}{\cos x^2}$ 10. $y'' + yy' = 0$

11. $\dfrac{dP}{dt} = -rP\left(1 - \dfrac{P}{K}\right)$ 12. $\dfrac{d^2 Q}{dt^2} + t\dfrac{dQ}{dt} = \sin t$

For the following exercises, verify that the given function is a solution of the ODE.

13. $y = 6x^2 + 24x,$ $(\frac{1}{2}x + 1)y' - y = 24$

14. $y = 3x^4 - 5x,$ $xy' - 4y = 15x$

15. $y = xe^{2x},$ $y'' - 2y' = 2e^{2x}$

16. $y = 5e^{3x} + 2x,$ $y'' - 3y' + 6 = 0$

17. $y = e^{5x^2 - 3x},$ $y'' - (10x - 3)y' - 10y = 0$

For the following exercises, verify that the given function is a solution of the IVP.

18. $y = 2e^{4x}$, $y' - 4y = 0$, $y(0) = 2$

19. $y = 5x^3 + 4x$, $xy' - 3y + 8x = 0$, $y(1) = 9$

20. $y = \cos(4x)$, $y'' + 16y = 0$, $y(0) = 1$, $y'(0) = 0$

21. $y = 3x^2 + 2$, $x^2 y'' - 2y + 4 = 0$, $y(0) = 2$, $y'(1) = 6$

For the following exercises, find the value(s) of the constant terms (i.e., C, c_1, c_2) such that the function is a solution of the IVP.

22. $y = Cx^4 + 3x$, $xy' - 4y = -9x$, $y(-2) = 2$

23. $y = Ce^{-2x+6}$, $y' + 2y = 0$, $y(3) = -5$

24. $y = c_1 \sin 5x + c_2 \cos 5x$, $y'' + 25y = 0$, $y(\pi) = 3$, $y'(\pi) = -12$

25. $y = c_1 e^{4x} + c_2 e^{3x}$, $y'' - 7y' + 12y = 0$, $y(0) = 2$, $y'(0) = 1$

26. $y = c_1 e^x + c_2 x e^x$, $y'' - 2y' + y = 0$, $y(0) = 7$, $y'(0) = 5$

2.2   Existence and Uniqueness of Solutions of Linear First-Order ODEs

ODEs without any initial values have a family of solutions. When an initial condition is included, we can eliminate some of the family of solutions from consideration. In general we are mostly concerned with finding a particular solution of an IVP. However, there are two questions you should be thinking about:

1. Does this IVP have a solution at all?

2. Is this solution the only possible one?

These questions together are asking about the *existence* and *uniqueness* of a solution of the IVP. Existence and uniqueness theorems are very common in mathematics, and as you may have guessed, we will look at the existence and uniqueness theorem for solutions of linear first-order ODEs.

Our goal is to determine whether or not there is a unique solution of an ODE without actually solving the ODE. Before we can determine if a solution is unique though, we must first determine if one even exists! The following theorem is both an existence and uniqueness theorem. It provides a way to find a specific interval on which a unique solution is guaranteed to exist.

Consider the IVP

$$y' + p(x)y = q(x), \qquad y(x_0) = y_0.$$

If $p(x)$ and $q(x)$ are continuous on an interval (a, b) containing the point $x = x_0$, then there exists a function $y = F(x)$ which is the unique solution of this initial value problem on the interval (a, b).

Example: We can use this theorem to determine an interval on which a solution of the IVP:

$$xy' - (x + 1)y = 5, \qquad y(1) = 4$$

exists. However, we need to rewrite the equation so that it is in the form:

$$y' + p(x)y = q(x).$$

Hence, we write:

$$y' - \frac{x+1}{x} y = \frac{5}{x}, \qquad y(1) = 4.$$

Now we can see that $p(x) = -\frac{x+1}{x}$ and $q(x) = \frac{5}{x}$. These two functions are both continuous everywhere except for $x = 0$, and thus both have domain $(-\infty, 0) \cup (0, \infty)$. Therefore, possible solutions may occur in this domain. However, there is a catch. Notice that this domain is not connected, so there is no way a continuous solution can exist on both intervals $(-\infty, 0)$ and $(0, \infty)$. As such, we must use our initial condition to decide on which of the two intervals our solution will exist. As the theorem states, there exists a function $y = F(x)$ which is the unique solution on the interval (a, b), **which contains** the point $x = x_0$ where $y(x_0) = y_0$ is the initial value. The below number line represents the domain on which both $p(x)$ and $q(x)$ are continuous.

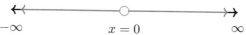

However, we are looking for the solution with initial value $y(1) = 4$, so $x_0 = 1$. This is in the interval $(0, \infty)$.

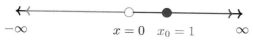

Therefore, by the theorem above, there exists a unique function that satisfies the IVP which is a solution on the interval $(0, \infty)$. This is the interval on which a solution exists because it is the largest interval containing $x_0 = 1$ on which both $p(x)$ and $q(x)$ are continuous. This interval corresponds with the blue region in the number line. Note that this interval contains x_0 and is the largest such interval.

○○

Example: Consider the IVP

$$y' - \ln(x-1)y = \sqrt{3-x}, \qquad y(2) = 1.$$

This is already in the form $y' + p(x)y = q(x)$ with $p(x) = -\ln(x-1)$ and $q(x) = \sqrt{3-x}$. We can see that the domain of $p(x) = -\ln(x-1)$ is $(1, \infty)$, and it is continuous on its entire domain.

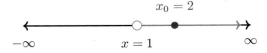

The domain of $q(x) = \sqrt{3-x}$ is $(-\infty, 3]$, so the largest open interval on which $q(x)$ is continuous is $(-\infty, 3)$.

Now, the intersection of these two open continuous domains is $(1, 3)$, and the point $x_0 = 2$ is contained within this interval.

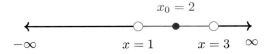

Therefore, as $(1, 3)$ is the largest open interval containing $x_0 = 2$ on which both functions are continuous, there exists a unique solution of this IVP on the interval $(1, 3)$.

○○

Example: Consider the IVP:

$$y' + \frac{1}{2x-3} y = \tan(x), \qquad y(-1) = 0.$$

The differential equation is already in the form $y' + p(x)y = q(x)$. In this case, we have $p(x) = \frac{1}{2x-3}$ and $q(x) = \tan(x)$. We know that $p(x) = \frac{1}{2x-3}$ is discontinuous at $x = \frac{3}{2}$ and can visualize its domain on a number line.

Also, $q(x) = \tan x$ is discontinuous at $x = \frac{\pi}{2} + n\pi$, where n is an integer. Focusing our attention near the given initial value of $x_0 = -1$, we can sketch the domain of $q(x)$ on a number line.

$$x_0 = -1$$

$$-\infty \quad x = -\frac{\pi}{2} \qquad\qquad x = \frac{\pi}{2} \quad \infty$$

The largest interval containing $x_0 = -1$ on which both $p(x)$ and $q(x)$ are continuous is $(-\frac{\pi}{2}, \frac{3}{2})$.

$$x_0 = -1$$

$$-\infty \quad x = -\frac{\pi}{2} \qquad\qquad x = \frac{3}{2} \quad \infty$$

Therefore, a unique solution of the above IVP exists on the interval $(-\frac{\pi}{2}, \frac{3}{2})$.

○○○○○○○○○○○○○○○○○○○○○○○○○○○○○○○○○○○○○○○

Example: Consider the IVP:

$$y' + \sqrt{x - 3}\, y - \frac{2}{5 - x} = 0, \qquad y(6) = 4.$$

The first step is to rewrite the equation in the form $y' + p(x)y = q(x)$. In this case, we have:

$$y' + \sqrt{x - 3}\, y = \frac{2}{5 - x}, \qquad y(6) = 4.$$

We can see that $p(x) = \sqrt{x - 3}$, which has domain $[3, \infty)$. As we consider only open intervals, we'll restrict this to $(3, \infty)$, which can be represented by the number line:

$$x_0 = 6$$

$$-\infty \qquad x = 3 \qquad\qquad \infty$$

Similarly, we can sketch a number line for the domain of $q(x) = \frac{2}{5-x}$, which is discontinuous at $x = 5$.

$$x_0 = 6$$

$$-\infty \qquad\qquad x = 5 \quad \infty$$

The largest open interval containing the given initial value $x_0 = 6$ on which both $p(x)$ and $q(x)$ are continuous, in this case, is $(5, \infty)$.

$$x_0 = 6$$

$$-\infty \qquad\qquad x = 5 \qquad\qquad \infty$$

Hence, we know that a unique solution of the IVP exists on the interval $(5, \infty)$.

2.2 Exercises

For the following exercises, determine the domain on which a solution exists for the IVP.

1. $y' + x^2 y = \sqrt{1+x}, \qquad y(0) = 1$

2. $y' + xy' = e^{x^2}, \qquad y(\pi) = 1$

3. $y' + \dfrac{y}{x} = \sqrt{x-1}, \qquad y(1) = \pi$

4. $y' + \dfrac{y}{x^2 - 4} = x, \qquad y(3) = 12$

5. $y' + y\tan x = \ln(x-1), \qquad y(\pi/3) = 2$

6. $y' - 4y = 2, \qquad y(0) = 1$

For the following exercises, determine the domain on which a solution exists for the IVP.

7. $x^2 y' + x^2 y = \sqrt{1+x}, \qquad y(0) = 1$

8. $(x-2)y' + (x^2 - 4)y = \sqrt{x^2 - 1}, \qquad y(-2) = 1$

9. $y' = \dfrac{xy - e^x}{\cos x}, \qquad y(1) = \pi$

10. $x^2 y' + xy - 1 = 0, \qquad y(5) = 1$

11. $(x^2 - 4)y' + \ln(x^2 - 1)y = 0, \qquad y(3/2) = 2$

12. $y' = \dfrac{x^2 y - y^2}{x^3 y + e^x y}, \qquad y(0) = 1$

2.3   Vector Fields

There may be instances when you are confronted with a differential equation whose solution you may not be able to or want to find. However, it is still possible to understand the behavior of these solutions. One method for understanding the behavior of solutions of differential equations without explicitly finding the solution is the vector field.

> **Definition:** The **vector field** (sometimes referred to as a slope field or direction field) of a differential equation $y' = f(x, y)$ assigns a line segment with slope $f(x, y)$ to every point (x, y) in the xy-plane. To visualize the vector field, we often draw these line segments at a collection of points uniformly distributed in the plane.

Note that the vector field assigns a line segment to *every* point in the plane. Trying to graph this whole object would certainly take a long time, and wouldn't be very helpful in visualizing the behavior of the differential equation. Often, we use mathematical graphing software to generate visualizations of vector fields, and depending on the software, we can sometimes choose with what frequency the line segments are displayed to aid in visualization. When we lack graphing software, we can calculate a handful of values of the vector field at important points to get an idea of the behavior of the differential equation.

○○

Example: Consider the differential equation $y' = y - 2$. By substituting a few values of (x, y) into the function $f(x, y) = y - 2$, we can obtain the slopes of some of the line segments in the vector field located at those points. For example, at the point $(0, 0)$, we have $y' = f(0, 0) = 0 - 2 = -2$, so we will draw a line segment of slope -2 at the point $(0, 0)$. Similarly, at the point $(0, 3)$ we will draw a line segment of slope 1, at $(3, 0)$ a segment of slope -2, at $(3, 3)$ a segment of slope 1, and at $(0, 4)$ a segment of slope 2.

Unfortunately, the line segments at these five points don't do very much to illustrate the behavior of possible solutions of this differential equation. However, if we calculate y' at each integer point (x, y), as above, and plot a line segment whose slope corresponds to the value of y' at that point, we generate a more dense vector field. Notice that because there is no x in the original equation, that these slopes can be easily calculated just by changing the y values (for example, $(0, 0)$, $(1, 0)$, and $(2, 0)$ will all have line segments with the same slope because $y = 0$ for each of these points, and similarly, $(0, 1)$, $(1, 1)$, and $(2, 1)$ will all have the same slope because $y = 1$ for each of these points). We can use this idea to sketch a more complete vector field, while only computing a few select values. Alternatively, we can turn to mathematical software to help us.

How does this graph correspond to the solutions that we are really interested in? If you start at a point (x_0, y_0) in the plane and follow the direction of the line segments in the vector field, the resulting curve is a solution of the ODE. A handy way to think about what is happening is to imagine a river or stream, with the vector field representing the direction of the currents. If you drop a rubber duck into the river at the point (x_0, y_0), the solution of the ODE corresponds to the path in which the duck follows through the river.

Plotting three solutions on the graph corresponding to initial conditions $(4, 3)$, $(0, 2)$, and $(-3, 1)$, we can see the three solutions of the ODE corresponding to these initial conditions and can make a statement about their **long-term behavior**, or in our rubber duck analogy, where in the river the duck will end up.

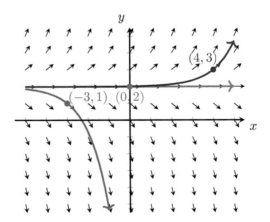

You may notice that for one of our solutions, it appears that the path is simply a certain value of y and does not deviate from that as x increases. This is because it is an **equilibrium solution**, which is a solution that corresponds to no change in y as x increases. From looking at the vector field and plotting the solution as above, we can see that the solution of the IVP:

$$y' = y - 2, \qquad (x_0, y_0) = (0, 2)$$

is an equilibrium solution. Despite not explicitly solving the ODE at all, we already know that $y = 2$ is a solution of the original differential equation, $y' = y-2$. We have also obtained two other solutions which have very different behaviors than the equilibrium solution, one of which grows without bound, and one of which decays without bound. We would characterize behavior like this as typical of an **unstable equilibrium**.

○○○

Example: Consider the differential equation $y' = xy$. This differential equation has some symmetry to it, which allows for more line segments in the vector field to be generated with fewer calculations. Note that, for example, we have $y' = f(x,y)$ where $f(x,y) = xy$, and $f(1,2) = f(2,1) = 2$. Therefore, at both the points $(1,2)$ and $(2,1)$ we should place a line segment of slope 2. Similarly, $f(-1,2) = f(2,-1) = -2$, so at both the points $(-1,2)$ and $(2,-1)$ we should place a line segment of slope -2.

We can also see that the first and third quadrants of this vector field coincide, so negating both x and y results in the same output in the vector field, and similarly for the second and fourth quadrants. Furthermore, we can see that $f(x,y) = -f(-x,y)$, so the values in the vector field in the second quadrant are the opposite of the values in the vector field in the first quadrant. With this information, we can see that by finding the value of one line segment in the first quadrant, we can sketch four line segments in the vector field!

For example, we can see that at the point $(3,4)$, we have $y' = f(3,4) = 12$, so we should draw a line segment of slope 12 at $(3,4)$. As the first and third quadrants coincide, at the point $(-3,-4)$ we should also draw a line segment of slope 12. As the second quadrant is the opposite of the first, we should draw a line segment of slope -12 at the point $(-3,4)$, and as the second and fourth quadrants coincide, we should draw a line segment of slope -12 at the point $(3,-4)$ as well. We can see in the graph of the vector field below that this behavior is realized.

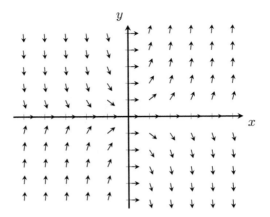

As we have shown, by computing the slope of the line segment at the point $(3, 4)$, we have obtained the slopes of the line segments at the points $(-3, 4)$, $(3, -4)$, and $(-3, -4)$ with minimal additional effort. This strategy of using the symmetry in the function $f(x, y)$ to more quickly sketch the vector field can be very useful in the absence of a graphing utility.

○○○

Example: Consider the differential equation $y' = y^2 - 1$. Note that we can rewrite this as $y' = (y + 1)(y - 1)$. An important thing to realize is that if we want y' to be equal to zero, as we would want were we to be searching for possible equilibrium solutions, then either $y + 1 = 0$ or $y - 1 = 0$. Thus, we can see that for all x, $y' = 0$ when $y = \pm 1$. As you may expect, with an initial condition of $(x_0, 1)$ or $(x_0, -1)$, for any x_0, the solution will remain constant over time. What happens when we do not start at a point already at the equilibrium value for y? In a previous example we saw that we had an unstable equilibrium. Let's look at the vector field for $y' = y^2 - 1$.

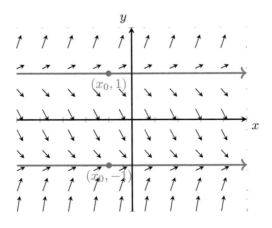

Notice that the arrows near $y = 1$ appear to be moving away from that equilibrium as in that previous example. However, the arrows near $y = -1$ appear to be moving towards that equilibrium. This behavior is characteristic of a **stable equilibrium**.

○○○

We can also consider systems of ODEs, for example, predator/prey relationships, which we will discuss in more detail in Chapter 5.

Example: Let's consider the interaction of species x and species y, which can be modeled by the equations:

$$\begin{cases} \dfrac{dx}{dt} = 1.2x - 0.01xy \\[2mm] \dfrac{dy}{dt} = -0.5y + 0.01xy \end{cases}$$

The vector field for the system of differential equations is shown below. Note that we can sketch this type of vector field in the same way as a vector field with only one equation by computing:

$$\frac{dy}{dx} = \frac{-0.5y + 0.01xy}{1.2x - 0.01xy}$$

and continuing with the process described earlier in this section.

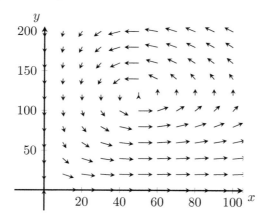

You may notice from the vector field that solutions circle around a point at $(50, 120)$. This is the equilibrium.

The main interest when confronted with a system of differential equations is in the behavior of x and y with respect to time. If the system can be solved analytically, the two graphs of x and y with respect to time can be plotted and give a picture of the future behavior of the variables. Let's focus on one particular solution, highlighted below, with initial condition $(80, 100)$.

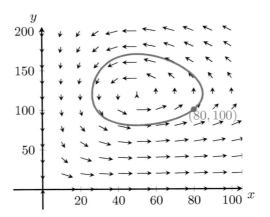

The direction of this solution is counterclockwise. If our vector field didn't have arrows to indicate direction, we could still determine this by evaluating the original derivative equations at the initial condition to see whether the solutions are initially increasing or decreasing.

Now that we have a vector field, we can make rough sketches of the solution for each species as functions of time. For example, we know that our initial condition is $(80, 100)$. After some amount of time, we can see that the populations reach the approximate levels of $(50, 170)$, then continue around the curve to approximately $(30, 120)$, and so on. If we graph only the x values, and then separately graph only the y values, as functions of t, then we will be able to see how these populations change over time.

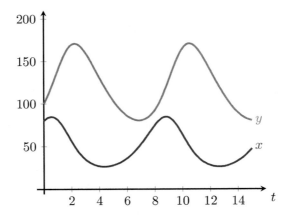

We can see that both populations have a periodic behavior over time. This corresponds to the circular shape that we saw in the vector field, and occurs because these populations coexist.

Example: Consider the interaction of species x and species y, which can be modeled by the equations:

$$\begin{cases} \dfrac{dx}{dt} &= 2x - 0.02x^2 + 0.1xy \\[2mm] \dfrac{dy}{dt} &= y - 0.1xy. \end{cases}$$

As before, we can sketch the vector field for this system by computing $\frac{dy}{dx} = \frac{y-0.1xy}{2x-0.02x^2+0.1xy}$. Doing so results in the following vector field.

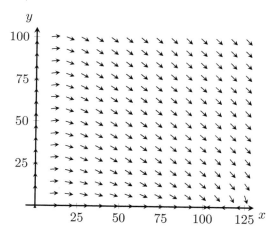

You may notice that all the arrows seem to be gravitating towards a point on the x-axis, which will ultimately be the equilibrium for this system. To get an idea of the possible behavior and interaction of these two species, we can plot one particular solution given, for example, the initial condition $(5, 65)$.

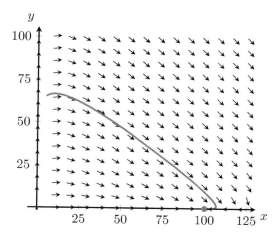

We can see that for this solution, the population of y initially increases just slightly, but quickly the growing population of x overpowers the growth rate of y. The population of y then starts decreasing rapidly as x increases. Eventually, there are almost no y remaining, but the population of x was growing so quickly as to exceed the natural carrying capacity of 100. As the population of y goes to zero, the population of x settles at the carrying capacity of 100, and we find the fixed point of $(100, 0)$. We can observe this behavior more easily if we plot the graphs of x and y with respect to time.

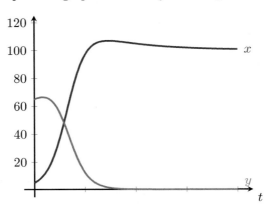

As we can see, our analysis from the solution curve is correct! The population of y increases to a maximum then quickly goes to zero while the population of x grows past its natural carrying capacity before settling at 100. This corresponds to the fixed point at $(100, 0)$ in our vector field. We will explore these types of models and carrying capacities further in Chapter 5.

2.3 Exercises

Sketch the vector fields for the following ODEs.

1. $y' = x^2 - 1$

2. $y' = y^2 + 4y + 4$

3. $y' = \sin x$

4. $y' = e^{-x}$

5. $y' = 2y - 2$

6. $y' = \frac{x}{y}$

7. $y' = y(y^2 - 4)$

8. $y' = (y - 2)(y + 3)^2$

For the problems below, match the differential equation with the corresponding vector field. Then, sketch solution curves corresponding to the initial conditions provided.

9. $y' = y^2 - 1$ with initial conditions $y(0) = 0$, $y(0) = 3$, and $y(2) = -2$.

10. $y' = y/x$ with initial conditions $y(1) = 0$ and $y(-2) = 3$.

11. $y' = (x - 1)(y + 1)$ with initial conditions $y(0) = 1$, $y(1) = 4$, and $y(0) = -2$.

12. $y' = x^2 - 1$ with initial conditions $y(0) = 3$ and $y(-2) = 0$.

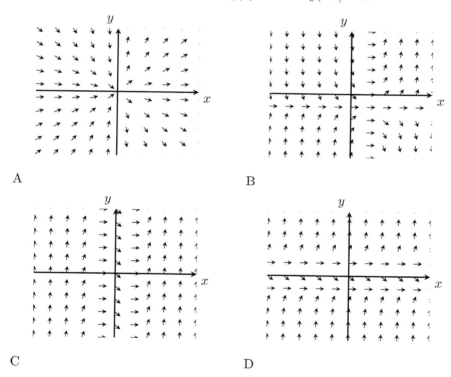

A
B
C
D

For the following problems, sketch the vector field with the aid of computer software and use it to determine the long-term behavior of the solution of the given IVP.

13. $y' = xy^2$, $y(0) = -2$

14. $y' = \sin y$, $y(0) = 1$

15. $y' = x + 3y$, $y(-2) = 5$

16. $y' = y - y^2$, $y(0) = 3$

17. $y' = x^2 - y^2$, $y(-2) = 4$

18. $y' = e^{-y} - 1$, $y(3) = -17$

Chapter 3

Modeling with First-Order ODEs

In the last chapter we used vector fields to understand the behavior of solutions of ODEs and learned to verify solutions of ODEs. Typically, when modeling with ODEs the goal is to make some sort of predictive model. For example, say we know the reproduction rate of rabbits and how many rabbits an area can sustain. It is natural that we would like to know at what point of time the rabbit population will reach a certain level. This may be useful information because once that level is reached, it might be beneficial to introduce population control measures.

Our methods from Chapter 2 have limited utility in this regard. Vector field analysis can give us a qualitative understanding of the long term behavior of a model, but cannot give us the types of explicit answers to questions that we sometimes require. To find more exact answers to these fundamental questions, it can behoove us to find explicit, closed form solutions of the differential equations we are using to model our biological systems.

Classifying ODEs, which we covered in Chapter 2, is a critical skill because the approach we take to solving and finding explicit solutions of ODEs varies depending on the classification. In this chapter, we will explore a few methods for finding the solutions of first order ODEs, i.e., ODEs of the form $y' = f(t, y)$ for a wide class of functions. Solutions of higher order ODEs and systems of ODEs will be covered in Chapters 4 and 5, respectively.

The methods covered in this chapter do not constitute an exhaustive list, and it is important to note that we are unable to find a general solution for every function of t and y. In cases where general solutions cannot be found analytically, numerical methods can be used to approximate the solutions. In the absence of using numerical methods, the tools discussed in Chapter 2 can be an excellent substitute for gleaning a qualitative understanding for the behavior of solutions of an ODE.

DOI: 10.1201/9781003298663-3

3.1   First-Order ODEs and Their Applications

While finding the solutions of differential equations is fulfilling in its own right, there are a number of real world problems which can be modeled using first-order ODEs. The solutions of these ODEs will then provide us with valuable information about the problem at hand.

We will begin by looking at exponential, migration, and logistic models to study population dynamics, while recognizing that these types of models can be used to study other types of growth and decay as well. We will then examine a similar model, Newton's Law of Cooling, that allows us to study how quickly an object changes temperature. We will then investigate mixing problems. These model, for example, the amount of chlorine in a pool provided that water with a certain concentration of chlorine is added at a given rate while water is drained at a different rate as effluent. These types of problems can also be used to describe the process of dialysis and other processes in biology. Finally, we will return to population dynamics in the study of interacting species. We use the tools found in this chapter to find the solution of the ODE and to answer questions about the behavior of the model for all of these applications.

In this section we will explore these different applications and how to construct first-order ODE models for them. We will then return to these models throughout the chapter as we learn the various methods which can be used to solve these ODEs. We will not solve any ODEs in this section, but merely outline the modeling process.

3.1.1   Exponential, Migration, and Logistic Models

The first of many applications of modeling with first-order ODEs, which we will explore is population models. These include finding the exponential growth of a population with no predation, finding the behavior of a population with migration, and finding the logistic growth of a population with a carrying capacity. For these population dynamics questions, the following models are very helpful.

The solutions of these models will take two main forms. The first type is exponential in which the population grows without bound. The rate of growth will depend on the intrinsic growth rate for that population. The second type is logistic in which the population grows in a similar fashion to an exponential model at the onset but then levels off once it approaches a carrying capacity for the population.

Let's look at the exponential model first!

Exponential Model

A population may grow at a rate proportional to the existing population. Under this assumption, an appropriate model is the **exponential model** for population growth:

$$\frac{dP}{dt} = rP,$$

where $r > 0$ is the growth constant. Note that we can also study decay using $r < 0$. Solutions of the exponential model behave as follows.

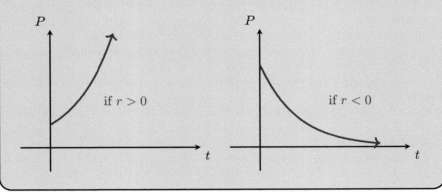

Example: A small population of 10 rabbits is introduced into a large geographic area with no known predators. Suppose that the population of rabbits doubles every four years. Construct an initial value problem (IVP) to model this situation.

From the above box you may suspect that we use $r = 2$ as the growth constant, and that we should use the initial value problem:

$$\frac{dP}{dt} = 2P, \quad P(0) = 10$$

to model this situation, but let's take a closer look at this equation and the units involved. Keeping track of the units in the model parameters can be a helpful way to ensure you develop the model correctly.

We have two variables in this model and one parameter. These are:

- P: population, which has rabbits as the unit.

- t: time, which has years as the unit.

- r: the intrinsic growth rate.

Let's presume we don't know the units of the intrinsic growth rate and see if we can arrive at them.

The left-hand side of this ODE is:

$$\frac{dP}{dt}.$$

Recall that in the differential, dP is denoting a very small step of the variable P while dt is denoting a very small step of the variable t. Therefore, the units on the left-hand side are:

$$\frac{dP \text{ rabbits}}{dt \text{ year}} = \frac{dP}{dt} \text{ rabbits/year}.$$

We should expect the right hand side to have the same units, and currently the right-hand side is:

$$(r \text{ units})(P \text{ rabbits}) = rP \text{ units} \times \text{ rabbits}.$$

As the units on the left- and right-hand side of the equation must agree, the unknown units for the intrinsic growth rate must be 1/year, or "per year." These are the standard units for an intrinsic growth rate. Naively we chose $r = 2$, but neglected to account for the fact that this is 2 per 4 years. Therefore, to convert to the correct units for the intrinsic growth rate, we must have $r = 0.5$ / year. Therefore, the correct equation for us should be:

$$\frac{dP}{dt} = 0.5P, \quad P(0) = 10.$$

○○○

In the above example we saw a model with a positive intrinsic growth rate. In other words, the population was *increasing* at a rate proportional to the amount of rabbits. There are other circumstances where there is a negative "growth" rate. These are cases where the variable *decreases* at a rate proportional to the amount present. For example, we can analyze the concentration of a drug in a patient's body using this exponential model.

Example: Suppose that 100 mg of a drug is administered to a patient. This drug decays with a decay rate of 0.15/hour.

We can thus set up a similar IVP as with the rabbit example. However, we must take care to realize that we have been given a decay rate which is positive. To account for the fact that the amount of the drug is *decreasing* at a rate proportional to the quantity, we must include a negative sign in the equation. Therefore, for this situation, the IVP:

$$\frac{dQ}{dt} = -0.15Q, \quad Q(0) = 100$$

is the correct model, where Q represents the quantity of the drug in mg and t is measured in hours.

○○○○○○○○○○○○○○○○○○○○○○○○○○○○○○○○○○○○○○○

Often, it is desirable to determine the *half-life* of a drug. That is, how long after a drug is administered that half of the drug has been metabolized. Alternatively, you may know the half-life of a drug and want to know how long after administration the amount of the drug present crosses below a therapeutic threshold. The ability to solve the above IVP, which we will develop later in this chapter, will allow us to answer these vitally important questions.

There are also times when the exponential model is insufficient. For example, suppose there is an intrinsic growth rate, but in addition, there is constant migration out of the area of study. This situation can be modeled as below.

Migration Model

We may know that a population grows exponentially on its own, but that k individuals migrate to another area each day. In this case we can modify the exponential differential equation to account for the migration:

$$\frac{dP}{dt} = rP - k.$$

with $r, k > 0$. Note that we can also use a similar model to study exponential decay with $r < 0$, or with immigration $k < 0$ rather than emigration $k > 0$. Also, solutions of this type of equation are merely transformations of the exponential model solutions.

Returning to our rabbit example, we now intend to capture a certain number of rabbits per year (in an idealized, continuous fashion) and release them in an animal sanctuary farm upstate. Suppose the example is as follows:

Example: A small population of 10 rabbits are introduced into a large geographic area with no known predators. Suppose that the population of rabbits doubles every four years. Furthermore, the local wildlife management agency captures four rabbits per year and releases them in a farm upstate. Construct an IVP to model this situation.

As in our previous example, our intrinsic growth rate is computed to be 0.5/year. As such, our equation with units included looks like this:

$$\frac{dP}{dt} \text{ rabbits/year} = (0.5/\text{year})(P \text{ rabbits}) - k.$$

What should the value of k be? Judging by the units in the equation, we should expect k to have rabbits/year as units. As we were given the number of rabbits removed per year rather than the number removed per month or otherwise, we can simply use $k = 4$. Therefore, the correct IVP to model this situation is:

$$\frac{dP}{dt} = 0.5P - 4, \quad P(0) = 10.$$

○○○

Surely, a population cannot grow without bound as could happen with an exponential model. Eventually there won't be enough food to sustain the growth, and the rate of change of the population will level off. The logistic model, outlined below, allows us to capture this phenomenon.

Logistic Model

One other common example is the **logistic growth model**. This models a population which grows proportional to the existing population at first, but which has some carrying capacity which limits the long term growth of the population. This situation is modeled by the differential equation:

$$\frac{dP}{dt} = rP\left(1 - \frac{P}{M}\right),$$

where M is the carrying capacity of the population and r is the intrinsic growth rate. Solutions of the logistic model behave differently, depending on the initial value.

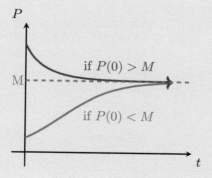

You should think of the term $\left(1 - \dfrac{P}{M}\right)$ as in a way scaling the otherwise exponentially growing rP term. When P is close to zero, this term is close to one, so the model will behave similarly to an exponential model. However, as the population approaches the carrying capacity, $\dfrac{P}{M}$ approaches one. This will force $\dfrac{dP}{dt}$ to go to zero, so the total population will level off at the carrying capacity.

Note that we could *combine* the logistic and migration models or in other ways generalize the model being used. However, for now we will focus on these three most general population models. Let's return once more to the rabbits.

○○○○○○○○○○○○○○○○○○○○○○○○○○○○○○○○○○○○○○○

Example: A small population of 10 rabbits are introduced into a large geographic area with no known predators. Suppose that the population of rabbits doubles every four years and that the carrying capacity of the area is 10,000 rabbits. Construct an IVP to model this situation.

As before, the intrinsic growth rate is 0.5/year. Now that there is a carrying capacity, we simply must append the term $(1-\frac{P}{M})$. Therefore, as $M = 10,000$, our IVP must be:

$$\frac{dP}{dt} = 0.5P\left(1 - \frac{P}{10,000}\right), \quad P(0) = 10.$$

3.1.2 Newton's Law of Cooling

Another very important application which we will outline here is finding the temperature of an object as it heats or cools due to the ambient temperature of its surroundings. A classical model for this situation is **Newton's Law of Cooling**.

Newton's Law of Cooling states that the rate of change of temperature of an object is proportional to the difference between the object's temperature and the surrounding temperature. This can be modeled using the differential equation:

$$\frac{dT}{dt} = -k(T - s),$$

where k is the heat transfer coefficient and s is the temperature of the surroundings.

You may note that the equation in the above box can be rewritten as:

$$\frac{dT}{dt} = -kT + ks$$

and think that this is very similar to a population model with an intrinsic decay rate but positive migration. While it is true that these are both linear first order ODEs with constant coefficients, these are being used to model different situations. In the migration model the idea is that the rate of change of the population is proportional to the current population, and that the population is being reduced at a constant rate. In contrast, Newton's Law of Cooling presumes that the rate of change of the temperature of an object is proportional not to the current temperature but the *difference* between the current temperature of the object and the ambient temperature. This is a reasonable assumption to make and falls in line with your real-world experience. Take two hot cups of coffee. Put one to cool in a sunlit windowsill. Put the other in the freezer. For which cup of coffee do you presume the rate of change of the temperature will be greatest?

Let's see how to set up one of these problems, which will allow us to get a handle on the units involved as well.

○○○

Example: Suppose a cup of freshly brewed coffee at 200°F is placed on a sunlit windowsill at 80°F. Set up an IVP to model this situation, where time is measured in minutes.

Note that there is missing information. We haven't been given the heat transfer coefficient. In these problems you will either be given the coefficient or asked to approximate it. You can do this in two ways. First, you may be given the initial rate of change of the temperature. Otherwise, you can take a second measurement after a certain amount of time has passed to approximate that initial rate of change. For now, let's set the problem up with what we know.

We know that the coffee is at 200° to begin. As we are going to construct an ODE to model T, the temperature of the coffee, this is our initial condition. Thus we have: $T(0) = 200$. We also know the temperature of the sunlit windowsill is $s = 80°$. Thus, using the Newton's Law of Cooling model formula, we have:

$$\frac{dT}{dt}\frac{°\text{ F}}{\text{minute}} = -(k \text{ units})(T - 80)°, \quad T(0) = 200.$$

As you may have suspected, the units for the heat transfer coefficient, k, is $1/\text{minute}$ or, in general, per timestep used for the model. We will see how to find an approximate value of k after we acquire the tools necessary to solve this type of ODE.

○○

Example: Suppose a sample of Ustilago maydis, a type of fungus which causes the disease corn smut, or Huitlacoche, is removed from an ear of corn at 55°F and immediately placed in a skillet held at 350°F.

Note that this is actually a heating problem! That is no problem, and we can continue as we did in the previous example. We know that our initial temperature is 55°, so we have that $T(0) = 55$. The skillet is held at 350°, so $s = 350$. Therefore, we have:

$$\frac{dT°}{dt}\frac{\text{ F}}{\text{minute}} = -(k \text{ units})(T - 350)°, \quad T(0) = 55.$$

3.1.3   Mixing Models

Another classic application is the mixing problem, sometimes referred to as tank problems. In problems of this type, we have a tank which contains some amount of a substance dissolved in its contents. For example, a tank may contain water of a certain salinity. Then, water with a different salinity is pumped into the tank at a given rate while the tank drains off its contents at a certain rate. There is generally a baseline assumption that the tank is "well mixed," meaning that once salt is added to the water, the salinity of all the water in the tank increases rather than just the area around the pipe. The question is what salinity level the tank will have at a certain time, or as t goes to infinity. The trick in these problems is to recognize that the object we are concerned with is not the amount of salt per gallon, but the total amount of salt in the tank.

Mixing Model

If we let A be the total amount of salt in the tank, then the differential equation we are interested in is:

$$\frac{dA}{dt} = R_{\text{in}} - R_{\text{out}},$$

where R_{in} is the rate at which salt is entering the tank and R_{out} is the rate at which it is leaving.

Note that this formula is much more general than those given for the population models or Newton's Law of Cooling. This is due to the fact that the rate in and rate out can vary wildly from problem to problem. For example, the concentration of the solution coming in may be constant, or it may be variable. The flow rate in and out may be the same, in which case the volume remains constant, or they may be different, in which case we must account for the change in the volume.

Let's consider a basic example to get a feeling for how to set these problems up.

○○

Example: A pond with volume 10^6 L currently contains a concentration of 2.5×10^{-8} mol/L of the hydronium ion, H_3O^+. This corresponds with a pH of approximately 7.6. The stream that feeds the pond has been polluted by an upstream chemical factory. The stream flows into the pond at a rate of 150 L/s and has a concentration of 3×10^{-2} mol/L of the hydronium ion. This corresponds with a pH of approximately 1.5. The pond drains via another stream that flows at a rate of 150 L/s as well. Find the amount of time before the pH of the pond is below 4.5, after which fish will begin to die off, given that pH is calculated as pH$=-\log(H_3O^+\text{mol/L})$.

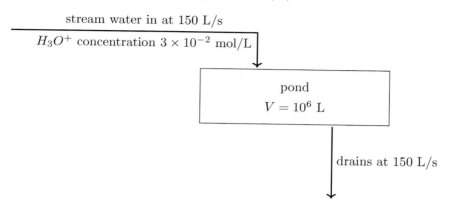

stream water in at 150 L/s

H_3O^+ concentration 3×10^{-2} mol/L

pond
$V=10^6$ L

drains at 150 L/s

Remember, we want A to represent the total amount of material in solution, not the concentration. As such, our units for A should be mol in this problem. To find the rate in and rate out, we use the information given. Note that here volume is measured in liters, time in seconds, and the amount measured in moles (mol), which corresponds to 6.02214076×10^{23} elemental units, here hydronium ions. Note also that the rate that the amount of material is flowing in or out is equal to the flow rate of the liquid times the concentration of the material. So far we have:

$$\frac{dA}{dt} = R_{in}-R_{out}$$

$$\frac{dA}{dt} = (\text{flow in})\times(\text{concentration in})-(\text{flow out})\times(\text{concentration out})$$

$$\text{mol/s} = \left(\frac{150\text{L}}{\text{s}}\right)\times\left(\frac{3\times 10^{-2}\text{mol}}{\text{L}}\right)-\left(\frac{150\text{L}}{\text{s}}\right)\times(\text{concentration out})$$

As you can see the liter units will cancel on the right-hand side, so we will match the mol/s units on the left-hand side. What remains to be seen is the concentration out. This is where the well-mixed assumption comes into play. We are going to assume that the concentration flowing out is equal to the amount in the pond divided by the volume of the pond, i.e.:

$$\text{pond concentration} = \frac{\text{total amount of material (mol)}}{\text{pond volume (L)}}$$

The pond volume is given to us as 10^6 L. The total amount of material is the variable A which we will eventually be solving the IVP and will find a solution. For now we will use A to represent this amount, so we have that:

$$\text{concentration out} = \frac{A\text{ mol}}{10^6\text{ L}}.$$

Returning to our ODE and leaving off the units, we now have:

$$\frac{dA}{dt} = \left(\frac{150L}{s}\right) \times \left(\frac{3 \times 10^{-2}\text{mol}}{L}\right) - \left(\frac{150\ L}{s}\right) \times \left(\frac{A\ \text{mol}}{10^6\ L}\right)$$

$$\frac{dA}{dt} = 4.5 - 0.00015A.$$

For our initial condition, we use the fact that the initial concentration is 2.5×10^{-8} mol/L to compute:

$$A(0) = \left(\frac{2.5 \times 10^{-8}\text{mol}}{L}\right)(10^6 L)$$

$$= 2.5 \times 10^{-2}\text{mol/L}$$

$$= 0.025 \text{ mol/L}.$$

Therefore, the IVP for this problem is:

$$\frac{dA}{dt} = 4.5 - 0.00015A, \quad A(0) = 0.025.$$

As mentioned, this is a simple example where the inflow and outflow rates are the same, and the inflow concentration is constant. In later sections we will explore setting up more complicated models once we are armed with the tools to solve the IVPs in question.

3.1.4   Interacting Species

When two species interact, their interaction can be modeled using a system of ODEs. However, there is a way to analyze the behavior of that system using a first-order ODE. Let's explore that idea.

Suppose there are two species, species A, which is prey, and species B, which is a predator. Suppose the population of species A grows exponentially in the absence of a predator with an intrinsic growth rate, a. The population of species B decays exponentially with an intrinsic decay rate of c in the absence of any prey (they will starve to death). If these species do not interact, we have the following system:

$$\begin{cases} \dfrac{dx}{dt} = ax \\ \dfrac{dy}{dt} = -cy \end{cases}$$

where x represents the quantity of species A and y represents the quantity of species B.

Now suppose there is interaction, then the rate of change of the population of species A also depends on the amount of predators which exist from species B. The more predators, the slower the growth rate of the population. Similarly, the rate of change of species B depends on the amount of prey, so the growth rate depends on the amount of species A that exists. We can use the *interaction terms* b and d here to capture this. Thus we have the system:

$$\begin{cases} \dfrac{dx}{dt} = ax - bxy \\ \dfrac{dy}{dt} = -cy + dxy. \end{cases}$$

Notice that there is a negative sign before b and a plus before d. Here all constants are positive. This system can be considered as a first-order ODE with respect to y and x by dividing $\frac{dy}{dt}$ by $\frac{dx}{dt}$. The result is:

$$\frac{\frac{dy}{dt}}{\frac{dx}{dt}} = \frac{-cy + dxy}{ax - bxy}$$

$$\frac{dy}{dx} = \frac{-cy + dxy}{ax - bxy}.$$

We will see in later sections that for some equations of this type, we can find a closed-form solution. Note that this is the simplest model for predator-prey. We could also consider a situation where there is a carrying capacity for one or both populations.

○○

Example: Suppose there is a population of 100 rabbits on an island. The rabbits have an intrinsic growth rate of 0.5/year and a carrying capacity on the island in the absence of predators of 10,000. A population of 5 wolves is introduced to the island. The wolves have an intrinsic decay rate of 1.5/year in the absence of prey and no carrying capacity. Suppose the interaction terms are $b = 0.5$ and $d = 0.3$ Construct an IVP to model this scenario.

Here we simply replace the exponential growth assumption on the population x with a logistic growth assumption. In the absence of interaction, we would have:

$$\begin{cases} \dfrac{dx}{dt} = 0.5x\left(1 - \dfrac{x}{10,000}\right) \\ \dfrac{dy}{dt} = -1.5y \end{cases}$$

However, we are given the interaction terms, and thus have:

$$\begin{cases} \dfrac{dx}{dt} = 0.5x\left(1 - \dfrac{x}{10,000}\right) - 0.5xy \\ \dfrac{dy}{dt} = -1.5y + 0.3xy \end{cases}$$

Dividing these two equations and distributing the $0.5x$ yields the first-order ODE:

$$\frac{dy}{dx} = \frac{-1.5y + 0.3xy}{0.5x - 0.00005x^2 - 0.5xy}.$$

It is important here to recall that y represents the wolf population and x represents the rabbit population. As such, this ODE describes the rate of change of the population of wolves with respect to the population of rabbits. As we initially have 100 rabbits and 5 wolves, our initial condition should be $y(100) = 5$. Therefore, we can model this scenario with the IVP:

$$\frac{dy}{dx} = \frac{-1.5y + 0.3xy}{0.5x - 0.00005x^2 - 0.5xy}, \quad y(100) = 5.$$

3.1 Exercises

For the following exercises determine if an exponential, migration, or logistic model would be the most appropriate to model the circumstance. Explain why. Do not construct an ODE.

1. A cluster of mold growing in a Petri dish.

2. A population of emu in Australia with no natural predators.

3. A population of emu in Australia with a constant amount being culled every year.

4. A population of guppies in a fish tank which can only sustain 500 fish.

For the following exercises construct, but do not solve, an IVP to model the given scenario.

5. A turkey is removed from the refrigerator at $37°F$ and immediately placed in an oven at $325°F$. Suppose the heat transfer coefficient is $k = 0.00327$. Construct an IVP to model the temperature of the turkey over time.

6. A population of 500 herring is introduced in a large sea. The herring has an intrinsic growth rather of $0.59/year$, there are no predators, and an unlimited number of herring can reside in the sea.

7. A 500 L tank of salt wanted with salinity 35 g/L is being filled at a rate of 2 L/min from a brackish water source with salinity 15 g/L. At the same time, the tank is being drained at a rate of 2 L/min. Construct an IVP to model the salinity of the water in the tank.

8. An elk dies of natural causes in the winter with a body temperature of 38.5°C. If the surroundings are a constant -10°C, construct a model for the body temperature of the carcass.

9. A bag of sweet potatoes is placed on a windowsill at a temperature of 40°F to cure for a week. The ambient temperature of the windowsill is variable, and can be modeled by the equation $s(t) = 70 + 20 \sin(\pi t/24)$. Construct an IVP to model the temperature of the bag of sweet potatoes over time.

10. A 27,000-gallon saltwater swimming pool has a salinity of 0.026 lb/gal. The pool is at full capacity when it starts raining, adding water to the pool at a constant rate of 5 gal/min. The pool owner begins draining the pool at the same rate in order to maintain the water level.

3.1 Project: Coronavirus Outbreak

In December 2019, the World Health Organization (WHO) was first notified of many cases of pneumonia with an unknown cause. This cause was finally identified as the coronavirus on January 7, 2020, and officials tried to follow the spread of infection ever since. Beginning on January 20, 2020, the WHO released a *Situation Report* explaining the current status of the coronavirus epidemic.[1] The table below summarizes the total number of cases (which may include individuals who have already recovered or died from infection) for the first 15 days of reporting.

Date of Report	t days	Total Cases
1/20/2020	0	47
1/21/2020	1	282
1/22/2020	2	314
1/23/2020	3	581
1/24/2020	4	846
1/25/2020	5	1,320
1/26/2020	6	2,014
1/27/2020	7	2,798
1/28/2020	8	4,593
1/29/2020	9	6,065
1/30/2020	10	7,818
1/31/2020	11	9,826
2/1/2020	12	11,953
2/2/2020	13	14,557
2/3/2020	14	17,391
2/4/2020	15	20,630

[1]Coronavirus disease (COVID-19) weekly epidemiological updates and monthly operational updates. (2023). World Health Organization. https://www.who.int/emergencies/diseases/novel-coronavirus-2019/situation-reports

1. Plot the total cases given in the table. What kind of equation do you think is appropriate for modeling this kind of epidemic?

2. Let $N(t)$ be the total number of coronavirus cases after t days. Write a differential equation to represent the change in total cases of coronavirus, assuming that the infection grows at a rate proportional to the total number of cases.

3. With the knowledge that differential equations of the form $\frac{dy}{dt} = ky$ have solutions $y = Ce^{kt}$, where C is a constant, solve the differential equation you obtained in Question 2 by finding C and k. Use the data from January 20 and January 25 as your initial conditions to find an expression for the total number of coronavirus cases at time t.

4. Estimate how long it will take for there to be 2,500 cases of coronavirus. Explain how this number compares to the given data.

5. Estimate the total cases after 4 days, 6 days, and 8 days and compare these results to the given data. Do you think this model is realistic in describing coronavirus at any point during the first 15 days of reporting? Do you think this model is realistic for the entire time period of the coronavirus pandemic? Explain your reasoning.

6. Interpret your model by using computer software to sketch the solution curve for your differential equation along with the data from the table in the same graph. Explain in words how your solution compares to the data from the table.

7. When creating our model, we assumed that the coronavirus infection spreads at a rate proportional to the total number of cases. Is this realistic? Explain any strengths and/or weaknesses of this type of model to describe the pandemic.

8. Is there a different model that may fit the data better? If so, find it! Sketch a vector field of the differential equation that you think would best fit the data. Make sure to include the solution going through the initial condition from the table, and plot the data from the table in the same graph. If the vector field does not fit the data, adjust your differential equation so that it fits better. Explain how you arrived at your final differential equation.

3.2 Autonomous Equations

A special case of first-order ODEs is the **autonomous** first-order ODE. The differential equations describing exponential growth and decay are autonomous first order ODEs, for example. There are several ways to qualitatively describe the behavior of solutions of autonomous first order ODEs which we will explore in this section.

> **Definition:** An **autonomous** first order ODE is an equation of the form $\frac{dy}{dt} = f(y)$.

Note that for autonomous ODEs, we have $y' = f(y)$ rather than $y' = f(t, y)$ as in a general first-order ODE. Therefore, $y' = y^2 + 2$ is an autonomous ODE, whereas $y' = y + t$ is not as there is a t term in the equation.

Example: Exponential growth $\frac{dy}{dt} = ry$, where r is the growth constant, is an autonomous first-order ODE. We can solve this ODE easily and will do so in the coming sections, but for now we would like to focus on what the possible solutions of this ODE look like without solving it. We can do this by examining the vector field as we did in the previous chapter, or we can conduct a **phase line analysis**.

> To conduct a **phase line analysis** for the ODE $\frac{dy}{dt} = f(y)$, we first sketch a graph of $f(y)$ versus y. It is important to note the values of y for which $f(y) = 0$, as these values, which we will call y^*, are the equilibria of the ODE.
>
> - If $f(y) > 0$, draw arrows pointing to the right on the horizontal axis.
>
> - If $f(y) < 0$, draw arrows pointing to the left on the horizontal axis.
>
> The phase line is the horizontal axis and shows the stability of the equilibria. If arrows on either side of an equilibrium, y^*, point towards that equilibrium, then we call y^* (asymptotically) **stable**. If both arrows point away, then we call y^* **unstable**. It is also possible that one arrow points towards and one arrow points away from the equilibrium. In this case, we call y^* **semistable**.

Example: Continuing with $\frac{dy}{dt} = ry$, we need to plot $f(y) = ry$ versus y. Assuming $r > 0$, we have:

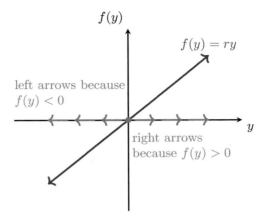

The equilibrium of the ODE occurs when $f(y) = ry = 0$, so we have equilibrium $y^* = 0$. The phase line is simply the horizontal axis in the above graph and can be drawn on its own:

Based on the phase line, we can determine the stability of the equilibrium $y^* = 0$. As all arrows point away from $y^* = 0$, we call $y^* = 0$ unstable.

Notice that the arrows in the phase line correspond to the instances when $f(y) > 0$ (or in other words, when $y' > 0$, corresponding to when y is increasing), or when $f(y) < 0$ (when y is decreasing). We can use this idea to sketch a graph of the possible solutions (as functions of t) of the ODE $\frac{dy}{dt} = ry$. In this process, imagine rotating the phase line so that it becomes the vertical axis in the graph below. We will typically graph the solutions for positive t only, as t generally represents time. Below, we focus on sketching the graph of a few representative solutions, one in each interval surrounding the equilibrium.

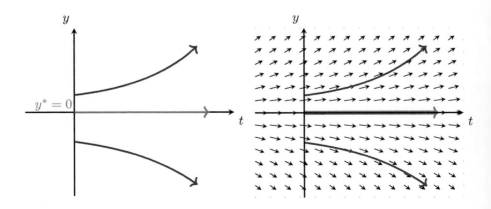

If we compare our solutions plot to the vector field for the ODE $\frac{dy}{dt} = ry$, we see a very similar result. The phase line analysis allows us to get an idea of what the solutions of an ODE look like without solving the ODE or the hassle of drawing a vector field.

○○○

We can also prove the stability of equilibria without graphing by using calculus.

Stability Theorem for Autonomous Equations: Let $\frac{dy}{dt} = f(y)$ have an equilibrium at $y = y^*$, then:

- if $f'(y^*) > 0$, y^* is an unstable equilibrium;

- if $f'(y^*) < 0$, y^* is an (asymptotically) stable equilibrium;

- if $f'(y^*) = 0$, the test is inconclusive.

Example: For the ODE $\frac{dy}{dt} = ry$, we previously found the equilibrium $y^* = 0$. We also know that $f(y) = ry$. Then $f'(y) = r$, so $f'(y^*) = f'(0) = r > 0$. Hence, by the Stability Theorem, $y^* = 0$ is an unstable equilibrium. This matches our result from the phase line analysis.

○○

Example: Let's conduct a phase line analysis for the ODE $y' = (y-4)(y+2)$ in order to determine the stability of all equlibria and sketch a solutions plot. We will also confirm our results using the Stability Theorem.

First, we need to plot $f(y) = (y-4)(y+2)$ versus y. We have:

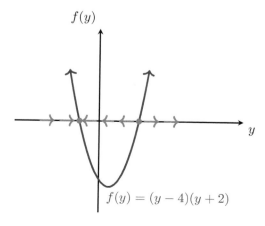

Recall that the left arrows are drawn when the function is below the axis and that we draw right arrows when the function is above the axis. Notice that the equilibria $y^* = -2$ and $y^* = 4$ occur when $f(y) = 0$. We can draw just the phase line as follows:

As we can see in our phase line, $y^* = -2$ is stable because all arrows nearby $y^* = -2$ point towards this equilibrium and $y^* = 4$ is unstable because all arrows nearby $y^* = 4$ point away from this equilibrium. Finally, we can use our understanding of this stability to sketch several solutions:

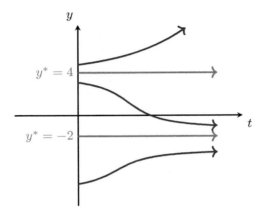

Just as before, we see that solutions will tend towards the stable equilibrium and away from the unstable equilibrium. We may also prove stability using the Stability Theorem. We know that $f(y) = (y - 4)(y + 2)$. Then $f'(y) = 2y - 2$ and we must evaluate this at both equilibrium solutions $y^* = -2$ and $y^* = 4$ (the solutions of $f(y) = 0$). We have $f'(-2) = -6 < 0$, so $y^* = -2$ is a stable equilibrium, and $f'(4) = 6 > 0$, so $y^* = 4$ is an unstable equilibrium.

$\circ\circ\circ\circ\circ\circ\circ\circ\circ\circ\circ\circ\circ\circ\circ\circ\circ\circ\circ\circ\circ\circ\circ\circ\circ\circ\circ\circ\circ\circ\circ\circ\circ\circ\circ\circ\circ\circ\circ\circ$

You may have noticed that some of the models we examined in the previous section happen to be autonomous equations. One such model was the logistic growth model. Let's conduct phase line analysis on a logistic model.

Example: A population of fruit flies is introduced into a large atrium. The fruit flies have an intrinsic growth rate of 0.15/day, and the atrium has a carrying capacity of 25,000 flies. Develop a model for the growth of the fruit fly population over time and conduct phase line analysis to determine the long-term behavior of the model.

We recognize that as there is a carrying capacity, this should be modeled as a logistic growth model. As such, we have the ODE:

$$\frac{dP}{dt} = 0.15P\left(1 - \frac{P}{25,000}\right).$$

Let's conduct phase line analysis on this ODE. We plot this in the $P, f(P)$ plane and draw the appropriate arrows, left when the function is below the axis and right when the function is above the axis. We have:

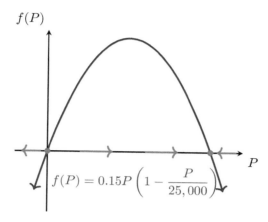

In this case, we have equilibria $P^* = 0$ and $P^* = 25,000$, which occur when $f(P) = 0$. We can sketch the phase line:

As we can see from both the $f(P)$ versus P graph and from the phase line itself, $P^* = 0$ is unstable because all arrows nearby $P^* = 0$ point away from this equilibrium. Similarly, $P^* = 25,000$ is stable because all arrows nearby $P^* = 25,000$ point towards this equilibrium.

Let's also prove stability using the Stability Theorem. We know that:

$$\begin{aligned}
f(P) &= 0.15P\left(1 - \frac{P}{25,000}\right) \\
&= 0.15P - \frac{3}{500,000}P^2 \quad. \\
f'(P) &= 0.15 - \frac{3}{250,000}P
\end{aligned}$$

For $P^* = 0$, we have $f'(0) = 0.15 > 0$. Then $P^* = 0$ is an unstable equilibrium, which agrees with our results from the phase line analysis. Similarly, for $P^* = 25,000$, we have $f'(25,000) = -0.15 < 0$, so $P^* = 25,000$ is a stable equilibrium. Finally, we can use this knowledge to sketch several solutions:

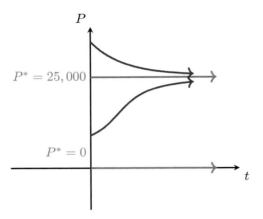

We made sure to graph solutions on and around the equilibria, with solutions tending towards the stable equilibrium and away from the unstable equilibrium. Note that because the original ODE represented the change in the fruit fly population, it only made sense to graph the solutions for positive time t and for positive population size P. We could have restricted our phase line analysis to positive values only in the same way; however, including the negative values in the phase line gave us a better understanding of the stability.

○○

Example: For the ODE $\frac{dy}{dt} = y^2(y - 1)$, determine the stability of the equilibria. We may decide to use the Stability Theorem first. In this case, $f(y) = y^2(y - 1)$, so that $f'(y) = 3y^2 - 2y$. We know that $y^* = 0$ and $y^* = 1$ are equilibria because these are solutions of $f(y) = y^2(y - 1) = 0$. Evaluating $f'(y)$ at our equilibria, we obtain $f'(0) = 0$, which is inconclusive, and $f'(1) = 1 > 0$, so $y^* = 1$ is an unstable equilibrium. However, because the theorem was inconclusive for $y^* = 0$, we must check the stability some other way, perhaps by sketching a phase line and/or by looking at the graph of several solutions.

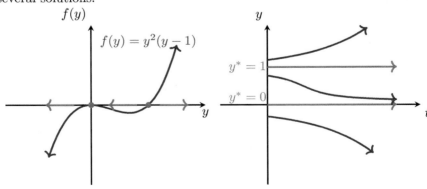

In either graph, we can see that $y^* = 1$ is unstable, just as we proved earlier, and that $y^* = 0$ is semistable due to solutions approaching $y^* = 0$ when the initial condition satisfies $0 < y_0 < 1$ and solutions tending away from $y^* = 0$ when $y_0 < 0$.

3.2 Exercises

1. For the ODE $y' = 7(y - 3)^2$:

   (a) Sketch the graph of $f(y)$ versus y and determine the equilibrium point(s).

   (b) Draw the phase line and use it to classify the stability of each equilibrium as stable, unstable, or semistable.

   (c) Graph several solutions in a single plot in the ty-plane.

   (d) Prove stability of the equilibrium point(s) using the Stability Theorem to confirm the results of your phase line.

2. For the ODE $y' = y^2(y^2 - 16)$:

   (a) Sketch the graph of $f(y)$ versus y and determine the equilibrium point(s).

   (b) Draw the phase line and use it to classify the stability of each equilibrium as stable, unstable, or semistable.

   (c) Graph several solutions in a single plot in the ty-plane.

   (d) Prove stability of the equilibrium point(s) using the Stability Theorem to confirm the results of your phase line.

3. For the ODE $y' = y(y - 2)(y - 5)$:

   (a) Sketch the graph of $f(y)$ versus y and determine the equilibrium point(s).

   (b) Draw the phase line and use it to classify the stability of each equilibrium as stable, unstable, or semistable.

   (c) Graph several solutions in a single plot in the ty-plane. For this problem, assume that the given ODE represents a changing population and thus do not consider negative initial conditions when sketching your graph.

   (d) Prove stability of the equilibrium point(s) using the Stability Theorem to confirm the results of your phase line.

4. For the ODE $y' = \sin(y)$:

   (a) Sketch the graph of $f(y)$ versus y and determine the equilibrium point(s).

   (b) Draw the phase line and use it to classify the stability of each equilibrium as stable, unstable, or semistable.

   (c) Graph several solutions in a single plot in the ty-plane. For this problem, assume that the given ODE represents a changing population and thus do not consider negative initial conditions when sketching your graph.

   (d) Prove stability of the equilibrium point(s) using the Stability Theorem to confirm the results of your phase line.

5. For the ODE $y' = y^2 - 6y + 5$:

   (a) Sketch the graph of $f(y)$ versus y and determine the equilibrium point(s).

   (b) Draw the phase line and use it to classify the stability of each equilibrium as stable, unstable, or semistable.

   (c) Graph several solutions in a single plot in the ty-plane. For this problem, assume that the given ODE represents a changing population and thus do not consider negative initial conditions when sketching your graph.

   (d) Prove stability of the equilibrium point(s) using the Stability Theorem to confirm the results of your phase line.

6. For the ODE $y' = e^y - y - 2$:

   (a) Sketch the graph of $f(y)$ versus y and determine the equilibrium point(s) (Note: you may need to use a graphing software to approximate the value of these equilibria).

   (b) Draw the phase line and use it to classify the stability of each equilibrium as stable, unstable, or semistable.

   (c) Graph several solutions in a single plot in the ty-plane. Do not disregard negative values.

   (d) Prove stability of the equilibrium point(s) using the Stability Theorem to confirm the results of your phase line.

7. For the ODE $y' = 2y - 8y^3$:

   (a) Sketch the graph of $f(y)$ versus y and determine the equilibrium point(s).

   (b) Draw the phase line and use it to classify the stability of each equilibrium as stable, unstable, or semistable.

   (c) Graph several solutions in a single plot in the ty-plane.

   (d) Prove stability of the equilibrium point(s) using the Stability Theorem to confirm the results of your phase line.

8. A 500 gallon tank of distilled water is filled at a rate of 5 gal/min with a solution of water which contains 20 g/gal of salt. The tank is drained at the same rate of 5 gal/min.

   (a) Construct an ODE to model this scenario.

   (b) Sketch the graph of $f(y)$ versus y to determine the equilibrium point(s).

   (c) Graph several solutions in a single plot in the ty-plane.

9. Assume $a, b > 0$. For the ODE $y' = ay - by^3$:

   (a) Sketch the graph of $f(y)$ versus y and determine the equilibrium point(s).

   (b) Draw the phase line and use it to classify the stability of each equilibrium as stable, unstable, or semistable.

   (c) Graph several solutions in a single plot in the ty-plane.

   (d) Prove stability of the equilibrium point(s) using the Stability Theorem to confirm the results of your phase line.

10. Consider a population of red-cockaded woodpeckers with an intrinsic growth rate of 0.03/year which reside in a small reservation which can support 300 woodpeckers.

    (a) Construct an ODE to model this scenario

    (b) Sketch the graph of $f(y)$ versus y to determine the equilibrium point(s).

    (c) Graph several solutions in a single plot in the ty-plane.

11. Consider a population of beetles which have infested a large garden. Suppose the intrinsic growth rate of the beetles in the current season is 2.1/month, and k beetles are removed per month on a continuous basis.

     (a) Construct an ODE to model this scenario.

     (b) Sketch the graph of $f(y)$ versus y to determine the equilibrium point(s).

     (c) Graph several solutions in a single plot in the ty-plane.

     (d) Suppose there is an initial infestation of 100 beetles. Determine the minimum value of k required so that the beetle population can be eliminated over the long term.

12. A 1000-gallon tank of distilled water is filled at a rate of 1 gal/min with a solution of water which contains 27 g/gal of anthocyanin which has been distilled from black currants. The tank is drained at the same rate of 1 gal/min.

     (a) Construct an ODE to model this scenario.

     (b) Sketch the graph of $f(y)$ versus y to determine the equilibrium point(s).

     (c) Graph several solutions in a single plot in the ty-plane.

3.3   Bifurcation Diagrams

Sometimes a differential equation may depend on some unknown value. For example, consider the ODE:

$$y' = 0.2y \left(1 - \frac{y}{500}\right) - ay.$$

This might be modeling the logistic growth of a population with intrinsic growth rate of 0.2/year in an environment with a carrying capacity of 500. The $-ay$ term may account for a certain amount of predation which is proportional to the population y. You may also have a similar ODE, for example:

$$y' = 0.2y \left(1 - \frac{y}{500}\right) - a,$$

where the extra term is just $-a$, which would correspond with a logistic growth model with constant migration.

The natural question is this: how much predation (or migration) can exist without the population failing to be able to reach carrying capacity? How can we approach this question? As we will see, these two similar looking problems have different types of **bifurcation diagrams**. To see why, let's consider some similar problems then return to these models.

ooooooooooooooooooooooooooooooooooooooooo

Consider the differential equation $y' = y^2 - 8y + a$, where $a \geq 0$. Depending on the value of a, is it possible to have different equilibria and/or different stability of those equilibria? If so, then the value of a for which the solution changes is called the **bifurcation value**.

To see how this works exactly, let's consider a few examples with our differential equation $y' = y^2 - 8y + a$. First, let's assume $a = 0$. Then:

$$
\begin{aligned}
y' &= y^2 - 8y + a \\
&= y^2 - 8y \\
&= y(y - 8),
\end{aligned}
$$

so the equilibria are $y^* = 0$ and $y^* = 8$. We will also do this for the a values of 16 and 20. If $a = 16$, then:

$$
\begin{aligned}
y' &= y^2 - 8y + a \\
&= y^2 - 8y + 16 \\
&= (y - 4)^2,
\end{aligned}
$$

so the equilibrium is at $y^* = 4$. If $a = 20$, then $y' = y^2 - 8y + 20$. To find the equilibrium, we could use the quadratic formula $y^* = \frac{8 \pm \sqrt{64 - 4(20)}}{2}$, but this produces only complex roots. We can now sketch a phase line for each of these three cases to determine stability.

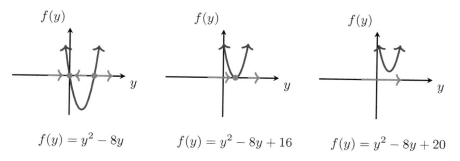

$$f(y) = y^2 - 8y \qquad\qquad f(y) = y^2 - 8y + 16 \qquad\qquad f(y) = y^2 - 8y + 20$$

In the first case of $a = 0$ which corresponds to $y' = y^2 - 8y$, we see that $y^* = 0$ is a stable equilibrium and $y^* = 8$ is an unstable equilibrium. In fact, for any $a < 16$, we will always have two equilibria, one of which will be stable and the other unstable, and the phase line analysis will look similar to the left graph above. When $a = 16$, which corresponds to $y' = y^2 - 8y + 16$, we have exactly one semistable equilibrium at $y^* = 4$. When $a = 20$, which corresponds to $y' = y^2 - 8y + 20$, we have no equilibria and because $y' > 0$ for all values of y, our solution is unbounded. In fact we will obtain a similar result for any $a > 16$.

Because $a < 16$ always has 2 equilibria and $a > 16$ always has 0 equilibria, the value $a = 16$ (which has exactly one equilibrium) is called the bifurcation value. It is the value of a for which the number and type of equilibria changes. We can illustrate this further by sketching a bifurcation diagram.

> **Definition: A bifurcation diagram** is a graph in which we plot the equilibria as a function of the bifurcation parameter. It is a curve in the plane with the bifurcation parameter on the horizontal axis. Generally, we draw a dotted line to indicate unstable sections of the curve and a solid line to indicate stable sections.

Example: Let's consider again the differential equation $y' = y^2 - 8y + a$. In order to sketch the bifurcation diagram, we first must determine the value(s) of y for which $y' = 0$ (in other words, find the equilibrium point(s) as a function of a). In this case, we have $0 = y^2 - 8y + a$ which can be solved using the quadratic formula so that:

$$
\begin{aligned}
y^* &= \frac{8 \pm \sqrt{8^2 - 4a}}{2} \\
&= \frac{8 \pm 2\sqrt{16 - a}}{2} \\
&= 4 \pm \sqrt{16 - a}.
\end{aligned}
$$

Thus our equilibria are $y^* = 4 + \sqrt{16 - a}$ and $y^* = 4 - \sqrt{16 - a}$. We will need to determine the stability of these equilibria before sketching the graph because ultimately we will sketch stable equilibria as solid lines and unstable equilibria as dashed (or dotted) lines. The stability of the equilibria typically depends on the value of a. For example, let's look at the first equilibrium of $y^* = 4 + \sqrt{16 - a}$. In order to determine the stability, we will use the Stability Theorem. For this example $f(y) = y^2 - 8y + a$, so that $f'(y) = 2y - 8$. Then:

$$
\begin{aligned}
f'(4 + \sqrt{16 - a}) &= 2(4 + \sqrt{16 - a}) - 8 \\
&= 2\sqrt{16 - a}.
\end{aligned}
$$

This quantity is always positive for $a < 16$, so $y^* = 4 + \sqrt{16 - a}$ is unstable when $a < 16$. Notice that if $a > 16$, this quantity is undefined and, in fact, this equilibrium does not exist when $a > 16$.

Similarly, let's consider the equilibrium $y^* = 4 - \sqrt{16 - a}$. Again, this equilibrium does not exist when $a > 16$. However, when $a < 16$, we can determine stability using the Stability Theorem. We have:

$$
\begin{aligned}
f'(4 - \sqrt{16 - a}) &= 2(4 - \sqrt{16 - a}) - 8 \\
&= -2\sqrt{16 - a}.
\end{aligned}
$$

This quantity is always negative when $a < 16$, so the second equilibrium of $y^* = 4 - \sqrt{16 - a}$ is stable when $a < 16$ and does not exist when $a > 16$.

Finally, we can sketch the graphs of $f(a) = 4 + \sqrt{16 - a}$ and $f(a) = 4 - \sqrt{16 - a}$, remembering to sketch stable equilibria using solid lines and unstable equilibria using dashed lines. Our bifurcation diagram gives a visual for the same results we have already stated:

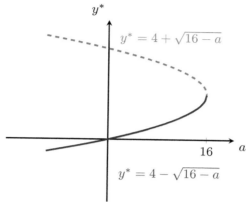

This bifurcation diagram shows a **saddle-node bifurcation**. When $a < 16$, we have two equilibria. The red dashed line that corresponds to the equilibrium $y^* = 4 + \sqrt{16 - a}$ represents the unstable equilibrium, while the blue solid line that corresponds to the equilibrium $y^* = 4 - \sqrt{16 - a}$ represents the stable one. Furthermore, we can see from the diagram that when $a > 16$ there are no equilibria. We call $a = 16$, where the number of equilibria changes from two to zero, the **bifurcation value** for this ODE.

Saddle-Node Bifurcation

On one side of the bifurcation value, there are no equilibria. On the other side of the bifurcation value, there are two equilibria, one of which is stable and the other is unstable.

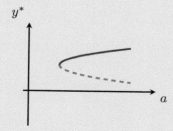

Example: Consider the population model:

$$y' = 0.2y \left(1 - \frac{y}{500}\right) - a,$$

which is modeling the growth of a population with intrinsic growth rate 0.2 per year, with carrying capacity 500. We want to find the bifurcation value for a which is a constant migration term. The question is this: how much negative migration can occur before the population fails to reach the carrying capacity?

We proceed as before by setting y' equal to zero to find the equilibria. We see that we have:

$$y' = 0.2y\left(1 - \frac{y}{500}\right) - a$$

$$y' = -0.0004y^2 + 0.2y - a$$

$$0 = -0.0004(y^2 - 500y + 2500a)$$

which, using the quadratic formula, we see has solutions:

$$y^* = 250 \pm 50\sqrt{25 - a}.$$

Prior to graphing the bifurcation diagram, we need to determine the stability of these equilibria. We do so using the Stability Theorem. We note that for $y' = f(y)$, we have $f'(y) = -0.0008y + 0.2$. Thus we see that:

$$f'(250 + 50\sqrt{25 - a}) = -0.0008(250 + 50\sqrt{25 - a}) + 0.2$$

$$= -0.2 - 0.04\sqrt{25 - a} + 0.2$$

$$= -0.04\sqrt{25 - a}$$

$$\leq 0$$

with equality only when $a = 25$. Thus, the equilibrium $y^* = 250 + 50\sqrt{25 - a}$ is a stable equilibrium when $a < 25$ (and doesn't exist when $a > 25$). Similarly, we see that:

$$f'(250 - 50\sqrt{25 - a}) = -0.0008(250 - 50\sqrt{25 - a}) + 0.2$$

$$= -0.2 + 0.04\sqrt{25 - a} + 0.2$$

$$= 0.04\sqrt{25 - a}$$

$$\geq 0$$

with equality only when $a = 25$. Thus, the equilibrium $y^* = 250 - 50\sqrt{25 + a}$ is an unstable equilibrium when $a < 25$ (and again, doesn't exist when $a > 25$). This results in the below bifurcation diagram.

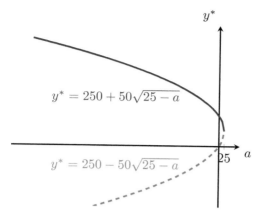

$$y^* = 250 + 50\sqrt{25 - a}$$

$$y^* = 250 - 50\sqrt{25 - a}$$

Thus we can see that we have a bifurcation value of $a = 25$. Note that if we have $a = 0$, we have two equilibrium points: $y^* = 0$ and $y^* = 500$. This corresponds with the unstable equilibrium at zero and the stable one at the carrying capacity.

In the event that the parameter $a = 9$, we have the two equilibrium solutions $y^* = 50$ and $y^* = 450$. This tells us that at that level of migration, we must have at least a population of 50 in order to not decay to zero, and the limit of our population is 450 due to the carrying capacity and migration.

At the bifurcation value, $a = 25$, these two solutions merge together into one equilibrium at $y^* = 250$ which turns out to be semi-stable. For any $a > 25$ there is no equilibrium at all.

This allows us to conclude that the critical migration value is $a = 25$ after which point the population cannot grow and will only die out.

○○○○○○○○○○○○○○○○○○○○○○○○○○○○○○○○○○○○○○○

The last example required us to use the quadratic formula and resulted in the parameter a being inside a square root, which determined the behavior of the bifurcation. Let's consider another autonomous differential equation with a quadratic function $f(y)$, however this time we will have a function which factors nicely into distinct real roots.

Example: Consider the differential equation $y' = (y - 5)(y - b)$, where b is a constant. This ODE has equilibria $y^* = 5$ and $y^* = b$. To determine stability, we consider $f(y) = (y - 5)(y - b)$. Then $f'(y) = 2y - 5 - b$.

For the equilibrium $y^* = 5$, we have $f'(5) = 5 - b < 0$ when $b > 5$. Thus $y^* = 5$ is stable when $b > 5$ and unstable when $b < 5$.

For the equilibrium $y^* = b$, we have $f'(b) = b - 5 < 0$ when $b < 5$. Hence $y^* = b$ is stable when $b < 5$ and unstable when $b > 5$. Visually, we can see these results in the bifurcation diagram:

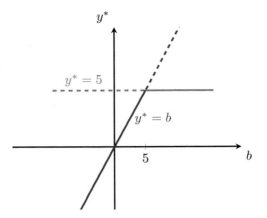

This is an example of a **transcritical bifurcation**. There are two equilibria before $b = 5$ and two equilibria after $b = 5$, however, the stability of these equilibria have switched. When $b < 5$, $y^* = 5$ is unstable and $y^* = b$ is stable, whereas when $b > 5$, $y^* = 5$ is stable and $y^* = b$ is unstable, which can be seen via the solid and dashed lines in the bifurcation diagram. The value $b = 5$ in this case is the bifurcation value because this is the value for which the stability changes.

Transcritical Bifurcation

There are the same number of equilibria before and after the bifurcation value (the intersection point in the graph), but the stability changes. An equilibrium that was stable before the bifurcation value becomes unstable after the bifurcation value, and vice versa.

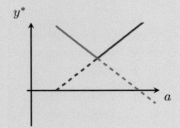

Example: Consider the population model:

$$y' = 0.2y \left(1 - \frac{y}{500}\right) - ay,$$

which is modeling the growth of a population with intrinsic growth rate 0.2 per year, with carry capacity 500. Unlike the earlier example, we also have the term $-ay$ which here is modeling a predation term which is proportional to the population, y.

We proceed by finding the equilibria, seeing that:

$$y' = 0.2y\left(1 - \frac{y}{500}\right) - ay$$

$$y' = -0.0004y^2 + 0.2y - ay$$

$$0 = -0.0004y(y - 500 + 2500a)$$

which we can see yields the equilibria $y^* = 0$ and $y^* = 500 - 2500a$. In order to graph the bifurcation diagram, we must first determine the stability of these equilibria. We again use the Stability Theorem and see that:

$$f'(y) = -0.0008y + 0.2 - a.$$

This yields

$$f'(0) = 0.2 - a$$

and

$$f'(500 - 2500a) = -0.4 + 2a + 0.2 - a$$

$$= a - 0.2.$$

For the equilibrium $y^* = 0$ we have that $f'(0) < 0$ when $a > 0.2$. Thus $y^* = 0$ is stable when $a > 0.2$ and unstable when $a < 0.2$.

For the equilibrium $y^* = 500 - 2500a$ we have that $f'(500 - 2500a) < 0$ when $a < 0.2$. Thus $y^* = 500 - 2500a$ is stable when $a < 0.2$ and unstable when $a > 0.2$. This results in a very similar bifurcation diagram as the preceding example:

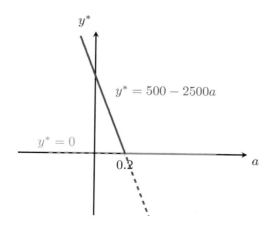

Thus, it is clear that at the bifurcation value $a = 0.2$ the two equilibria switch stability. The interpretation here is that while $a < 0.2$, there is a positive, stable equilibrium and an unstable equilibrium at zero, so at that level of predation the population will stabilize at $y^* = 500 - 2500a$. However, at a predation level of $a > 0.2$, there is now no positive stable equilibrium and the population will stabilize at $y^* = 0$.

$$\circ\circ\circ\circ\circ\circ\circ\circ\circ\circ\circ\circ\circ\circ\circ\circ\circ\circ\circ\circ\circ\circ\circ\circ\circ\circ\circ\circ\circ\circ\circ\circ\circ\circ\circ\circ\circ\circ\circ\circ$$

Let's look at another type of bifurcation problem whose bifurcation diagram takes a different shape than the two we have seen so far.

Example: Consider the differential equation $y' = (y^2 - 2y + k + 1)(1 - y)$, where k is a constant. This ODE has an equilibrium at $y^* = 1$. To find the remaining equilibria, we must solve $0 = y^2 - 2y + k + 1$, and we can do so using the quadratic formula. Then $y^* = \frac{2 \pm \sqrt{4 - 4(k+1)}}{2} = 1 \pm \sqrt{-k}$. To determine stability, we need to find $f'(y)$. Note that:

$$
\begin{aligned}
f(y) &= (y^2 - 2y + k + 1)(1 - y) \\
f(y) &= -y^3 + 3y^2 - (3 + k)y + k + 1 \\
f'(y) &= -3y^2 + 6y - 3 - k.
\end{aligned}
$$

For the equilibrium $y^* = 1$, we have $f'(1) = -k < 0$ when $k > 0$. Thus $y^* = 1$ is stable when $k > 0$ and unstable when $k < 0$. For the equilibria $y^* = 1 \pm \sqrt{-k}$, we have:

$$
\begin{aligned}
f'(1 + \sqrt{-k}) &= -3(1 + \sqrt{-k})^2 + 6(1 + \sqrt{-k}) - 3 - k \\
&= 2k < 0
\end{aligned}
$$

when $k < 0$, and

$$
\begin{aligned}
f'(1 - \sqrt{-k}) &= -3(1 - \sqrt{-k})^2 + 6(1 - \sqrt{-k}) - 3 - k \\
&= 2k < 0
\end{aligned}
$$

when $k < 0$. Hence both of these equilibria are stable when $k < 0$. However, it is important to notice that $y^* = 1 \pm \sqrt{-k}$ do not exist for $k > 0$. Finally, we can sketch the bifurcation diagram to illustrate our results.

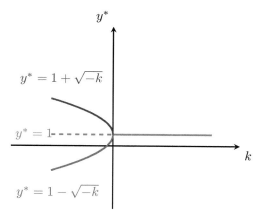

This is an example of a **supercritical bifurcation**. There are three equilibria before $k = 0$, two stable and one unstable. After $k = 0$, there is only one stable equilibrium. The equilibria $y^* = 1 \pm \sqrt{-k}$ do not exist after $k = 0$, giving the diagram a pitchfork shape. It's also important to note that the stability of the remaining equilibrium at $y^* = 1$ changes at this same point when $k = 0$. Thus $k = 0$ is the bifurcation value for this ODE.

Pitchfork Bifurcation

This type of bifurcation looks like a pitchfork, hence the name. There are two kinds of pitchfork bifurcations: supercritical and subcritical. The supercritical pitchfork bifurcation shows one stable equilibrium becoming unstable and the appearance of two new stable equilibria as the "branches" of the pitchfork. In contrast, the subcritical pitchfork bifurcation has three equilibria (one stable and two unstable) before the bifurcation value and only one unstable equilibrium after.

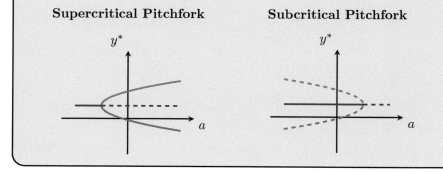

Of course, we can also identify bifurcation values and describe the quantity and stability of equilibria for a differential equation just by looking at the bifurcation diagram.

Example: Consider the bifurcation diagram below:

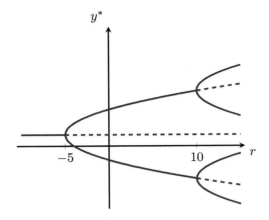

From this bifurcation diagram, we can tell that r is our bifurcation parameter and there are two bifurcation values of $r = -5$ and $r = 10$. There is a supercritical pitchfork bifurcaton at $r = -5$, where we have one stable equilibrium when $r < -5$ and three equilibria (the original stable equilibrium is now unstable and there are two new stable branches) when $-5 < r < 10$. At $r = 10$, we see two more supercritical pitchfork bifurcations. When $r > 10$, we have seven equilibria (four stable and three unstable).

3.3 Exercises

For each of the ODEs below: draw a bifurcation diagram, identify any bifurcation value(s), and determine the type of bifurcation for each value.

1. $y' = y^2 - 3y + a$

2. $y' = y - k^2 + 2$

3. $y' = y^3 - ay^2 + ay - a^2$

4. $y' = (y - 5)(y - a)(y + a)$

5. $y' = y^3 - ry$

6. $y' = (y - 4)(y - b^2)$

7. $y' = (y - 2)(y^2 - a)$

8. $y' = y^3 + y^2 - 2ay$

9. $y' = y^2 - a$

10. $y' = (k - y)(y^2 - k)$

11. $y' = y^2 + y - a$

12. $y' = y^4 + y^3 - ay^2 - ay$

13. A population of cats on an abandoned island has an intrinsic growth rate of $r = 0.3$ and the island has a carrying capacity of 5,000 cats. In order to control the growing population, naturalists decide to introduce a hunting program where volunteers are allowed to hunt a certain amount of cats per year.

   (a) Construct a differential equation to model this situation which includes a parameter, a, to account for the amount of hunting allowed.

   (b) Draw a bifurcation diagram identifying any bifurcation values.

   (c) Advise the naturalists on what value of a they should aim for if they want to eliminate the cat population (i.e., ensure there is not a positive stable equilibrium).

   (d) Advise the naturalists on what value of a they should aim for the they want to ensure that the cat population stabilizes at 3000 cats.

   (e) Is it possible to stabilize the cat population at 500 cats?

14. A population of mice in a nature preserve has an intrinsic growth rate of $r = 1.2$ and there is a carrying capacity of 25,000 mice at the preserve. In order to control the population, scientists have decided to introduce a predator which will kill a certain percentage of the population every year.

   (a) Construct a differential equation to model this situation including a parameter r to account for the predation.

   (b) Draw a bifurcation diagram identifying any bifurcation values. What kind of bifurcation is this?

   (c) Suppose the preserve wants to stabilize the mouse population at 20,000 mice. Is this possible? If so, what value of r would be required?

   (d) What value of a must be used to ensure the mouse population dies out?

15. A population of aphids in a garden has an intrinsic growth rate of $r = 1.2$ and the garden has a carrying capacity of 5000 aphids. In order to control the growing population, a gardener decide to introduce a population of ladybugs which will consume a given percentage of the aphids every day.

   (a) Construct a differential equation to model this situation which includes a parameter, a to account for the predation.

   (b) Draw a bifurcation diagram identifying any bifurcation values.

   (c) Advise the gardener on how many ladybugs he needs to introduce if 100 ladybugs correspond with a parameter of $a = 0.05$, 200 ladybugs correspond with a parameter of $a = 0.10$, and in general n hundred ladybugs corresponds with a parameter of $a = 0.05n$.

16. A population of wild hogs in a farming community has an intrinsic growth rate of $r = 0.7$ and there is a carrying capacity of 250 hogs in the community. In order to control the population, the game warden has decided to introduce a program in which a certain number of hogs can be killed every month.

   (a) Construct a differential equation to model this situation including a parameter r to account for the predation.

   (b) Draw a bifurcation diagram identifying any bifurcation values. What kind of bifurcation is this?

   (c) What value of r must be used to ensure the hog population dies out?

3.4   Separable Equations

So far in this chapter we've focused our attention on qualitatively describing the solutions of ODEs. We have used the phase line and the Stability Theorem to determine the stability of equilibria. This has helped us in sketching a graph of potential solutions, much like the solutions that would result from analyzing the vector field of the ODE. Then we discussed bifurcation diagrams in an effort to determine when the number of equilibria changes. In the following sections, we are more interested in finding the explicit equation for the solution of the ODE. In other words, it is great that we could graph a possible solution in previous sections without knowing the equation of the solution, but now we would like to actually solve the ODE to obtain the equation. We will examine a few different types of ODEs, beginning in this section with separable equations.

> **Definition:** An ODE is called **separable** if it can be written in the form
> $$M(x) \ dx + N(y) \ dy = 0.$$
> In other words, the x and y terms can be separated from each other so that we can solve the differential equation by simply integrating.

Example: The ODE $-x \ dx + y^4 \ dy = 0$ is separable. To solve this differential equation, we separate the variables and integrate:

$$-x \ dx + y^4 \ dy \ = \ 0$$

$$y^4 \ dy \ = \ x \ dx$$

$$\int y^4 \ dy \ = \ \int x \ dx$$

$$\tfrac{1}{5}y^5 \ = \ \tfrac{1}{2}x^2 + C$$

$$y^5 \ = \ \tfrac{5}{2}x^2 + C$$

$$y \ = \ \left(\tfrac{5}{2}x^2 + C\right)^{\frac{1}{5}},$$

where C is a constant. You may have noticed that when we integrated, we only included the "$+C$" on the right side of the equation, whereas normally this should have been included on both sides. However, as we don't know the actual value of C and would end up subtracting these two values when solving for y anyway, it is much easier and more concise to only include the "$+C$" on the right side of the equation. Furthermore, when multiplying our equation by 5, we technically have $5C$ on the right side of the equation. However, as we don't know the value of C, the 5 can be absorbed into this quantity.

Hence all solutions of $-x \ dx + y^4 \ dy = 0$ are of the form $y = \left(\tfrac{5}{2}x^2 + C\right)^{\frac{1}{5}}$.

To further understand this solution, let's graph it! The curves $y = \left(\tfrac{5}{2}x^2 + C\right)^{\frac{1}{5}}$ with $C = -20, 0, 20$ are in the graph below. We will also sketch the vector field with these solutions overlaid to remind us how these graphs are related to each other.

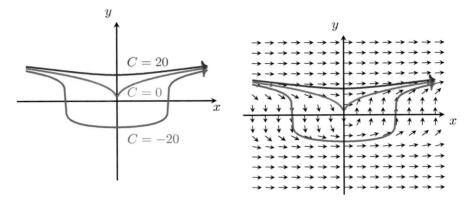

We see that all three solutions we have chosen to graph, $y = \left(\frac{5}{2}x^2 - 20\right)^{\frac{1}{5}}$, $y = \left(\frac{5}{2}x^2\right)^{\frac{1}{5}}$, and $y = \left(\frac{5}{2}x^2 + 20\right)^{\frac{1}{5}}$, all follow the paths in the vector field for the ODE $-x\,dx + y^4\,dy = 0$. It is also important to note that we could not have used phase line analysis for this example, and for some of the many examples that follow in this and future sections, because the ODE $-x\,dx + y^4\,dy = 0$ is not autonomous. Thus, finding the explicit equation for the solution of the ODE $y = \left(\frac{5}{2}x^2 + C\right)^{\frac{1}{5}}$ is invaluable.

○○

Example: Consider the exponential model $y' = 2y$. We can rewrite this differential equation as $\dfrac{dy}{dx} = 2y$ so that it is more obvious how to separate and integrate:

$$\frac{dy}{dx} = 2y$$

$$\frac{1}{y}\,dy = 2\,dx$$

$$\int \frac{1}{y}\,dy = \int 2\,dx$$

$$\ln|y| = 2x + C_1$$

$$y = \pm e^{2x + C_1}$$

$$y = \pm e^{2x} e^{C_1}$$

$$y = Ce^{2x},$$

where $C = \pm e^{C_1}$ is a constant. We will continue to use this "trick" to simplify expressions, like we did here: $e^{ax+C} = Ce^{ax}$. Of course, the C on each side of the equation simply represents a constant and these C values are not technically equal. As we don't yet know the correct value of C, rewriting the equation as $y = Ce^{2x}$ could make the calculation of C easier if we had been given more information.

○○

Of course, given a differential equation and an initial value, we can solve for the constant C. Let's look at an example that gives us this extra piece of information.

Example: To solve the IVP:

$$\frac{dy}{dx} = \frac{y^2 x^3}{x^4 + 1}, \qquad y(0) = \frac{1}{2},$$

we first must separate x and y. Then we have:

$$\begin{aligned}
\frac{dy}{dx} &= \frac{y^2 x^3}{x^4 + 1} \\
\frac{1}{y^2}\, dy &= \frac{x^3}{x^4 + 1}\, dx \\
\int y^{-2}\, dy &= \int \frac{x^3}{x^4 + 1}\, dx \\
-y^{-1} &= \tfrac{1}{4} \ln\left(x^4 + 1\right) + C \quad \text{by } u\text{-sub, where } u = x^4 + 1 \\
y^{-1} &= -\tfrac{1}{4} \ln\left(x^4 + 1\right) + C \\
y &= \left(-\tfrac{1}{4} \ln\left(x^4 + 1\right) + C\right)^{-1}.
\end{aligned}$$

As we were given initial condition $y(0) = \frac{1}{2}$, we must also solve for C. So we have $\frac{1}{2} = \left(-\frac{1}{4} \ln(0 + 1) + C\right)^{-1}$. This gives $\frac{1}{2} = C^{-1}$, so that $C = 2$. Hence the solution of the given IVP is $y = \left(-\frac{1}{4} \ln\left(x^4 + 1\right) + 2\right)^{-1}$.

○○

As separable equations by their nature separate into two functions that can be integrated independently, many of the integration techniques you may have learned in previous courses now may become useful again. One technique that can be extremely useful is the method of *partial fractions*.

Example: Consider the first-order ODE:

$$\frac{dy}{dx} = y^2 - 1.$$

We take our usual approach and separate x and y. Then we have:

$$\frac{dy}{dx} = y^2 - 1$$

$$\frac{dy}{y^2 - 1} = dx$$

$$\frac{dy}{(y-1)(y+1)} = dx.$$

It is here that the method of partial fractions will be helpful. We want to integrate $(y^2 - 1)^{-1}$ but there is nothing useful in the numerator to allow us to try u substitution. Therefore, we seek to split the denominator instead with the goal of arriving at something we can integrate nicely. Note that if we let

$$\frac{1}{(y-1)(y+1)} = \frac{A}{y-1} + \frac{B}{y+1},$$

and simplify the right-hand side by creating a common denominator, we have:

$$\frac{1}{(y-1)(y+1)} = \frac{A(y+1)}{(y-1)(y+1)} + \frac{B(y-1)}{(y-1)(y+1)}$$

$$= \frac{Ay + A}{(y-1)(y+1)} + \frac{By - B}{(y-1)(y+1)}$$

$$= \frac{(A+B)y + (A-B)}{(y-1)(y+1)}.$$

Then we see that for the numerators to agree, the equations

$$1 = A - B \quad \text{(from the constant term)}$$

$$0 = A + B \quad \text{(from the coefficients for the } y \text{ term)}$$

must be satisfied. Solving this system we find that $A = \frac{1}{2}$ and $B = -\frac{1}{2}$. Thus, returning to the differential equation, we have:

$$\frac{dy}{(y-1)(y+1)} = dx$$

$$\frac{1}{2}\left(\frac{dy}{y-1}\right) - \frac{1}{2}\left(\frac{dy}{y+1}\right) = dx$$

$$\frac{1}{2}\int \frac{dy}{y-1} - \frac{1}{2}\int \frac{dy}{y+1} = \int dx$$

$$\frac{1}{2}\left(\ln|y-1| - \ln|y+1|\right) = x + C$$

$$\ln\left|\frac{y-1}{y+1}\right| = 2x + C$$

$$\frac{y-1}{y+1} = Ce^{2x}$$

$$y - 1 = (y+1)Ce^{2x}$$

$$y - 1 = Ce^{2x}y + Ce^{2x}$$

$$y - Ce^{2x}y = Ce^{2x} + 1$$

$$y(1 - Ce^{2x}) = Ce^{2x} + 1$$

$$y = \frac{Ce^{2x} + 1}{1 - Ce^{2x}}.$$

ooooooooooooooooooooooooooooooooooooooooo

Example: Consider an invasive insect species with no natural predators and an effectively unlimited supply of food. Initially there are 1000 insects and the species grows at a rate proportional to its size, with an intrinsic growth rate of $r = 0.05$ per day. In how many days will the population reach one million insects?

This is an example of exponential growth because there are no predators, and as there is effectively no limit to the food supply, there is no carrying capacity. As such, we can set up the IVP:

$$\frac{dP}{dt} = 0.05P, \quad P(0) = 1000.$$

This equation is separable, so we have that:

$$\frac{dP}{dt} = 0.05P$$

$$\frac{1}{P} dP = 0.05 \ dt$$

$$\int \frac{1}{P} dP = \int 0.05 \ dt$$

$$\ln |P| = 0.05t + C$$

$$P = Ce^{0.05t}.$$

Using the initial condition, we have that $C = 1000$. Then the solution of the IVP is $P = 1000e^{0.05t}$. In order to determine how many days it will take for the population to reach one million, we must solve:

$$1000000 = 1000e^{0.05t}$$

$$1000 = e^{0.05t}$$

$$\ln 1000 = 0.05t$$

$$t = 20 \ln 1000 \approx 138.16 \text{ days.}$$

Hence, it takes about 138.16 days for the insect population to reach one million insects. But, more importantly, we have a solution, $P = 1000e^{0.05t}$, that allows us to calculate the number of insects at any point in time.

○○

Example: Consider a colony of bacteria in a Petri dish. The population will grow proportional to its size, but due to the physical constraints of the Petri dish and the finite amount of sugar in the dish, there is a carrying capacity of 20 million bacteria cells. If there are initially 4 million bacteria cells in the dish, how many cells will there be after 10 days if the intrinsic growth rate is $r = 0.02$ per day?

This problem follows the logistic growth model with carrying capacity $M = 20$ (million) and growth rate $r = 0.02$ per day. As such, we have the IVP:

$$\frac{dP}{dt} = 0.02P\left(1 - \frac{P}{20}\right), \quad P(0) = 4.$$

This ODE is separable and can be rewritten as:

$$\frac{1}{P\left(1 - \frac{P}{20}\right)} \, dP = 0.02 \, dt.$$

In order to integrate the left-hand side, we will need to use partial fractions to express it as two fractions which we will be able to integrate. We have that:

$$\frac{1}{P\left(1 - \frac{P}{20}\right)} = \frac{A}{P} + \frac{B}{1 - \frac{P}{20}},$$

so

$$1 = A\left(1 - \frac{P}{20}\right) + BP.$$

Solving for A and B, we obtain $A = 1$, and $B = \frac{1}{20}$. Thus, we can rewrite our ODE and integrate:

$$
\begin{aligned}
\frac{1}{P\left(1 - \frac{P}{20}\right)} \, dP &= 0.02 \, dt \\
\left(\frac{1}{P} + \frac{\frac{1}{20}}{1 - \frac{P}{20}}\right) dP &= 0.02 \, dt \\
\int \left(\frac{1}{P} + \frac{\frac{1}{20}}{1 - \frac{P}{20}}\right) dP &= \int 0.02 \, dt \\
\ln|P| - \ln\left|1 - \frac{P}{20}\right| &= 0.02t + C \\
\ln\left|\frac{P}{1 - \frac{P}{20}}\right| &= 0.02t + C
\end{aligned}
$$

$$\frac{P}{1 - \frac{P}{20}} = Ce^{0.02t}$$

$$P = Ce^{0.02t}\left(1 - \tfrac{P}{20}\right)$$

$$P = Ce^{0.02t} - C\tfrac{P}{20}e^{0.02t}$$

$$P + C\tfrac{P}{20}e^{0.02t} = Ce^{0.02t}$$

$$P\left(1 + \tfrac{C}{20}e^{0.02t}\right) = Ce^{0.02t}$$

$$P = \frac{Ce^{0.02t}}{1 + \tfrac{C}{20}e^{0.02t}}.$$

Now, using the initial condition $P(0) = 4$, we have that:

$$4 = \frac{C}{1 + \frac{C}{20}}$$

$$4\left(1 + \tfrac{C}{20}\right) = C$$

$$4 + \tfrac{C}{5} = C$$

$$4 = \tfrac{4}{5}C$$

$$5 = C.$$

Therefore, our solution is:

$$P = \frac{5e^{0.02t}}{1 + \tfrac{1}{4}e^{0.02t}},$$

and after 10 days, we can find the population to be

$$P = \frac{5e^{0.2}}{1 + \tfrac{1}{4}e^{0.2}} \approx 4.68.$$

Thus, after 10 days there will be approximately 4.68 million bacteria in the Petri dish.

○○

Let's consider the following example using Newton's Law of Cooling.

Example: A cast iron skillet is removed from an oven at a temperature of 175° C and placed on the counter of a kitchen held at 25° C to cool. After 5 minutes, the temperature of the skillet is 150° C. How long must the cook wait to store the skillet if it must be at a temperature of no more than 50° C?

Note that in the above problem statement we are presented with both an initial condition and a secondary temperature reading. Why is this? Well, Newton's law of cooling only states that the rate of change is proportional to the difference in temperature, but not the actual value for that proportion (known as the *heat transfer coefficient*). To set up and solve the differential equation, we have that:

$$\frac{dT}{dt} = -k(T - s)$$

$$\frac{dT}{dt} = -k(T - 25)$$

$$\frac{1}{T - 25} dT = -k \, dt$$

$$\int \frac{1}{T - 25} dT = \int -k \, dt$$

$$\ln |T - 25| = -kt + C$$

$$T - 25 = Ce^{-kt}$$

$$T = 25 + Ce^{-kt}.$$

Then, using the initial condition $T(0) = 175°$, we have $175 = 25 + C$, so that $C = 150$. Then $T = 25 + 150e^{-kt}$. However, we still need to solve for k. We do this using the second given condition that $T(5) = 150$. We have:

$$150 = 25 + 150e^{-5k}$$

$$125 = 150e^{-5k}$$

$$\frac{125}{150} = e^{-5k}$$

$$\ln \left(\frac{125}{150}\right) = -5k$$

$$-\frac{1}{5} \ln \left(\frac{125}{150}\right) = k.$$

Hence $k = -\frac{1}{5} \ln \left(\frac{125}{150}\right) \approx 0.036$. Let's use the rounded value for k in our equation so that the solution of our differential equation is:

$$T = 25 + 150e^{-0.036t}.$$

Finally, we must answer the original question: how long until the temperature of the skillet is $50°C$? Then:

$$50 = 25 + 150e^{-0.036t}$$

$$25 = 150e^{-0.036t}$$

$$\frac{25}{150} = e^{-0.036t}$$

$$\ln \left(\frac{25}{150}\right) = -0.036t$$

$$-\frac{1}{0.036} \ln \left(\frac{25}{150}\right) = t.$$

Thus, we should wait $t = -\frac{1}{0.036} \ln \left(\frac{25}{150}\right) \approx 49.77$ minutes.

○○

Example: Consider the same cast iron skillet as above. The skillet has cooled completely to 50°C, and is now placed back in the oven held at a temperature of 175°C. If the heat transfer coefficient is still $k = -0.036$, how many minutes must the skillet remain in the oven to reach a temperature of 120°C?

To set up and solve the differential equation, we have that:

$$\frac{dT}{dt} = -k(T - s)$$

$$\frac{dT}{dt} = -0.036(T - 175)$$

$$\frac{1}{T - 175}\,dT = -0.036\,dt$$

$$\int \frac{1}{T - 175}\,dT = \int -0.036\,dt$$

$$\ln |T - 175| = -0.036t + C$$

$$T - 175 = Ce^{-0.036t}$$

$$T = 175 + Ce^{-0.036t}.$$

Using the initial condition $T(0) = 50$, we have $50 = 175 + C$. Then $C = -125$, so the solution is $T = 175 - 125e^{-0.036t}$. We want to find the number of minutes until the skillet reaches a temperature of 120°C, so we need to solve for t when $T = 120$. Therefore, we have:

$$120 = 175 - 125e^{-0.036t}$$

$$-55 = -125e^{-0.036t}$$

$$\frac{11}{25} = e^{-0.036t}$$

$$\ln\left(\frac{11}{25}\right) = -0.036t$$

$$-\frac{1}{0.036}\ln\left(\frac{11}{25}\right) = t.$$

Therefore, we must wait $t = -\frac{1}{0.036}\ln\left(\frac{11}{25}\right) \approx 22.81$ minutes before the skillet reaches the required temperature.

3.4 Exercises

Find the general solution of the given ODE.

1. $y' = \dfrac{x^3 + 2x + 1}{e^y}$

2. $y' = (x^2 + 2)(y + 1)$

3. $y' - 2y \csc x = 0$

4. $(y^2 + 1)^2 y' = \dfrac{4x}{(5 + x^2)y}$

5. $y' = x^2 y^2 - x^2$

6. $xy' = y^2$

7. $y' + 4xy^2 = 0$

8. $x^3 y^2 \sin y\, y' = x^4 y e^{x^2}$

9. $(yx^2 e^x - 3y)dx + (xy^2 - y \sin y)dy = 0$

10. $y' = \dfrac{y^2 - 5y + 6}{x^2}$

11. $xy' + x^2 \sin x^2 y = 0$

12. $y' = \dfrac{y^2 - 5y + 14}{x^2 - 6x + 8}$

Find the solution of the given IVP.

13. $y' + \dfrac{\cos 2x}{y} = 0, \quad y(0) = -5$

14. $y' = \dfrac{10x - 3x^3 + e^{x-3}}{y^3}, \quad y(3) = 2$

15. $(y^4 - \sin y)dx + (e^x - x^2)dy = 0, \quad y(0) = 1$

16. $y' = \dfrac{xe^x}{y^3 - y^2 + y - 1}, \quad y(1) = 2$

17. $y + x^2 y = 0, \quad y(0) = 15$

18. $(x + 1)y' = \dfrac{e^y}{y}, \quad y(0) = 1$

19. $\dfrac{dy}{dx} = \dfrac{x^3 - x^2 - 7}{y \sin y^2}, \quad y(1) = 0.$

20. The Emerald Ash Borer is an invasive insect species that has caused millions of dollars of damage to ash trees across North America. The Ash Borer has no native predators, and its population can grow when introduced into an unaffected ash forest at a rate proportional to its existing population. If 2750 Ash Borers are introduced into an unaffected ash forest, how many Ash Borers will be in the forest after a year (365 days) if the intrinsic growth rate is $r = 0.075$ per day?

21. An ant-keeper has decided to start a new colony with 250 ants. The formicarium has the capacity to hold 20000 ants. If the colony's intrinsic growth rate is $r = 0.03$ per day, how many ants will the colony contain after 100 days?

22. In the mid-1800s, 24 rabbits were released in Australia for the purposes of sport hunting. The rabbits have an intrinsic growth rate of $r = 0.05$ per day, have no natural predators, and their population is not impacted by hunting. How long will it take for the rabbit population to reach one million rabbits? How many rabbits will be present in Australia within a year?

23. Lenny, the lemon tree, was left outside overnight and the temperature unexpectedly dropped below freezing. When he was brought inside, the temperature of Lenny's soil was 33°F. His caretaker checked the temperature of his soil after 1 hour and found it to be 37°F. If the temperature of the room Lenny was placed in was 72°F, how long will it take for Lenny's soil to reach the minimum recommended temperature for lemon trees of 50°F?

24. A container of ice cream was left in a deep freezer with a temperature of -25°F. The ice cream is taken out and placed on a counter to thaw slightly so that it is easier to scoop. The kitchen counter is at a temperature of 70°F. Presuming the ice cream has a heat transfer coefficient of $k = 0.573$. After how long will the ice cream reach the desired temperature of 10°F?

25. A watermelon is brought inside after being picked from a garden. The temperature in the garden was 100°F at the time the melon was picked. The watermelon is brought inside and placed in the refrigerator at a temperature of 37°F. After five minutes the temperature of the watermelon has dropped to 80°F. After how long will the watermelon reach the desired temperature of 45°F?

3.5   Integrating Factors

Not every differential equation you may encounter in the wild will be a separable equation. We have already explored how to classify these differential equations, how to determine existence and uniqueness of their solutions, and how to use vector fields and phase line analysis to obtain qualitative data about these solutions. However, what if you are approached by the following differential equation?

$$x^3 \frac{dy}{dx} + 3x^2 y = 6$$

This equation is not separable, and while you may obtain some information about the behavior of its solutions by the methods at your disposal so far, you may want to find the closed-form solution of this equation. What can you do?

You may look at this equation, see the term $3x^2 y$, and think to yourself that $3x^2$ is the derivative of x^3, and there is already an x^3 term elsewhere in the equation. Thinking that this may be your way out of the woods, you take the derivative of $x^3 y$, just to see. Doing so requires the product rule, and you find that:

$$\frac{d}{dx}\left[x^3 y\right] = x^3 \frac{dy}{dx} + 3x^2 y.$$

This is exactly what you have on the left-hand side of your equation. So then you can see that:

$$
\begin{aligned}
x^3 \frac{dy}{dx} + 3x^2 y &= 6 \\
\frac{d}{dx}\left[x^3 y\right] &= 6 \\
\int \frac{d}{dx}\left[x^3 y\right] dx &= \int 6 \, dx \\
x^3 y &= 6x + C \\
y &= \frac{6x + C}{x^3}
\end{aligned}
$$

Success! But what if you don't recognize part of your equation as the derivative of a known function? We want to use the idea we had above to solve ODEs of the general form:

$$\frac{dy}{dx} + p(x)y = q(x),$$

by the method of **integrating factors**.

To execute the method of **integrating factors**, we follow the below steps:

(1) Write the ODE in standard form:

$$\frac{dy}{dx} + p(x)\, y = q(x)$$

Find the integrating factor $\psi(x) = e^{\int p(x)\ dx}$

(2) Multiply every term in the ODE by $\psi(x)$

(3) Use ψ to write the left hand side as the derivative $\dfrac{d}{dx}\,[\psi(x)y]$

(4) Integrate!

Before we move onto some examples, let's examine why this works. What makes this tick is a combination of both the product rule and the chain rule applied to the exponential function. Notice first that if:

$$\psi(x) \;=\; e^{\int p(x)dx}$$

$$\psi'(x) \;=\; p(x)e^{\int p(x)dx}.$$

Now if we take our standard form linear first-order ODE and multiply every term by $\psi(x)$, we have:

$$y' + p(x)y \;=\; q(x)$$

$$e^{\int p(x)dx}y' + p(x)e^{\int p(x)dx}y \;=\; e^{\int p(x)dx}q(x)$$

$$\frac{d}{dx}\left[e^{\int p(x)dx}y\right] \;=\; e^{\int p(x)dx}q(x).$$

Note that the function on the right-hand side of the equation may be difficult to integrate, but this method allows us to successfully obtain an expression for y in terms of x.

○○○○○○○○○○○○○○○○○○○○○○○○○○○○○○○○○○○○○○○

Example: Let's use the method of integrating factors to solve the ODE: $\frac{dy}{dx} = 3x^2y$.

(1) We start by writing the ODE in standard form $\frac{dy}{dx} - 3x^2y = 0$ and

then find the integrating factor $\psi(x)$. In standard form the equation has $p(x) = -3x^2$ and $q(x) = 0$. Then we have:

$$\psi(x) = e^{\int -3x^2 \, dx} = e^{-x^3}.$$

② Multiply every term in the ODE by $\psi(x)$. Doing so gives us

$$e^{-x^3} \frac{dy}{dx} - 3x^2 e^{-x^3} y = 0.$$

③ Use ψ to write the left-hand side as a derivative. We notice that

$$\frac{d}{dx}\left[e^{-x^3} y\right] = e^{-x^3} \frac{dy}{dx} - 3x^2 e^{-x^3} y,$$

so we have

$$\frac{d}{dx}\left[e^{-x^3} y\right] = 0.$$

④ Integrate!

$$\int \frac{d}{dx}\left[e^{-x^3} y\right] dx = \int 0 \, dx$$

$$e^{-x^3} y = C$$

$$y = Ce^{x^3}.$$

You may have noticed that this ODE was in fact separable. You may find it fun to try to find the solution by separating and integrating instead and checking that you arrive at the same result.

○○

Let's look at an example involving an ODE that is not separable.

Example: We can use the method of integrating factors to solve the ODE: $t\frac{dy}{dt} + y = 3t^2$ where $t > 0$.

① We start by writing the ODE in standard form $\frac{dy}{dt} + \frac{1}{t} y = 3t$ and then find the integrating factor $\psi(t)$. In standard form the equation has $p(t) = \frac{1}{t}$ and $q(t) = 3t$. Then we have:

$$\psi(t) = e^{\int \frac{1}{t} \, dt} = e^{\ln t} = t.$$

② Multiply every term in the ODE by $\psi(t)$. Doing so gives us:

$$t\frac{dy}{dt} + y = 3t^2.$$

Notice that this is the ODE we started with but now we know the strategy to solve from here.

③ Use ψ to write the left-hand side as a derivative. We notice that:

$$\frac{d}{dt}[ty] = t\frac{dy}{dt} + y,$$

so we have

$$\frac{d}{dt}[ty] = 3t^2.$$

④ Integrate!

$$\int \frac{d}{dt}[ty]\,dt = \int 3t^2\,dt$$

$$ty = t^3 + C$$

$$y = \frac{t^3 + C}{t}.$$

○○

Here's another example where the use of the integrating factor is critical in finding the solution.

Example: To solve the differential equation:

$$y' + 2xy = 8x,$$

we first notice that the ODE is in standard form. Thus our first real step is to find $\psi(x)$. We note that $p(x) = 2x$ and compute:

$$\psi(x) = e^{\int p(x)\,dx}$$

$$= e^{\int 2x\,dx}$$

$$= e^{x^2},$$

so $\psi(x) = e^{x^2}$ is our integrating factor. We now multiply all terms of the ODE by $\psi(x)$ and see that we have:

$$y' + 2xy = 8x$$

$$e^{x^2}y' + 2xe^{x^2}y = 8xe^{x^2}$$

$$\int \frac{d}{dx}\left[e^{x^2}y\right]dx = \int 8xe^{x^2}\,dx$$

$$e^{x^2}y = 4e^{x^2} + C$$

$$y = 4 + Ce^{-x^2}.$$

○○

The following example is a bit more complicated. It requires us to recall a somewhat obscure antiderivative and to perform integration by parts. However, despite this, the ODE is still a linear first order differential equation and as such, the method of integrating factors will allow us to solve it.

Example: Consider the IVP

$$y' + \cot(t)y = t, \qquad y(\frac{\pi}{2}) = 0.$$

Noting that we are given an ODE in standard form, we compute $\psi(t)$ by:

$$
\begin{aligned}
\psi(x) &= e^{\int p(t)dt} \\
&= e^{\int \cot(t)dt} \\
&= e^{\ln|\sin(t)|} \\
&= \sin(t).
\end{aligned}
$$

Now that we have obtained $\psi(t) = \sin(t)$, we multiply all terms of the ODE by ψ and find that we have:

$$
\begin{aligned}
y' + \cot(t)y &= t \\
\sin(t)y' + \cos(t)y &= t\sin(t) \\
\int \frac{d}{dt}[\sin(t)y]dt &= \int t\sin(t)dt.
\end{aligned}
$$

Recalling integration by parts, we can see that:

$$\int t\sin(t)dt = \sin(t) - t\cos(t) + C.$$

Therefore, we have:

$$
\begin{aligned}
\int \frac{d}{dt}[\sin(t)y]dt &= \int t\sin(t)dt \\
\sin(t)y &= \sin(t) - t\cos(t) + C \\
y &= 1 - t\cot(t) + C\csc(t).
\end{aligned}
$$

Using our initial condition of $y(\frac{\pi}{2}) = 0$ we see that:

$$
\begin{aligned}
0 &= 1 - \frac{\pi}{2} \cot\left(\frac{\pi}{2}\right) + C \csc\left(\frac{\pi}{2}\right) \\
0 &= 1 - \frac{\pi}{2}(0) + C(1) \\
C &= -1.
\end{aligned}
$$

Therefore, the solution of this IVP is $y = 1 - t\cot(t) - \csc(t)$.

○○○

Example: Consider a group of 1000 clownfish on a reef. The population grows in proportion to their size, with an intrinsic growth rate of $r = 0.005$ per day, except there is migration of the young clownfish away from the reef at a constant 4 clownfish per day. How many days will it take for the population of clownfish on the reef to double?

We can set up the IVP using this information as:

$$
\frac{dP}{dt} = 0.005P - 4, \quad P(0) = 1000.
$$

This ODE can be solved using the method of integrating factors. Putting it in standard form, we have:

$$
\frac{dP}{dt} - 0.005P = -4,
$$

which has integrating factor:

$$
\psi(t) = e^{\int -0.005 dt} = e^{-0.005t}.
$$

Multiplying the ODE by $\psi(t)$ and solving, we have:

$$
\begin{aligned}
\frac{d}{dt}\left[e^{-0.005t}P\right] &= -4e^{-0.005t} \\
\int \frac{d}{dt}\left[e^{-0.005t}P\right] dt &= \int -4e^{-0.005t} dt \\
e^{-0.005t}P &= 800e^{-0.005t} + C \\
P &= 800 + Ce^{0.005t}.
\end{aligned}
$$

Using the initial condition $P(0) = 1000$, we have that $1000 = 800 + C$, so $C = 200$. Hence, the solution is:

$$
P = 800 + 200e^{0.005t}.
$$

In order to solve the problem, which asked when the population of the clownfish would double, we must remember that we started with 1000 clownfish on the reef. So we are trying to find out when the population reaches 2000 clownfish. Then we have:

$$2000 = 800 + 200e^{0.005t}$$

$$1200 = 200e^{0.005t}$$

$$6 = e^{0.005t}$$

$$\ln 6 = 0.005t$$

$$t = 200 \ln 6 \approx 358.35 \text{ days.}$$

Therefore, it will take about 358.35 days (about a year) for the population of clownfish on the reef to double.

○○

Next we will utilize this method to find the solutions for a wide array of mixing problems. There are a number of different circumstances that can occur in these models. The examples that follow include models with: constant rate in and volume, variable volume, and two tanks.

Example: Consider an Olympic-sized swimming pool (a fun tank!) of volume 2500 m^3, which contains chlorine at a concentration of 1.5 g/m^3. This concentration is too low, so we pump water at a rate of 1 m^3/hr which contains chlorine at a concentration of 200 g/m^3 into the pool. The pool is well mixed and drains at a rate of 1 m^3/hr. If we want the pool to have a chlorine concentration of 2 g/m^3, how long will we need to pump and drain it?

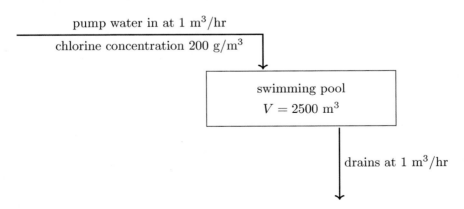

Notice that we did not label the initial concentration of chlorine (1.5 g/m^3) in our diagram. This is because ultimately the concentration of chlorine is changing over time. However, it will be important to remember this initial concentration of chlorine as it will be useful later. Additionally, the swimming pool is labeled $V = 2500 \text{ m}^3$ because we know that the amount of liquid in the pool is not changing over time, simply because we are pumping liquid in and out of the pool at the same rate.

To begin, we need to establish R_{in} and R_{out}. As the water being pumped into the pool is pumped in at a rate of $1 \text{ m}^3/\text{hr}$ and it contains chlorine at 200 g/m^3, we can see that $R_{in} = 1\frac{\text{m}^3}{\text{hr}} \cdot 200\frac{\text{g}}{\text{m}^3} = 200 \text{ g/hr}$.

Finding R_{out} is a bit different. We don't know what the concentration of chlorine will be at any given time, but we are letting A equal the total amount of chlorine, so we know that the concentration of chlorine is $\frac{A \text{ g}}{2500 \text{ m}^3}$ as 2500 m^3 is the volume of the pool. Then, as the outflow rate is $1 \text{ m}^3/\text{hr}$, we know that $R_{out} = 1\frac{\text{m}^3}{\text{hr}} \cdot \frac{A \text{ g}}{2500 \text{ m}^3} = \frac{A}{2500} \text{ g/hr}$.

Finally, we need to determine the initial conditions. At $t = 0$, we are told that the concentration of chlorine is 1.5 g/m^3. Given the pool's volume, we can calculate that $A(0) = 1.5 \text{ g/m}^3 \cdot 2500 \text{ m}^3 = 3750 \text{ g}$. Thus, we have set up our IVP to be:

$$\frac{dA}{dt} = 200 - \frac{A}{2500}, \quad A(0) = 3750.$$

After rearranging, we recognize that this can be written in standard form and solved using the method of integrating factors. In standard form this becomes:

$$\frac{dA}{dt} + \frac{1}{2500}A = 200, \quad A(0) = 3750.$$

The integrating factor for this ODE is:

$$\psi(t) = e^{\int \frac{1}{2500}\,dt} = e^{t/2500},$$

so we can rewrite this and integrate as follows:

$$e^{t/2500}\frac{dA}{dt} + e^{t/2500}\frac{1}{2500}A = 200e^{t/2500}$$

$$\frac{d}{dt}\left[e^{t/2500}A\right] = 200e^{t/2500}$$

$$\int \frac{d}{dt}\left[e^{t/2500}A\right]dt = \int 200e^{t/2500}\,dt$$

$$e^{t/2500}A = 500000e^{t/2500} + C$$

$$A = 500000 + Ce^{-t/2500}.$$

Using the initial condition $A(0) = 3750$, we see that $3750 = 500000 + C$, so $C = -496250$. Thus our solution of the IVP is:

$$A = 500000 - 496250e^{-t/2500}.$$

Now, our question is how long to run the pump so that there is a chlorine concentration of 2 g/m^3. This would correspond to a total amount of chlorine of 2 g/m^3 · 2500 m^3 = 5000 g. Therefore, we want to solve the following equation for t:

$$5000 = 500000 - 496250e^{-t/2500}.$$

This can be done, and we find that:

$$5000 = 500000 - 496250e^{-t/2500}$$

$$-495000 = -496250e^{-t/2500}$$

$$\frac{396}{397} = e^{-t/2500}$$

$$\ln\left(\frac{396}{397}\right) = -\frac{t}{2500}$$

$$t = -2500\ln\left(\frac{396}{397}\right) \approx 6.31 \text{ hours}.$$

Hence, it would take approximately 6.31 hours to increase the chlorine concentration to 2 g/m^3.

○○

Example: In this example, let's consider a tank in which the inlet rate and outlet rate are different.

Consider a tank of brine which has a maximum capacity of 1000 gallons, but which currently only contains 250 gallons of brine concentrated to 1 lb/gal. In order to refill the tank with a lower concentration of brine, seawater is pumped into the tank at a rate of 2 gal/min and a concentration of 0.3 lb/gal. To speed up the dilution process, brine is pumped out of the well-mixed tank at a rate of 1 gal/min. What will be the concentration of salt when the tank is full?

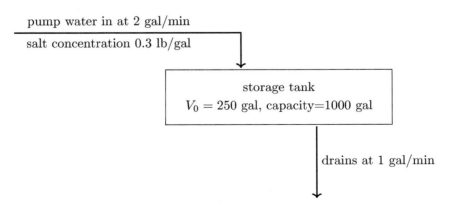

pump water in at 2 gal/min

salt concentration 0.3 lb/gal

storage tank
$V_0 = 250$ gal, capacity$=1000$ gal

drains at 1 gal/min

As before, we have that $\frac{dA}{dt} = R_{in} - R_{out}$. The computation for R_{in} is similar to the last problem. We have that:

$$R_{in} = \text{(rate of inflow)} \cdot \text{(concentration)} = 2\frac{\text{gal}}{\text{min}} \cdot 0.3\frac{\text{lb}}{\text{gal}} = 0.6 \text{ lb/min.}$$

However, the computation for R_{out} here is different. As before,

$$R_{out} = \frac{\text{(rate of outflow)(amount of salt)}}{\text{volume of tank}}.$$

Here the volume of the tank is variable because the inflow and outflow rate differ. The volume of the tank follows the equation:

$$V(t) = \text{initial volume} + \text{(inflow rate} - \text{outflow rate)}t.$$

So for our problem, $V(t) = 250 + (2 - 1)t = 250 + t$. Therefore, we have $R_{out} = \frac{A}{250+t}$, and

$$\frac{dA}{dt} = 0.6 - \frac{A}{250 + t}, \quad A(0) = 250.$$

This can be solved using the method of integrating factors. In standard form this becomes:

$$\frac{dA}{dt} + \frac{A}{250 + t} = 0.6, \quad A(0) = 250.$$

The integrating factor for this ODE is:

$$\psi(t) = e^{\int \frac{1}{250+t}\, dt} = e^{\ln(250+t)} = 250 + t.$$

Note that we have assumed $t > 0$, which is common in applied problems in which we are concerned with what happens as time progresses into the future. We can rewrite our original standard form equation and integrate as follows:

$$(250 + t)\frac{dA}{dt} + A = 0.6(250 + t)$$

$$\frac{d}{dt}\left[(250 + t)A\right] = 0.6(250 + t)$$

$$\int \frac{d}{dt}\left[(250 + t)A\right] dt = \int 0.6(250 + t)\, dt$$

$$(250 + t)A = 0.3(250 + t)^2 + C$$

$$A = 0.3(250 + t) + \frac{C}{250 + t}.$$

Using the initial condition $A(0) = 250$, we see that $250 = 75 + \frac{C}{250}$. Then $C = 43750$, and our solution of the IVP is:

$$A = 0.3(250 + t) + \frac{43750}{250 + t}.$$

Now, the question asked what the concentration of brine would be when the tank is at the full capacity of 1000 gallons. Our first step is to find at what time the volume reaches this point. Luckily, we already computed $V(t) = 250 + t$, and it is an easy computation to find that $V(t) = 1000$ at time $t = 750$.

In order to find the final concentration, we must remember that a concentration is the ratio of salt to solution in the tank. We know that at $t = 750$, there are $A(750) = 343.75$ lbs of salt in the tank. Of course, there are 1000 gallons of solution at this time. Thus, the concentration of salt in the tank when the tank is full is $\frac{343.75 \text{ lb}}{1000 \text{ gal}} = 0.34375$ lb/gal.

○○

This final tank example is a bit different. Before we have seen a problem with a single tank with fixed volume, and a single tank with variable volume. This next problem's focus is an apparatus with two tanks.

Example: The first tank is a 200 liter tank which contains fresh water. Water that has been treated with fluoride at a concentration of 0.15 mg/L is pumped into the tank at a rate of 1 L/min. This tank outflows at a rate of 1 L/min into a second tank which contains 100 L of water with a fluoride concentration of 1 mg/L. This second tank also drains at the rate 1 L/min. What is the minimum concentration of fluoride in the second tank?

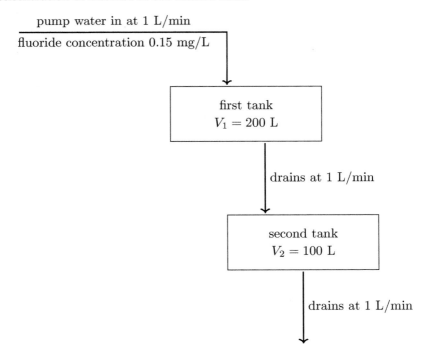

pump water in at 1 L/min

fluoride concentration 0.15 mg/L

first tank
$V_1 = 200$ L

drains at 1 L/min

second tank
$V_2 = 100$ L

drains at 1 L/min

There are a number of methods for solving a situation like this one, but the way we are going to approach this is to solve two tank problems. First, we will model the fluoride content in the first tank, and then use that to find the function defining the concentration of fluoride flowing into the second tank.

The first tank can be modeled similarly as before with $\frac{dA}{dt} = R_{in} - R_{out}$. We can find that:

$$R_{in} = 0.15\frac{mg}{L} \cdot 1\frac{L}{min} = 0.15mg/min$$

and

$$R_{out} = \frac{1L/min \cdot Amg}{200L} = \frac{A}{200}mg/min.$$

We also have initial condition $A(0) = 0$ because the first tank contains only fresh water (i.e., there is no fluoride in the tank initially). Therefore, our IVP is:

$$\frac{dA}{dt} = 0.15 - \frac{A}{200}, \quad A(0) = 0.$$

Rewriting the ODE in standard form, we have:

$$\frac{dA}{dt} + \frac{A}{200} = 0.15.$$

The integrating factor for this ODE is:

$$\psi(t) = e^{\int \frac{1}{200} \, dt} = e^{t/200},$$

so we can rewrite and solve the ODE as follows:

$$\frac{d}{dt}\left[e^{t/200}A\right] = 0.15e^{t/200}$$

$$\int \frac{d}{dt}\left[e^{t/200}A\right] dt = \int 0.15e^{t/200}dt$$

$$e^{t/200}A = 30e^{t/200} + C$$

$$A = 30 + Ce^{-t/200}.$$

Using the initial condition $A(0) = 0$, we see that $0 = 30 + C$. Then $C = -30$, and

$$A(t) = 30 - 30e^{-t/200}.$$

Now that the first tank has been addressed, we can set up the ODE for the second tank. To find R_{in}, we need to know the concentration of fluoride that is coming into the second tank. The amount of fluoride in the first tank is $A(t) = 30 - 30e^{-t/200}$ mg, and the volume of the first tank is $V_1 = 200$ L, so

$$R_{in} = 1\frac{L}{min} \cdot \frac{30 - 30e^{-t/200} \ mg}{200 \ L} = 0.15 - 0.15e^{-t/200} \ mg/min.$$

We also know that:

$$R_{out} = 1\frac{L}{min} \cdot \frac{Q \text{ mg}}{100 \text{ L}} = \frac{Q}{100}\text{mg/min},$$

where Q is the amount (or quantity) of fluoride in the second tank. Additionally, there is an initial fluoride concentration of 1 mg/L in the second tank. Then our initial condition is $Q(0) = 1$ mg/L$\cdot100$ L $= 100$ mg. Therefore, we have the IVP:

$$\frac{dQ}{dt} = 0.15 - 0.15e^{-t/200} - \frac{Q}{100}, \quad Q(0) = 100.$$

This ODE can be solved using integrating factors as well. Rewriting in standard form we have:

$$\frac{dQ}{dt} + \frac{Q}{100} = 0.15 - 0.15e^{-t/200},$$

with integrating factor

$$\psi(t) = e^{\int \frac{1}{100} \, dt} = e^{t/100}.$$

Therefore, we can rewrite and solve this ODE as follows:

$$\frac{d}{dt}\left[e^{t/100}Q\right] = 0.15e^{t/100} - 0.15e^{t/200}$$

$$\int \frac{d}{dt}\left[e^{t/100}Q\right] dt = \int \left(0.15e^{t/100} - 0.15e^{t/200}\right) dt$$

$$e^{t/100}Q = 15e^{t/100} - 30e^{t/200} + C$$

$$Q = 15 - 30e^{-t/200} + Ce^{-t/100}.$$

Now, using the initial condition $Q(0) = 100$, we find that $100 = 15 - 30 + C$, so that $C = 115$. Therefore, the equation describing the amount of fluoride in the second tank is:

$$Q(t) = 15 - 30e^{-t/200} + 115e^{-t/100}.$$

We need to find the minimum amount of fluoride in the second tank, then divide by $V_2 = 100$ L, in order to obtain the fluoride concentration in the tank. To find the minimum, we can find the time when the derivative $Q'(t)$

equals 0. Thus:

$$
\begin{aligned}
Q'(t) &= 0.15e^{-t/200} - 1.15e^{-t/100} \\
0 &= 0.15e^{-t/200} - 1.15e^{-t/100} \\
1.15e^{-t/100} &= 0.15e^{-t/200} \\
e^{-t/100} &= \tfrac{0.15}{1.15}e^{-t/200} \\
e^{-t/200} &= \tfrac{0.15}{1.15} \\
-\tfrac{t}{200} &= \ln\tfrac{0.15}{1.15} \\
t &= -200\ln\tfrac{0.15}{1.15} \approx 407.38.
\end{aligned}
$$

Therefore, the minimum amount of fluoride is in the second tank at time $t = 407.38$ min. We can find the minimum quantity of fluoride by finding

$$Q(407.38) = 15 - 30e^{-407.38/200} + 115e^{-407.38/100} \approx 13.043 \text{ mg.}$$

As the minimum amount of fluoride is 13.043 mg and the second tank has volume 100 L, the minimum concentration of fluoride in the second tank is $\frac{13.043 \text{ mg}}{100 \text{ L}} = 0.13043$ mg/L.

3.5 Exercises

Find the general solution of the following ODEs.

1. $xy' + 3y = 5x^2 - 9$
2. $y' + 5y = e^{-5t} + 3$
3. $y' - 4y = e^{7x} - 2x$
4. $(x-1)y' + 2y = xe^x$
5. $y' + x^2 y = x^2$
6. $y' + 4xy = 2x$
7. $xy' + y = x^2 + 2x$
8. $xy' + y = x\sin x$
9. $y' + 2y = e^{5x}$
10. $y' + 5y = 2x$
11. $y' + \dfrac{y}{x} = e^{x^2}$
12. $x^2 y' + xy = x^2 \sin x + xe^x + 8$
13. $(x^2 + x^3)y' + (2x + 3x^2)y = e^x$

Find the solutions of the following IVPs and state the interval on which they exist.

14. $x^2 y' - 2xy = 4x^4 + 3x^3, \quad y(1) = 10$

15. $y' = 3t^2 y + t^2, \quad y(0) = 1$

16. $y' + \tan(x)y = x \sec(x), \quad y(0) = 2$

17. $ty' - y = \sin(t), \quad y(t) = 1$

18. $y' = (1 + y)\sin t, \quad y(\pi/2) = 1.$

19. $xy' + y = \sin x + \cos x, \quad y(\pi/2) = 1$

20.   A dairy farm has a 500-gallon vat of heavy cream, which has a milk fat concentration of 0.31 lb/gal. They want to sell 500 gallons of half and half, which has a milk fat concentration of 0.2 lb/gal. In order to do so, the farmer pumps 2% milk with a fat concentration of 0.1 lb/gal into the vat at a rate of 2 gal/min and pumps the contents of the vat out at a rate of 2 gal/min. How long will it take for the well-mixed vat of cream to have the milk fat concentration of half and half?

21.   A pond has an initial population of 1500 guppies. Every day 100 guppies swim downstream and never return to the pond. If the guppies in the pond have an intrinsic growth rate of $r = 0.08$ per day, how long will it take for the population of guppies in the pond to reach 10000?

22.   A litter of 10 vole are nested on a large log on the shore of a bay. The log becomes displaced and floats to sea, coming ashore on a small island with no existing vole population. The vole has an intrinsic growth rate of $r = 0.11$ per day and the island has a carrying capacity of 7250 vole. How many days will it take for the island's vole population to reach 3125?

23.   An oil refinery is preparing a shipment of gasoline. However, the gasoline in their shipment tank currently contains too small of a concentration of an antiknock agent additive. In order to meet specifications, the refinery begins pumping gasoline from a 1000-liter storage tank into the 200 liter shipment tank at a rate of 2 liters per minute. Gasoline is pumped out of the shipment tank at the same rate of 2 liters per minute, and gasoline is pumped into the storage tank at a rate of 2 liters per minute as well. The gasoline pumped into the storage tank has an additive concentration of 5 grams per liter, the gasoline in the storage tank has a concentration of 3 grams per liter, and the gasoline in the shipment tank contains no additives. What will the additive concentration in the shipment tank be after 300 minutes?

24. A batch of cookies was removed from the oven, and each cookie had a temperature of 170°F. The cookies were left to cool on a wire rack in a kitchen at the temperature of 68°F. After 5 minutes, the cookies were 150°F. If the ideal temperature at which to eat a cookie is 90°F, how long must the cookies be out of the oven before the baker can taste test his cookies?

25. A chemical factory has a tank with a maximum capacity of 200 gallons that is currently filled with 100 gallons of iodine solution. The tank initially contains a concentration of 3 mg/gal of iodine. The factory pumps a solution with iodine concentration of 15 mg/gal into the tank at a rate of 5 gal/min while draining the tank at the rate of 3 gal/min. What is the concentration of iodine in the tank at the moment the tank reaches capacity?

26. A 40,000-liter-capacity Koi pond is filled with 20,000 liters of water. The pond drains at a rate of 5 liters per minute. Suppose there are currently 5.0 grams per liter of total dissolved solids in the pond. Flood water starts filling the pond at a rate of 8 liters per minute with water containing 18.5 grams per liter of total dissolved solids. What will the concentration of total dissolved solids in the pond be when it reaches its capacity?

3.6 Exact Equations

In Section 3.4 we considered separable ODEs of the form:

$$M(x)dx + N(y)dy = 0.$$

What was nice about ODEs of this form was that the terms containing x and y could be separated and integrated, which then allowed us to find the general solution of the ODE. Any ODE of the form $M(x)dx + N(y)dy = 0$ could be solved using this method, though depending on the functions $M(x)$ and $N(y)$, the integration might be difficult or impossible to do without assistance from computational software or numerical methods.

It is natural to ask about ODEs similar to separable equations, but where the functions M and N are functions of both x and y. Can we still solve ODEs of this form? Is there a restriction on the functions $M(x, y)$ and $N(x, y)$ for which we can solve it? It turns out that ODEs of the form:

$$M(x, y) + N(x, y)y' = 0$$

with functions M and N satisfying a specific restriction are a special class of ODEs called exact equations.

Definition: The ODE $M(x,y) + N(x,y)\, y' = 0$, where M and N are continuous functions, is an **exact** differential equation if $M_y = N_x$ where M_y denotes the derivative of M with respect to y and N_x denotes the derivative of N with respect to x.

Example: The differential equation $4xy^2 + 5 + (4x^2y + y)y' = 0$ is exact because:

$$\left.\begin{array}{l} M = 4xy^2 + 5 \implies M_y = 8xy \\[2mm] N = 4x^2y + y \implies N_x = 8xy \end{array}\right\} M_y = N_x, \text{ so it's exact!}$$

Theorem for Exact Equations: If an ODE is exact, then there exists a function $f(x,y)$ satisfying $f_x(x,y) = M$ and $f_y(x,y) = N$. Furthermore, solutions of the ODE are of the form $f(x,y) = C$, where C is a constant.

Example: To solve the exact equation from the previous example,

$$4xy^2 + 5 + (4x^2y + y)y' = 0,$$

we need to use the above theorem about exact equations. This exact equation is not separable and is not linear (so we cannot use integrating factors). So instead, we try to come up with a function f that satisfies:

$$f_x = M = 4xy^2 + 5$$

and

$$f_y = N = 4x^2y + y.$$

We start by integrating $f_x = M = 4xy^2 + 5$ with respect to x:

$$\begin{aligned} f_x &= 4xy^2 + 5 \\[2mm] f &= \int \left(4xy^2 + 5\right)\, dx \\[2mm] f &= 2x^2y^2 + 5x + C(y), \end{aligned}$$

where $C(y)$ is a constant function in x (but may depend on y). Then, taking the derivative with respect to y, we have $f_y = 4x^2y + C'(y)$. We know that $f_y = N = 4x^2y + y$. Then $4x^2y + C'(y) = 4x^2y + y$, so that $C'(y) = y$. Then $C(y) = \int y\, dy = \frac{1}{2}y^2$. Hence solutions of the ODE $4xy^2 + 5 + (4x^2y + y)y' = 0$ have the form $2x^2y^2 + 5x + \frac{1}{2}y^2 = C$.

○○

Example: Let's solve the differential equation:

$$(6x^2y + xy^2)dx + (2x^3 + x^2y + 5)dy = 0.$$

Notice that this equation is not quite in the form needed to determine if the ODE is exact. However, we can rewrite our ODE as:

$$6x^2y + xy^2 + (2x^3 + x^2y + 5)\frac{dy}{dx} = 0,$$

or in other words,

$$6x^2y + xy^2 + (2x^3 + x^2y + 5)y' = 0.$$

Now we can see that this ODE is exact because:

$$\left.\begin{array}{l} M = 6x^2y + xy^2 \quad \implies \quad M_y = 6x^2 + 2xy \\[2mm] N = 2x^3 + x^2y + 5 \quad \implies \quad N_x = 6x^2 + 2xy \end{array}\right\} M_y = N_x, \text{ so it's exact!}$$

Then according to the Theorem for Exact Equations, we can find the solution of the ODE by identifying the function f with $f_x = M$ and $f_y = N$. Then we have:

$$f_x = M$$

$$f_x = 6x^2y + xy^2$$

$$f = \int (6x^2y + xy^2)\ dx$$

$$f = 2x^3y + \tfrac{1}{2}x^2y^2 + C(y),$$

where $C(y)$ is a function of y (and is constant in x). Then, taking the derivative with respect to y, we have $f_y = 2x^3 + x^2y + C'(y)$. We also know that:

$$f_y = N = 2x^3 + x^2y + 5,$$

so that

$$2x^3 + x^2y + C'(y) = 2x^3 + x^2y + 5.$$

Hence $C'(y) = 5$. To solve for $C(y)$, we integrate to get $C(y) = \int 5\ dy = 5y$. Therefore, solutions of the ODE $(6x^2y + xy^2)dx + (2x^3 + x^2y + 5)dy = 0$ have the form $2x^3y + \tfrac{1}{2}x^2y^2 + 5y = C$.

There are situations where a given ODE is not exact, but that we can use integrating factors to make it exact and then solve the ODE using the above method. In this case, there are two possibilities of integrating factors that could work.

1. If $p = \dfrac{M_y - N_x}{N}$ is a function of x alone, then we can use the integrating factor $\psi = e^{\int p \, dx}$.

2. If $p = \dfrac{N_x - M_y}{M}$ is a function of y alone, then we can use the integrating factor $\psi = e^{\int p \, dy}$.

Example: Consider the ODE:

$$3x^2y + x^3 - 1 + (3x^3 + 2x^2)y' = 0.$$

Notice that:

$$
\left.
\begin{array}{ll}
M = 3x^2y + x^3 - 1 & \implies M_y = 3x^2 \\[2mm]
N = 3x^3 + 2x^2 & \implies N_x = 9x^2 + 4x
\end{array}
\right\} M_y \neq N_x, \text{ so it's NOT exact.}
$$

We must try using an integrating factor to see if we can make the equation exact, knowing that sometimes this does not work either. Let's try:

$$
\begin{aligned}
p &= \frac{M_y - N_x}{N} \\[2mm]
&= \frac{3x^2 - (9x^2 + 4x)}{3x^3 + 2x^2} \\[2mm]
&= \frac{-6x^2 - 4x}{3x^3 + 2x^2} \\[2mm]
&= \frac{-2(3x^2 + 2x)}{x(3x^2 + 2x)} \\[2mm]
&= -\frac{2}{x}.
\end{aligned}
$$

Because this is a function of x alone (there are no y in the equation for p), we may proceed with the integrating factor:

$$\psi = e^{\int p \, dx} = e^{\int -\frac{2}{x} \, dx} = e^{-2 \ln |x|} = x^{-2}.$$

If we multiply our original ODE by the integrating factor $\psi = x^{-2}$, we obtain:

$$x^{-2}(3x^2y + x^3 - 1) + x^{-2}(3x^3 + 2x^2)y' = 0,$$

or simplifying,
$$3y + x - x^{-2} + (3x + 2)y' = 0.$$
We can now check to see if this new ODE is exact or not. We have:

$$\left. \begin{array}{lll} M = 3y + x - x^{-2} & \implies & M_y = 3 \\[2mm] N = 3x + 2 & \implies & N_x = 3 \end{array} \right\} M_y = N_x, \text{ so it's exact!}$$

Now we can solve this new ODE in the same way as before knowing that both ODEs will have the same solution. In this case, we have:

$$\begin{aligned} f_x &= M \\[2mm] f_x &= 3y + x - x^{-2} \\[2mm] f &= \int \left(3y + x - x^{-2}\right) \, dx \\[2mm] f &= 3xy + \tfrac{1}{2}x^2 + x^{-1} + C(y), \end{aligned}$$

where $C(y)$ is a constant function in x. Then, taking the derivative with respect to y, we have $f_y = 3x + C'(y)$. We know that $f_y = N = 3x + 2$, so that $3x + C'(y) = 3x + 2$. Then $C'(y) = 2$, so that $C(y) = \int 2 \, dy = 2y$. Thus, solutions of the ODE $3x^2y + x^3 - 1 + (3x^3 + 2x^2)y' = 0$ have the form $3xy + \tfrac{1}{2}x^2 + x^{-1} + 2y = C$.

○○

Example: Let's consider the ODE:
$$y^2 + (xy + 5y^2 - 2)y' = 0,$$
where $x > 0$ and $y > 0$. Then:

$$\left. \begin{array}{lll} M = y^2 & \implies & M_y = 2y \\[2mm] N = xy + 5y^2 - 2 & \implies & N_x = y \end{array} \right\} M_y \neq N_x, \text{ so it's NOT exact.}$$

So we consider the integrating factor:
$$p = \frac{M_y - N_x}{N} = \frac{2y - y}{xy + 5y^2 - 2} = \frac{y}{xy + 5y^2 - 2}.$$

However, this is not a function of x alone, so this will not help us to solve the ODE. Let's instead try:
$$p = \frac{N_x - M_y}{M} = \frac{y - 2y}{y^2} = \frac{-y}{y^2} = -\frac{1}{y}.$$

This is a function of y alone, so we can use the integrating factor:

$$\psi = e^{\int p\ dy} = e^{\int -\frac{1}{y}\ dy} = e^{-\ln y} = y^{-1}.$$

We can now multiply our original ODE by this integrating factor to get:

$$y^{-1}y^2 + y^{-1}(xy + 5y^2 - 2)y' = 0,$$

or simplifying,

$$y + (x + 5y - 2y^{-1})y' = 0.$$

Then:

$$\left.\begin{array}{lll} M = y & \implies & M_y = 1 \\[2mm] N = x + 5y - 2y^{-1} & \implies & N_x = 1 \end{array}\right\} M_y = N_x, \text{ so it's exact!}$$

We are now able to proceed as in the previous example. We start by integrating $f_x = M = y$ with respect to x:

$$f_x = y$$

$$f = \int y\ dx$$

$$f = xy + C(y).$$

Then, taking the derivative with respect to y, we have $f_y = x + C'(y)$. We know that $f_y = N = x + 5y - 2y^{-1}$. Then we have $x + C'(y) = x + 5y - 2y^{-1}$, so that $C'(y) = 5y - 2y^{-1}$. Then:

$$C(y) = \int \left(5y - 2y^{-1}\right)\ dy = \frac{5}{2}y^2 - 2\ln|y|.$$

Hence solutions of the given ODE have the form $xy + \frac{5}{2}y^2 - 2\ln|y| = C.$

3.6 Exercises

For the following questions, determine whether each of the ODEs is exact. If it is exact, find the solution. If it is not exact, find an integrating factor to make the ODE exact and then find the solution.

1. $2xy^2 + 7y + (7x + 2x^2y)y' = 0$

2. $3y + 2xy + y^2 + 4 + (x^2 + 3x + 2xy)y' = 0$

3. $x^2 + \dfrac{4y}{x} + 4y' = 0$

4. $(2x + 2xy + y)dx + (x^2 + x + y^2)dy = 0$

5. $3x^2 + 4xy + (4x^2 + 5xy)y' = 0$

6. $3x^2e^{2y} + x^3 + 2 + (2x^3e^{2y} + \cos y + 3)y' = 0$

7. $(2x + 4y)dx + (4x + y^2)dy = 0$

8. $(3x^2 + 6xy\cos(x^2y)) + (3x^2\cos(x^2y) - e^y)y' = 0$

9. $(4x^3y + 2xy^2e^{x^2} - 2x) + (x^4 + 2ye^{x^2} + 1)y' = 0$

10. $(7x^6 + 2xye^{x^2y} + y\cos(xy)) + (x^2e^{x^2y} + x\cos(xy))y' = 0$

Solve the following IVPs. In the event that the equation is not exact, use an integrating factor to solve.

11. $(2x\sin y + ye^{xy})dx + (x^2\cos y + xe^{xy})dy = 0, \quad y(0) = 1$

12. $\left(\frac{1}{x} + y^2\right) + (2xy + 3y^2)y' = 0, \quad y(1) = 2$

13. $(y^2 + 2xy^3)dx + (2xy + 3x^2y^2)dy = 0, \quad y(1) = 2$

14. $y' = \dfrac{2xy}{5y^4 - x^2}, \quad y(1) = 2$

15. $(2xy - \sin x) + (x^2 + y^{-1})y' = 0, \quad y(0) = 4$

16. $\left(y\cos xy + \frac{e^x}{y}\right) + (x\cos xy + e^x\ln y)y' = 0, \quad y(\pi) = 1$

17. $(2x - 3x^2)dx + 2ydy = 0, \quad y(2) = 1$

18. $y' = \dfrac{x\sin x - 1}{2xy}, \quad y(1) = 2$

19. $y' = \dfrac{y\sin xy - 2xy - 1}{x^2 - x\sin xy - 4y^3}, \quad y(0) = 2$

20. $\left(\frac{x}{x^2+y^2} + 2xy^3e^{x^2y^3}\right) + \left(\frac{y}{x^2+y^2} + 3x^2y^2e^{x^2y^3}\right)y' = 0, \quad y(1) = 0$

3.6 Project: Euler's Method Lab

This project can be completed using any computer software which can graph functions, vector fields, and sequences of numbers defined using a for loop. MATLAB, Mathematica, Maple, free software such as Python, R, or online tools such as Desmos can all be used.

Euler's Method is a numerical method to approximate the value of the solution of a differential equation at a specific point. Given an IVP:

$$y' = f(x, y), \qquad y(x_0) = y_0,$$

Euler's method uses linearization to approximate the solution. Given a step size h, it defines recursive equations:

$$\begin{cases} x_{n+1} & = & x_n + h \\ \\ y_{n+1} & = & y_n + hf(x_n, y_n) \end{cases}$$

where x_0 and y_0 are defined in the IVP. In effect, we are after y_N for some chosen N. For example, consider the IVP:

$$y' = 2y, \qquad y(0) = 1.$$

You can easily find that $y = e^{2x}$ is the solution, but let's try to use Euler's method. If we are interested in approximating the value of y when $x = 1$, we can use step size $h = 0.5$ and have an approximation after two iterations.

1. Find an approximation for $y(1)$ using a step size of $h = 0.5$ by hand. Compare this approximation to the actual value of $y(1)$.

2. Using your code, find approximations for $y(1)$ with $h = 0.5$, $h = 0.25$, and $h = 0.1$. Compare these approximations with the actual value of $y(1)$.

3. Plot your Euler's method results against the vector field for $y' = 2y$ and analyze how well they align.

○○

Let's try this with a more interesting IVP. Consider a population of herring which has an intrinsic growth rate of 1.2 per day. Suppose the sound in which they reside can support a maximum of 10 million herring. The herring are harvested periodically so that they can grow naturally over the winter and are harvested over the summer. In January, there are one million Herring in the sound. This can be modeled with the following differential equation:

$$\frac{dy}{dx} = 1.2y \left(1 - \frac{y}{10}\right) - 0.5 \left(\sin\left(\frac{\pi(t-6)}{12}\right) + 1\right) y, \qquad y(0) = 1.$$

This does not have an easy to write down solution, so we'll use Euler's method to approximate the behavior.

4. Suppose that at $t = 0$ there are 3 million herring in the sound. For h=2,1, and 0.5, approximate the population of herring in the sound after 24 months, at $t = 24$.

5. Graph your approximation against the vector field for this differential equation.

○○○

Finally, let's consider a predator-prey model. We saw before that the system of first-order ODEs:

$$\begin{cases} \dfrac{dx}{dt} = 1.2x - 0.01xy \\ \dfrac{dy}{dt} = -0.5y + 0.01xy \end{cases}$$

can model a predator-prey scenario. This system is an example of the Lotka-Volterra equations. We saw that this system can be thought of as a single first-order ODE:

$$\frac{dy}{dx} = \frac{-0.5y + 0.01xy}{1.2x - 0.01xy}$$

and that we can look at the vector field in the x, y-plane rather than the x, t and y, t-planes. It's not obvious that we can even find a solution of this, so Euler's method looks much more attractive here.

6. Compare the results of Euler's Method with a step size of $h = 10, 5, 1$, and 0.5 to approximate $y(80)$ for the IVP:

$$\frac{dy}{dx} = \frac{-0.5y + 0.01xy}{1.2x - 0.01xy}, \qquad y(30) = 100.$$

Amazingly, you can actually solve this IVP. Let's try using exact equations!

7. Write the ODE:

$$\frac{dy}{dx} = \frac{-0.5y + 0.01xy}{1.2x - 0.01xy}$$

in standard $M(x,y)dx + N(x,y)dy = 0$ form and check if it is exact.

You should have found that it is NOT exact. If you attempt to use integrating factors like we do in this section, you will find that the approach will not work. However, for this specific type of differential equation, the integrating factor $\psi(x, y) = x^{-1}y^{-1}$ will make this an exact equation.

8. Using the integrating factor $\psi(x, y) = x^{-1}y^{-1}$, solve the IVP:

$$\frac{dy}{dx} = \frac{-0.5y + 0.01xy}{1.2x - 0.01xy}, \qquad y(30) = 100.$$

9. This equation actually is also a separable equation. Using separation of variables, find the solution and confirm the solution you just got using exact equations.

10. Plot your approximation for $y(80)$ with step size $h = 1$ along with the solution curve on the vector field.

11. Find the actual value of $y(80)$ and compare the Euler's Method approximations from earlier to this actual value.

Chapter 4

Modeling with Second-Order ODEs

It is a natural question to ask "we now know how to solve some first order ODEs, can we solve some higher order ones?" The answer is of course yes, and it is this problem that we intend to confront in this chapter. We would like to develop a single method of solving all higher-order ordinary differential equation (ODEs), but this is asking too much. We might then ask for a single method to solve all linear ODEs, of any order. Something like:

$$P_n(t)y^{(n)} + P_{n-1}(t)y^{(n-1)} + \cdots + P_1(t)y' + P(t)y = G(t).$$

This too is asking too much, at least in full generality.

We will limit ourselves to linear homogeneous differential equations with constant coefficients to begin. We will also limit ourselves to second-order ODEs. This is not because higher-order ODEs are more challenging; in fact, the methods used in this chapter extend naturally. Rather, we are limiting ourselves to second-order ODEs for clarify of presentation and so that certain linear algebra processes, such as finding the determinant of a matrix, can be dealt with in the easier two-dimensional setting.

Our main object of study therefore are ODEs of the type below.

> **Definition:** A second order ODE is called linear and homogeneous if it is of the form
> $$P(t)y'' + Q(t)y' + R(t)y = 0.$$

This differential equation is called **homogeneous** because it is of the form $F(y, y', y'') = 0$. In the event that $P(t)$, $Q(t)$, and $R(t)$ are all constant functions, for example:

$$y'' + 5y' + 6y = 0$$

we say that it is a *linear homogeneous second-order ODE with constant coefficients*. It is this limited subset of higher-order ODEs that we will now develop a method for solving.

DOI: 10.1201/9781003298663-4

4.1   The Wronskian and the Fundamental Set

We would like to be handed a linear homogeneous second-order ODE with constant coefficients and return a general solution, as we did with first-order ODEs. Furthermore, armed with suitable initial values we would like to return a *unique* solution. There is some work to do before we can arrive at that end state.

Our first question is how can we possibly try to solve one of these ODEs? You might be tempted to just make an educated guess and follow your nose, hoping you arrive at a solution. Let's give that a try! You may have noticed that we had a lot of exponential functions pop up in the solutions of first-order ODEs. Maybe it wouldn't hurt to just guess that a solution is $y = e^{rt}$ and just see if we find a restriction on r. It wouldn't hurt to try; let's see what happens!

oooooooooooooooooooooooooooooooooooooooo

Example: Can we find values of r such that $y = e^{rt}$ is a solution of the differential equation $y'' + 5y' + 6y = 0$?

We can find the derivatives of y, and substitute them into the equation to determine which values might work. We have:

$$y = e^{rt}$$

$$y' = re^{rt}$$

$$y'' = r^2 e^{rt}.$$

Substituting these values into the equation we have that:

$$y'' + 5y' + 6y = 0$$

$$r^2 e^{rt} + 5re^{rt} + 6e^{rt} = 0$$

$$e^{rt}(r^2 + 5r + 6) = 0$$

$$e^{rt}(r + 3)(r + 2) = 0,$$

and this equation only holds when either $r = -2$ or $r = -3$. Therefore, both $y = e^{-2t}$ and $y = e^{-3t}$ are solutions of the ODE.

oooooooooooooooooooooooooooooooooooooooo

This example may create more questions than answers for you. So far we have found unique solutions of ODEs, but here we seem to have found two! Also, we guessed that $y = e^{rt}$ is a solution, but how can you guarantee that this solution always exists? The following theorem provides the "existence" part of the existence and uniqueness question.

Let

$$y'' + p(t)y' + q(t)y = g(t), \qquad y(t_0) = y_0, \quad y'(t_0) = y_0',$$

where p, q, and g are continuous on an open interval I containing the point t_0. Then there is exactly one solution, which exists throughout the interval I.

Note that this theorem holds even for linear second-order ODEs which are nonhomogeneous and don't have constant coefficients. Additionally, it requires two initial conditions, one stating the initial value $y(t_0)$ and another the initial value of the derivative $y'(t_0)$. If we think of y as modeling the position of a particle, this boils down to knowing the initial position and velocity.

It is worth noting that this theorem extends to higher dimensions, in which case an nth-order ODE requires n initial conditions $y(t_0), y'(t_0), \ldots, y^{n-1}(t_0)$ to have a unique solution. The following example will illustrate the usefulness of this theorem.

○○

Example: Let's find the largest interval on which a solution of the initial value problem:

$$ty'' + 2y' - e^t y = 0, \qquad y(1) = 1, \quad y'(1) = 4$$

exists. In order to use the above theorem, we need to rewrite the initial value problem in the correct form, so we write:

$$y'' + \frac{2}{t}y' - \frac{e^t}{t}y = 0, \qquad y(1) = 1, \quad y'(1) = 4.$$

Now there is a discontinuity at $t = 0$, and as $t_0 = 1$, the largest interval on which a solution exists is $(0, \infty)$.

○○

Note that in the event that $p(t)$ and $q(t)$ are constant functions and $g(t) = 0$, so we have a linear homogeneous second order ODE with constant coefficients, the functions $p(t)$ and $q(t)$ are everywhere continuous. Therefore, the unique solution of ODEs of this type will be everywhere continuous, which is nice.

Your fears about existence assuaged, let us focus now on the question of uniqueness. In our first example we found two solutions, $y = e^{-2t}$ and $y = e^{-3t}$, but it would be nice if there was a single unique solution. Can't we combine them somehow? Yes, it turns out, you can.

If y_1 and y_2 are two solutions of the ODE

$$L[y] = y'' + p(t)y' + q(t)y = 0,$$

then the linear combination $c_1y_1 + c_2y_2$ is also a solution for any constants c_1 and c_2.

Proof: Assume that y_1 and y_2 are solutions of $L[y] = 0$ and let c_1 and c_2 be constants. We will show that $L[c_1y_1 + c_2y_2] = 0$, which will show that $c_1y_1 + c_2y_2$ is a solution of $L[y] = 0$. Notice that:

$$
\begin{aligned}
L[c_1y_1 + c_2y_2] &= [c_1y_1 + c_2y_2]'' + p(t)[c_1y_1 + c_2y_2]' + q(t)[c_1y_1 + c_2y_2] \\
&= c_1y_1'' + c_2y_2'' + p(t)c_1y_1' + p(t)c_2y_2' + q(t)c_1y_1 + q(t)c_2y_2 \\
&= c_1\left(y_1'' + p(t)y_1' + q(t)y_1\right) + c_2\left(y_2'' + p(t)y_2' + q(t)y_2\right) \\
&= c_1L[y_1] + c_2L[y_2] \\
&= 0.
\end{aligned}
$$

Hence, the linear combination $c_1y_1 + c_2y_2$ is also a solution.   □

○○○

We have another thing to consider here. Notably, this theorem tells us that because $y_1 = e^{-2t}$ and $y_2 = e^{-3t}$ were both solutions of the ODE that $y = c_1e^{-2t} + c_2e^{-3t}$ is as well. However, the theorem also would tell us that with solutions $y_3 = e^{-3t}$ and $y_4 = 3e^{-3t}$, the pretty trivial linear combination $y = c_1e^{-3t} + c_2(3e^{-3t})$ is a solution as well. I think you can easily see that the solution is different from the first in a pretty meaningful way. In the first one, y_2 is not simply a multiple of y_1, while $y_4 = 3y_3$. We want a way to test this, to determine if the solutions y_1 and y_2 are *linearly independent*.

Definition: A collection of functions $y_1, y_2, \ldots, y_n$ is **linearly independent** if no function y_k in the collection can be written as a linear combination of the others, i.e. there exists no $c_1, c_2, \ldots, c_n$, such that:

$$y_k = c_1y_1 + c_2y_2 + \cdots + c_{k-1}y_{k-1} + c_{k+1}y_{k+1} + \cdots + c_{n-1}y_{n-1} + c_ny_n.$$

Example: The functions $y_1 = e^{2t}$ and $y_2 = \sin 2t$ are linearly independent because one is not a constant multiple of the other. However, the set y_1, y_2, y_3 with y_1, y_2 as above and $y_3 = e^{2t} + 3\sin 2t$ is linearly *dependent* because we can write $y_3 = y_1 + 3y_2$.

○○○

We would like a way to detect if a set of functions is linearly independent or not. Luckily we have such a tool: The Wronskian.

Definition: The **Wronskian** is the determinant of the matrix $\begin{pmatrix} y_1 & y_2 \\ y_1' & y_2' \end{pmatrix}$. Recall that the determinant of a 2×2 matrix $A = \begin{pmatrix} a & b \\ c & d \end{pmatrix}$ is $\det(A) = |A| = ad - bc$, so the Wronskian of y_1 and y_2 is:

$$W(y_1, y_2) = \begin{vmatrix} y_1 & y_2 \\ y_1' & y_2' \end{vmatrix} = y_1 y_2' - y_2 y_1'.$$

In the event that the Wronskian, $W(y_1, y_2)$, is non-zero, the functions y_1, y_2 are linearly independent.

For nth-order ODEs, the Wronskian is the determinant of the $n \times n$ matrix formed by columns of each solution and its first $n-1$ derivatives. We will limit ourselves to the 2×2 case.

○○

Example: Let's find the Wronskian of $y_1 = e^{2t}$ and $y_2 = e^{5t}$. We have that:

$$
\begin{aligned}
W(y_1, y_2) &= \begin{vmatrix} y_1 & y_2 \\ y_1' & y_2' \end{vmatrix} \\
&= \begin{vmatrix} e^{2t} & e^{5t} \\ 2e^{2t} & 5e^{5t} \end{vmatrix} \\
&= e^{2t} \cdot 5e^{5t} - e^{5t} \cdot 2e^{2t} \\
&= 5e^{7t} - 2e^{7t} \\
&= 3e^{7t}.
\end{aligned}
$$

○○

This tool is useful to us due to the following theorem:

We say that y_1 and y_2 form a **fundamental set of solutions** if and only if $W(y_1, y_2) \neq 0$.

Furthermore, if y_1 and y_2 form a fundamental set of solutions, then every solution of the ODE can be written as the linear combination $y = c_1 y_1 + c_2 y_2$, for c_1 and c_2 constants.

You'll notice that y_1 and y_2 form a fundamental set of solutions if and only if y_1 and y_2 are linearly independent. Let's see an example of the use of this theorem.

Example: We will verify that the functions $y_1 = e^{5x}$ and $y_2 = xe^{5x}$ form a fundamental set of solutions for the ODE $y'' - 10y' + 25y = 0$.

First, we must verify that $y_1 = e^{5x}$ is actually a solution of the ODE. We know that $y_1' = 5e^{5x}$ and $y_1'' = 25e^{5x}$, so that:

$$y_1'' - 10y_1' + 25y_1 = 25e^{5x} - 10(5e^{5x}) + 25(e^{5x}) = 0.$$

Thus y_1 is, in fact, a solution of the ODE.

Next, we confirm that $y_2 = xe^{5x}$ is a solution of the ODE. We know that $y_2' = e^{5x} + 5xe^{5x}$ and $y_2'' = 10e^{5x} + 25xe^{5x}$. Then:

$$y_2'' - 10y_2' + 25y_2 = 10e^{5x} + 25xe^{5x} - 10(e^{5x} + 5xe^{5x}) + 25(xe^{5x}) = 0.$$

Thus y_2 is also a solution of the ODE.

Finally, we verify that y_1 and y_2 form a fundamental set. We have:

$$
\begin{aligned}
W(y_1, y_2) &= W(e^{5x}, xe^{5x}) \\
&= \begin{vmatrix} e^{5x} & xe^{5x} \\ 5e^{5x} & e^{5x} + 5xe^{5x} \end{vmatrix} \\
&= e^{5x}\left(e^{5x} + 5xe^{5x}\right) - xe^{5x}\left(5e^{5x}\right) \\
&= e^{10x} + 5xe^{10x} - 5xe^{10x} \\
&= e^{10x} \\
&\neq 0.
\end{aligned}
$$

Hence, y_1 and y_2 form a fundamental set and the general solution of the ODE can be written as $y = c_1 y_1 + c_2 y_2 = c_1 e^{5x} + c_2 xe^{5x}$, where c_1 and c_2 are constants.

ooooooooooooooooooooooooooooooooooooooooo

Example: We will verify that the functions $y_1 = e^{-2t}$ and $y_2 = e^{-3t}$ form a fundamental set of solutions for the ODE $y'' + 5y' + 6y = 0$.

We have already found earlier in this section that y_1 and y_2 are solutions of the ODE, so we just need to verify that y_1 and y_2 form a fundamental set. We have:

$$
\begin{aligned}
W(y_1, y_2) &= W(e^{-2t}, e^{-3t}) \\
&= \begin{vmatrix} e^{-2t} & e^{-3t} \\ -2e^{-2t} & -3e^{-3t} \end{vmatrix} \\
&= e^{-2t}\left(-3e^{-3t}\right) - e^{-3t}\left(-2e^{-2t}\right) \\
&= -3e^{-5t} + 2e^{-5t} \\
&= -e^{-5t} \\
&\neq 0.
\end{aligned}
$$

Hence, y_1 and y_2 form a fundamental set and the general solution of the ODE can be written as $y = c_1 y_1 + c_2 y_2 = c_1 e^{-2t} + c_2 e^{-3t}$, where c_1 and c_2 are constants.

4.1 Exercises

For the following functions, find the Wronskian.

1. $y_1 = e^{2x}$, $y_2 = e^{-2x}$

2. $y_1 = e^{4x}$, $y_2 = e^{-x}$

3. $x_1 = t^2 e^{t/2}$, $x_2 = te^{t/2}$

4. $y_1 = 5e^{3t}$, $y_2 = e^{-t}$

5. $u_1 = t$, $u_2 = t^2$

6. $y_1 = e^{2t}$, $y_2 = te^{2t}$

7. $u_1 = \sin t$, $u_2 = \cos t$

8. $y_1 = e^{7x}$, $y_2 = e^{3x}$

9. $x_1 = e^{2t} \sin 3t$, $x_2 = e^{2t} \cos 3t$

10. $y_1 = t^2 + 3t$, $y_2 = 2t^2 + 6t$

Verify that the given functions are solutions of the differential equation provided. Then verify that they form a fundamental set of solutions. Finally, write down the general solution of the ODE.

11. $y_1 = e^t$, $y_2 = e^{-2t}$ and $y'' + y' - 2y = 0$

12. $y_1 = e^{-t}$, $y_2 = e^{3t}$ and $y'' - 2y' - 3y = 0$

13. $u_1 = e^{-t/2} \sin t$, $u_2 = e^{-t/2} \cos t$ and $u'' + u' + 1.25u = 0$

14. $y_1 = e^{-2x}$, $y_2 = xe^{-2x}$ and $y'' + 4y' + 4y = 0$

15. $y_1 = e^{-2t}$, $y_2 = e^t$ and $y'' - y' - 2y = 0$

16. $u_1 = t^2$, $u_2 = t^3$ and $t^2 u'' - 4tu' + 6u = 0$

17. $y_1 = \sin 2x$, $y_2 = \cos 2x$ and $y'' + 4y = 0$.

18. $y_1 = e^{2t} \cos t$, $y_2 = e^{2t} \sin t$ and $y'' - 4y' + 5y = 0$

4.2   The Characteristic Equation and Solutions of Linear Homogeneous Second-Order ODEs

From our example in Section 4.1 of showing that $y = e^{-2t}$ and $y = e^{-3t}$ are solutions of the ODE:

$$y'' + 5y' + 6y = 0,$$

you may have noticed that we boiled the problem down to simply solving the equation $r^2 + 5r + 6 = 0$ for r. This is called the **characteristic equation** of this ODE. Here's an idea for how to solve ODEs of this form:

1. Suppose the ODE has a solution of the form $y = e^{rt}$.

2. Solve the characteristic equation of the ODE.

3. Use the roots you find to construct a general solution of the ODE.

This idea is, in fact, the right one. However, we need to break this plan up into cases, depending on the nature of the roots of the characteristic equation. Specifically, we must consider the cases when we have distinct real roots, repeated roots, and complex roots.

4.2.1   Distinct Real Roots

The linear homogeneous equation:

$$ay'' + by' + cy = 0$$

has characteristic equation $ar^2 + br + c = 0$. If r_1, r_2 are distinct real roots of the characteristic equation, then:

$$y = c_1 e^{r_1 t} + c_2 e^{r_2 t}$$

is the general solution of the ODE.

Example: Consider the ODE:

$$y'' + 8y' + 12y = 0.$$

We find the characteristic polynomial to be:

$$r^2 + 8r + 12 = 0$$

which factors as:

$$(r + 6)(r + 2) = 0.$$

The roots of the characteristic polynomial therefore are $r = -2, -6$. Thus, we can conclude that the general solution of this ODE is:

$$y = c_1 e^{-2t} + c_2 e^{-6t}.$$

○○

Example: Consider the ODE:

$$y'' - y' - y = 0.$$

This has characteristic polynomial $r^2 - r - 1 = 0$. This does not factor, so we must use the quadratic formula. We can see that the roots of this polynomial are $r = \frac{1}{2} \pm \frac{\sqrt{5}}{2}$. These are distinct, real roots despite the fact that they are not integers. That is okay! We continue as before and have general solution:

$$y = c_1 e^{\left(\frac{1+\sqrt{5}}{2}\right)t} + c_2 e^{\left(\frac{1-\sqrt{5}}{2}\right)t}.$$

You could simplify this slightly by recalling that $e^{a+b} = e^a e^b$ and factor out a common $e^{\frac{t}{2}}$ to obtain:

$$y = e^{\frac{t}{2}} \left[c_1 e^{\frac{\sqrt{5}t}{2}} + c_2 e^{\frac{-\sqrt{5}t}{2}} \right].$$

○○

Next, let's consider an initial value problem (IVP). Note that the process for finding the general solution is identical. However, with an IVP, we can then solve for the arbitrary constants c_1 and c_2.

Example: Consider the IVP:

$$y'' - 7y' - 8y = 0, \qquad y(0) = 0, \quad y'(0) = 9.$$

You may notice that we have two initial conditions, one for y and one for y'. As we will have two constants, c_1 and c_2, in the general solution, we need two initial conditions to find their values.

The characteristic equation of this ODE is:

$$r^2 - 7r - 8 = 0$$

which can be factored as

$$(r+1)(r-8) = 0,$$

giving $r = -1$ and $r = 8$. Therefore, the general solution will be

$$y = c_1 e^{-t} + c_2 e^{8t}.$$

In order to find the values of c_1 and c_2, we start with the first initial condition, $y(0) = 0$, and see that:

$$0 = c_1 e^{-0} + c_2 e^{8(0)}$$

$$0 = c_1 + c_2.$$

To utilize the second initial condition, $y'(0) = 9$, we must first differentiate the general solution to see that:

$$y' = -c_1 e^{-t} + 8c_2 e^{8t}.$$

We then use the initial condition to find that:

$$9 = -c_1 e^{-0} + 8c_2 e^{8(0)}$$

$$9 = -c_1 + 8c_2.$$

Now we have a system of linear equations:

$$\begin{cases} c_1 + c_2 = 0 \\ -c_1 + 8c_2 = 9 \end{cases}$$

which we can solve, finding the values of c_1 and c_2 to be $c_1 = -1$ and $c_2 = 1$. Therefore, the solution of the IVP in question is:

$$y = -e^{-t} + e^{8t}.$$

ooooooooooooooooooooooooooooooooooooooo

Example: Consider the IVP:

$$y'' - y' - 2y = 0, \qquad y(1) = 0, \quad y'(1) = 2.$$

The characteristic equation of this ODE is:

$$r^2 - r - 2 = 0$$

which can be factored as

$$(r+1)(r-2) = 0,$$

giving $r = -1$ and $r = 2$. Therefore, the general solution will be

$$y = c_1 e^{-t} + c_2 e^{2t}.$$

To find the values of c_1 and c_2, we can start with the first initial condition, $y(1) = 0$, and see that:

$$0 = c_1 e^{-1} + c_2 e^{2(1)}$$

$$0 = c_1 e^{-1} + c_2 e^2.$$

Note that because our initial value $t_0 = 1$ and not zero, our equation doesn't simplify as nicely as our previous IVP example. That is okay!

To utilize the second initial condition, $y'(1) = 2$, we must first differentiate the general solution to see that:

$$y' = -c_1 e^{-t} + 2c_2 e^{2t}.$$

We then use the initial condition to find that:

$$2 = -c_1 e^{-1} + 2c_2 e^{2(1)}$$

$$2 = -c_1 e^{-1} + 2c_2 e^2.$$

Now we have a system of linear equations:

$$\begin{cases} e^{-1}c_1 + e^2 c_2 = 0 \\ -e^{-1}c_1 + 2e^2 c_2 = 2 \end{cases}$$

which we can solve (elimination might be a good strategy), finding the values of c_1 and c_2 to be $c_1 = -\frac{2e}{3}$ and $c_2 = \frac{2}{3e^2}$. Therefore, the solution of the IVP in question is:

$$y = -\frac{2e}{3} e^{-t} + \frac{2}{3e^2} e^{2t}.$$

4.2.2 Repeated Roots

While these examples are great there is a limitation: we are only considering characteristic equations with distinct real roots. For example, this framework does not allow us to solve the following differential equation:

$$y'' - 14y' + 49y = 0.$$

This is a pretty elementary-looking equation. Why does our framework fail? Finding the characteristic equation, we have:

$$r^2 - 14r + 49 = 0$$

$$(r-7)(r-7) = 0$$

$$(r-7)^2 = 0,$$

so we have one root, $r = 7$, but it is a repeated root. If we were to put $y_1 = e^{7t}$, $y_2 = e^{7t}$ into the Wronskian, we would find that we don't have a fundamental set. We must have to develop a slightly augmented theory for dealing with these ODEs with repeated roots.

The fix to the above quagmire is to set $y_1 = e^{7t}$ as before and set $y_2 = te^{7t}$. This might seem like a random choice, however, we can check easily that y_2 is a solution of the ODE as:

$$y_2' = 7te^{7t} + e^{7t}$$

$$y_2'' = 49te^{7t} + 14e^{7t},$$

and then

$$y_2'' - 14y_2' + 49y_2 = 49te^{7t} + 14e^{7t} - 14(7te^{7t} + e^{7t}) + 49(te^{7t})$$

$$= 49te^{7t} + 14e^{7t} - 98te^{7t} - 14e^{7t} + 49te^{7t}$$

$$= 0.$$

We can readily check that this will always give us a fundamental set of solutions. Letting $y_1 = e^{7t}$ and $y_2 = te^{7t}$ we can see that:

$$W(e^{7t}, te^{7t}) = \begin{vmatrix} e^{7t} & te^{7t} \\ 7e^{7t} & 7te^{7t} + e^{7t} \end{vmatrix}$$

$$= e^{7t}(7te^{7t} + e^{7t}) - te^{7t}(7e^{7t})$$

$$= 7te^{14t} + e^{14t} - 7te^{14t}$$

$$= e^{14t}$$

$$\neq 0,$$

so we have a fundamental set. As such, we can alter the statement from before as follows:

The linear homogeneous equation:

$$ay'' + by' + cy = 0$$

has characteristic equation $ar^2 + br + c = 0$.

If r_1, r_2 are distinct real roots of the characteristic equation, then:

$$y = c_1 e^{r_1 t} + c_2 e^{r_2 t}$$

is the general solution of the ODE.

If r is a repeated root of the characteristic equation, then:

$$y = c_1 e^{rt} + c_2 t e^{rt}$$

is the general solution of the ODE.

For our example, as $r = 7$, we have that the general solution is:

$$y = c_1 e^{7t} + c_2 t e^{7t}.$$

Now that we understand the idea behind solutions of repeated root ODEs, let's see an example of a problem where we go top to bottom from the statement of the problem to the unique solution.

○○

Example: Consider the IVP:

$$y'' + 6y' + 9y = 0, \qquad y(0) = 1, \quad y'(0) = 3.$$

The characteristic equation is:

$$r^2 + 6r + 9 = 0$$

which has repeated root $r = -3$. Therefore the general solution is:

$$y = c_1 e^{-3t} + c_2 t e^{-3t}.$$

With the initial condition $y(0) = 1$, we have:

$$1 = c_1 e^{-3(0)} + c_2 (0) e^{-3(0)}$$

$$1 = c_1.$$

Taking the derivative of the general solution and substituting in the initial condition $y'(0) = 3$, we have:

$$y' = -3c_1 e^{-3t} + c_2 e^{-3t} - 3c_2 t e^{-3t}$$

$$3 = -3c_1 e^{-3(0)} + c_2 e^{-3(0)} - 3c_2(0)e^{-3(0)} \cdot$$

$$3 = -3c_1 + c_2$$

Therefore our initial conditions give us the system of equations:

$$\begin{cases} c_1 & = & 1 \\ -3c_1 & + & c_2 & = & 3 \end{cases}$$

which has solutions $c_1 = 1$, $c_2 = 6$. Therefore the solution is:

$$y = e^{-3t} + 6te^{-3t}.$$

○○

Example: Consider the IVP:

$$y'' + 10y' + 25y = 0, \qquad y(0) = 1, \quad y'(0) = 0.$$

The characteristic equation is:

$$r^2 + 10r + 25 = 0$$

which has repeated root $r = -5$. Therefore the general solution is:

$$y = c_1 e^{-5t} + c_2 t e^{-5t}$$

and its derivative is

$$y = -5c_1 e^{-5t} + c_2 e^{-5t} - 5c_2 t e^{-5t}.$$

Therefore our initial conditions give us the system of equations:

$$\begin{cases} c_1 & = & 1 \\ -5c_1 & + & c_2 & = & 0 \end{cases}$$

which has solutions $c_1 = 1$, $c_2 = 5$. Therefore the solution is:

$$y = e^{-5t} + 5te^{-5t}.$$

4.2.3   Complex Roots

With repeated roots the theorem we desire which describes the general solution of an arbitrary linear homogeneous second-order differential equation with constant coefficients is still incomplete. For example, the ODE:

$$y'' - 2y' + 2y = 0$$

has characteristic equation

$$r^2 - 2r + 2 = 0$$

which has roots

$$r = \frac{2 \pm \sqrt{4 - 4(2)}}{2} = 1 \pm i.$$

Now, naively, you might think that we could then just have a general solution:

$$y = c_1 e^{(1+i)t} + c_2 e^{(1-i)t}$$

and be done! You might even simplify the expression and write

$$y = e^t \left(c_1 e^{it} + c_2 e^{-it} \right).$$

Your head is in the right place, but our problem is that now we have *complex* solutions of a real-valued ODE. That won't do!

What should we do? We would like to find some way to write out the e^{it} and e^{-it} terms with real-valued functions. You might recall Euler's formula, which states that:

$$e^{it} = \cos t + i \sin t.$$

If you do not recall this formula, there are a few ways to realize it. One is by considering the Taylor series for $i \sin x$, $\cos x$, and e^{ix}. If you add the Taylor series for $i \sin x$ and $\cos x$ together you will recover the Taylor series for e^{ix}. This is a worthwhile exercise to do just one single time in your life. It is very cool that it works out so nicely, but it is a bit of a chore.

Another more elegant way to realize this is by considering the two ways to describe a complex number. One way is to describe:

$$z = a + bi$$

where a describes the real part of z and b describes the imaginary part of z. Another way is to describe the complex number by stating its distance from the origin and the angle at which you should proceed to reach the point. These should be thought of as the Cartesian and polar parametrizations of the number.

From basic trigonometry it is apparent that if $z = re^{i\theta}$ then the real part of z is $r \cos \theta$ while the imaginary part is $r \sin \theta$.

Now armed with Euler's formula, the idea here is to let:

$$y_1 = e^{(1+i)t} = e^t(\cos t + i \sin t)$$

and

$$y_2 = e^{(1-i)t} = e^t(\cos t - i \sin t).$$

As linear combinations of solutions are themselves solutions, then

$$y_1 + y_2 = e^t(\cos t + i \sin t) + e^t(\cos t - i \sin t) = 2e^t \cos t$$

and

$$y_1 - y_2 = e^t(\cos t + i \sin t) - e^t(\cos t - i \sin t) = 2ie^t \sin t$$

are solutions as well! Then our general solution is:

$$y = c_1 e^t \cos t + c_2 e^t \sin t.$$

You might be unsure about our last step. After all, $y_1 - y_2$ still has an imaginary number in it. While this is true, it can be shown that both c_1 and c_2 are real valued.

This approach works in general, and we can now finally rewrite the theorem from before to be as complete as possible.

> The linear homogeneous equation $ay'' + by' + cy = 0$ has characteristic equation $ar^2 + br + c = 0$.
>
> - If r_1, r_2 are distinct real roots of the characteristic equation, then the general solution is: $y = c_1 e^{r_1 t} + c_2 e^{r_2 t}$.
>
> - If r is a repeated root of the characteristic equation, then the general solution is: $y = c_1 e^{rt} + c_2 t e^{rt}$.
>
> - If $r = a \pm bi$ are complex roots of the characteristic equation, then the general solution is: $y = c_1 e^{at} \cos bt + c_2 e^{at} \sin bt$.

It is worth mentioning that while we are limiting ourselves to linear second-order ODEs here, this theory does naturally extend to higher-order ODEs. Some care needs to be taken with repeated roots with multiplicity higher than two and repeated complex roots, but the general strategy of multiplying by t still turns out to be the right approach.

A top-to-bottom example will be illuminating.

ooooooooooooooooooooooooooooooooooooooo

Example: Consider the ODE:

$$y'' - 8y' + 17y = 0.$$

We can see that the characteristic equation is:

$$r^2 - 8r + 17 = 0$$

which has roots $r = 4 \pm i$. Therefore, the general solution is:

$$y = c_1 e^{4t} \cos t + c_2 e^{4t} \sin t.$$

oooooooooooooooooooooooooooooooooooooo

Let's see an example now where we start with an IVP and need to solve for the arbitrary constants, c_1 and c_2.

Example: Consider the IVP

$$y'' - 4y' + 13y = 0, \qquad y(0) = 1, \quad y'(0) = -1.$$

Using the quadratic formula, we can see that the characteristic equation:

$$r^2 - 4r + 13 = 0$$

has roots $r = 2 \pm 3i$. Therefore, our general solution is:

$$y = c_1 e^{2t} \cos 3t + c_2 e^{2t} \sin 3t.$$

In order to solve for the arbitrary constants we need to find the derivative, y'. Note that in this case, we will require the use of the product rule and the chain rule. We find that we have:

$$y' = 2c_1 e^{2t} \cos 3t - 3c_1 e^{2t} \sin 3t + 2c_2 e^{2t} \sin 3t + 3c_2 e^{2t} \cos 3t.$$

Using our initial conditions of $y(0) = 1$ and $y'(0) = -1$, we see that we have the system of equations:

$$\begin{cases} 1 &= & c_1 \\ -1 &= & 2c_1 & + & 3c_2 \end{cases}$$

which has solutions $c_1 = 1$, $c_2 = -1$. As such, the unique solution of this IVP is:

$$y = e^{2t} \cos 3t - e^{2t} \sin 3t.$$

oooooooooooooooooooooooooooooooooooooo

A natural question might be: what happens when we have purely imaginary roots? The answer is that we utilize the same framework, but it's worth looking at the different shapes that the solution takes.

Example: Consider the ODE:

$$y'' + 16y = 0.$$

This has characteristic equation:

$$r^2 + 16 = 0,$$

which has purely imaginary roots $r = \pm 4i$. Using our same framework, the general solution should be

$$y = c_1 e^{0t} \cos 4t + c_2 e^{0t} \sin 4t$$

as our roots are $r = 0 \pm 4i$. In a more simplified manner, we would typically write

$$y = c_1 \cos 4t + c_2 \sin 4t.$$

As you can see, the solution of an ODE with purely imaginary roots has no exponential terms in it and is a function purely of sines and cosines.

○○○○○○○○○○○○○○○○○○○○○○○○○○○○○○○○○○○○○○○

Example: Consider the IVP:

$$y'' + 2y' + 26y = 0, \qquad y(0) = 2, \quad y'(0) = 0.$$

Using the quadratic formula, we can see that the characteristic equation:

$$r^2 + 2r + 26 = 0$$

has roots $r = -1 \pm 5i$. Therefore, our general solution is:

$$y = c_1 e^{-t} \cos 5t + c_2 e^{-t} \sin 5t.$$

In order to solve for the arbitrary constants we need to find the derivative, y'. As before, we will require the use of the product and chain rules. We find that we have:

$$y' = -c_1 e^{-t} \cos 5t - 5c_1 e^{-t} \sin 5t - c_2 e^{-t} \sin 5t + 5c_2 e^{-t} \cos 5t.$$

Using our initial conditions $y(0) = 2$ and $y'(0) = 0$, we see that we have the system of equations:

$$\begin{cases} 2 &=& c_1 \\ 0 &=& -c_1 + 5c_2, \end{cases}$$

which has solutions $c_1 = 2$, $c_2 = \frac{2}{5}$. As such, the unique solution of this IVP is:

$$y = 2e^{-t} \cos 5t + \frac{2}{5} e^{-t} \sin 5t.$$

4.2 Exercises

Find the general solution of the following differential equations.

1. $y'' - 4y' + 53 = 0$ 2. $y'' + 2y' + 5y = 0$

3. $y'' - 4y' + 40y = 0$ 4. $y'' + 6y' + 58y = 0$

5. $y'' + 8y' + 17y = 0$ 6. $4y'' - 4y' + 5y = 0$

7. $2y'' - 6y' + 5y = 0$ 8. $y'' - 4y' + 29y = 0$

9. $y'' - 15y' + 50y = 0$ 10. $y'' - 8y' + 16y = 0$

Find the solution of the following initial value problems.

11. $y'' - 6y' + 9y = 0,$ $y(0) = 1,$ $y'(0) = 2$

12. $y'' + 2y' - 15y = 0,$ $y(0) = 0,$ $y'(0) = 1$

13. $y'' + 25y = 0,$ $y(\pi) = 3,$ $y'(\pi) = -12$

14. $y'' - 7y' + 12y = 0,$ $y(0) = 2,$ $y'(0) = 1$

15. $y'' - 2y' + y = 0,$ $y(0) = 7,$ $y'(0) = 5$

16. $y'' - 13y' + 42y = 0,$ $y(0) = 1,$ $y'(0) = 2$

17. $y'' - 4y' + 2y = 0,$ $y(0) = 2,$ $y'(0) = 0$

18. $y'' + 2y' + 3 = 0,$ $y(0) = 1,$ $y'(0) = 2$

19. $y'' - 8y' + 16y = 0,$ $y(1) = 1,$ $y'(1) = 0$

20. $y'' - 8y' + 25y = 0,$ $y(0) = 3,$ $y'(0) = 2$

21. $y'' - 3y' - 28 = 0,$ $y(1) = 0,$ $y'(1) = 1$

22. $y'' - 4y' + 29y = 0,$ $y(1) = 2,$ $y'(1) = 0$

4.3   Mechanical and Electrical Vibrations

Second-order ODEs are fantastic tools for modeling different types of oscillation or vibration in the physical world. Two quintessential examples are mechanical vibrations, for example a spring-mass system, and electrical vibrations, such as the current across a capacitor in a standard RLC circuit, where RLC represents the resistor, inductor, and capacitor in a circuit. As you may be wondering, it is common practice to use an L to represent the inductor.

While the RLC circuit is a very simple circuit diagram, understanding how second-order ODEs can be used to model it can help us understand how more complex circuits, such as those which model the behavior of neurons, can be modeled in a similar fashion.

4.3.1   Mechanical Vibrations: Unforced Springs

One classic problem which can be well modeled by second-order ODEs is the **spring-mass system**. In this class of problems, there is a free-hanging spring which is then stretched by a weight attached to the end of the spring. This mass is then moved to either elongate or compress the spring, and released. The challenge is to model the motion of the mass as time progresses. The following equation models the motion of the mass over time.

We can derive this model by considering the forces acting on the mass. There is the force due to gravity, force applied by the spring when compressed or stretched, and force applied by damping, for example by the spring moving through a liquid or simply air resistance.

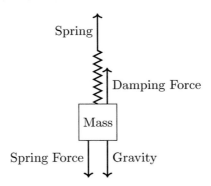

We proceed by considering all the forces applied to the mass.

- **Gravity:** There is constant acceleration due to gravity applied to the mass. Noting that force is mass times acceleration, we have:

$$F_g = ma = mu''(t)$$

  where $u(t)$ is the position of the spring. Note that we choose the convention that the positive $u(t)$ direction is downward while the negative $u(t)$ direction is upward. As such, you can think of $u(t)$ as representing the elongation of the spring from the equilibrium.

- **Spring Constant:** The spring constant is derived from Hooke's Law, which states that $F_s = kL$ where F is the weight, or force, and L is how much the spring is stretched. L is always measured in either feet or meters. Thus we have $F_s = ku(t)$.

- **Damping:** Damping force is force applied to the mass which is resistance against the movement of the spring. This force is assumed to be proportional to the velocity of the spring (think of how much force you feel on your hand when you move it underwater slowly compared to when you move it fast). As such, we have $F_d = \gamma u'(t)$ where γ is the damping coefficient.

Summing these forces, we have the following equation of motion for the mass:

Equation of Motion of Mass

Suppose a mass with weight w (kilograms or pounds) is attached to a spring in equilibrium, which stretches the spring by a length L (meters or feet) before reaching equilibrium again. Suppose a damping force is applied with damping coefficient γ. Then the equation of motion of the mass is given by:
$$mu'' + \gamma u' + ku = 0,$$
where:

m is mass: $m = \frac{w}{g}$
γ is the damping coefficient,
k is the spring constant, $k = \frac{w}{L}$.
For Earth, we will use $g = 32$ ft/s^2 or $g = 9.81$ m/s^2.

Example: A mass weighing 10 lbs stretches a spring 24 inches and comes to rest. The mass is then pulled down an additional 6 inches. The spring is suspended in a vacuum chamber so there is no damping. We are asked now to find the position, the frequency, period, and amplitude of the motion of the mass.

We can find $m = \frac{w}{g} = \frac{10}{32}$. As there is no damping $\gamma = 0$, and the spring constant can be found to be $k = \frac{w}{L} = \frac{10}{24/12} = 5$. Note that we converted from inches to feet so that the units match with that of gravity. Therefore, our equation:
$$mu'' + \gamma u' + ku = 0$$
becomes:
$$\frac{10}{32}u'' + 5u = 0.$$
Our initial conditions are easy to deduce from the information given. The spring is pulled down an extra 6 inches, so $u(0) = \frac{6}{12} = \frac{1}{2}$, and because the spring is at rest when it is released, $u'(0) = 0$. Therefore, we have the initial value problem:
$$\frac{10}{32}u'' + 5u = 0, \qquad u(0) = \frac{1}{2}, \quad u'(0) = 0.$$

This can be solved using our existing methods. We see that we have characteristic equation:

$$\frac{10}{32} r^2 + 5 = 0$$

$$\frac{10}{32} r^2 = -5$$

$$r^2 = -16$$

$$r = \pm 4i.$$

Therefore, $u = c_1 \cos 4t + c_2 \sin 4t$. From the initial condition $u(0) = \frac{1}{2}$, we can see that $\frac{1}{2} = c_1$. Taking the derivative of the general solution and substituting $u'(0) = 0$, we have:

$$u = \tfrac{1}{2} \cos 4t + c_2 \sin 4t$$

$$u' = -2 \sin 4t + 4 c_2 \cos 4t$$

$$0 = 4 c_2$$

$$0 = c_2,$$

so we have that

$$u = \frac{1}{2} \cos 4t.$$

This, in fact, is an example of **Simple Harmonic Motion**.

Simple Harmonic Motion can be described by the equation:

$$u = R \cos(\omega t - \delta),$$

where

- ω is the **natural frequency**
- R is the **amplitude**
- δ is the **phase angle**

and the **period** is $T = \frac{2\pi}{\omega}$.

Note that the phase shift is typically found while studying trigonometric functions, and this is NOT the same as the phase angle. While the phase angle is just the parameter δ in the equation $\cos(\omega t - \delta)$, the phase shift of $\frac{\delta}{\omega}$ can be seen by rewriting this equation as $\cos\left(\omega\left(t - \frac{\delta}{\omega}\right)\right)$.

○○

Example: Continuing our previous example, then, our solution is $u = \frac{1}{2} \cos 4t$.

- The frequency is $\omega = 4$ rad/sec.

- The period is $T = \frac{2\pi}{\omega} = \frac{2\pi}{4} = \frac{\pi}{2}$ sec.

- The amplitude is $R = \frac{1}{2}$ ft.

○○○○○○○○○○○○○○○○○○○○○○○○○○○○○○○○○○○○○○

It is important to note that simple harmonic motion is only achieved by undamped spring-mass systems. As you can imagine, in the presence of damping, the weight will not continue to move forever.

You may happen upon a spring-mass system (suspended in a vacuum, of course) whose solution is of the form:

$$u = A \cos \omega t + B \sin \omega t.$$

This is not in the form of simple harmonic motion, but we can still find the amplitude and phase angle of this spring.

If the motion of the spring-mass system is described by:

$$u = A \cos \omega t + B \sin \omega t$$

- Frequency is ω

- Amplitude is $R = \sqrt{A^2 + B^2}$.

- Phase angle is $\delta = \tan^{-1}\left(\frac{B}{A}\right)$

and the solution can be rewritten as

$$u = R \cos(\omega t - \delta).$$

Example: A 16 lb weight stretches a spring 6 inches and comes to rest. The weight is pushed up 2 inches above the equilibrium point and then set in motion with an initial upward velocity of 2 ft/sec. There is no damping.

Here we have:

$$m = \frac{w}{g} = \frac{16}{32} = \frac{1}{2}$$

$$\gamma = 0$$

$$k = \frac{w}{L} = \frac{16}{6/12} = 32.$$

Therefore, our equation is:

$$\frac{1}{2}u'' + 32u = 0, \qquad u(0) = -\frac{2}{12}, \qquad u'(0) = -2.$$

Our characteristic equation is:

$$\frac{1}{2}r^2 + 32 \;=\; 0$$

$$r^2 + 64 \;=\; 0$$

$$r \;=\; \pm 8i.$$

Therefore, $u = c_1 \cos 8t + c_2 \sin 8t$. From the initial conditions we can see that $c_1 = -\frac{1}{6}$ and:

$$u \;=\; -\tfrac{1}{6}\cos 8t + c_2 \sin 8t$$

$$u' \;=\; \tfrac{4}{3}\sin 8t + 8c_2 \cos 8t$$

$$-2 \;=\; 8c_2$$

$$-\tfrac{1}{4} \;=\; c_2.$$

Then we have:

$$u(t) = -\frac{1}{6}\cos 8t - \frac{1}{4}\sin 8t.$$

We can then find the frequency, period, amplitude, and phase angle:

$$\text{Frequency} \;=\; \omega \;=\; 8 \text{ rad/sec}$$

$$\text{Period} \;=\; T \;=\; \tfrac{2\pi}{\omega} = \tfrac{\pi}{4} \text{ sec}$$

$$\text{Amplitude} \;=\; R \;=\; \sqrt{\left(-\tfrac{1}{6}\right)^2 + \left(-\tfrac{1}{4}\right)^2} = \tfrac{\sqrt{13}}{12} \text{ ft}$$

$$\text{Phase Angle} \;=\; \delta \;=\; \tan^{-1}\left(\frac{-\frac{1}{4}}{-\frac{1}{6}}\right) = \tan^{-1}\left(\tfrac{3}{2}\right) \approx 4.1244 \text{ rad}$$

Note that in the calculation of the phase angle, we need to know in which quadrant our answer lies. Thinking back to trigonometry, $\tan(\delta) = \frac{y}{x}$, so that $\delta = \tan^{-1}\left(\frac{y}{x}\right)$. We can use the signs for x and y to determine the quadrant for our angle. For this problem, $y = -\frac{1}{4} < 0$ and $x = -\frac{1}{6} < 0$, so our answer should be in Quadrant III. How does this affect our answer? When typing $\tan^{-1}\left(\frac{3}{2}\right)$ into a calculator, you get the Quadrant I answer of approximately 0.9828 radians. In order to get the correct answer, which is in Quadrant III, we compute $0.9828 + \pi = 4.1244$ radians.

Therefore, we see that we can rewrite the solution:

$$u(t) = -\frac{1}{6}\cos 8t - \frac{1}{4}\sin 8t$$

as

$$u(t) = \frac{\sqrt{13}}{12}\cos(8t - 4.1244).$$

○○

In the presence of damping, we fail to achieve simple harmonic motion. This is in line with your everyday experience: springs don't stay in motion forever! The following example will illustrate this.

Example: A 2 lb weight stretches a spring 6 inches and comes to rest. The weight is pulled 2 inches below the equilibrium point and released. There is a damping coefficient of $\frac{1}{2}$.

Here we have:

$$m = \frac{w}{g} = \frac{2}{32} = \frac{1}{16}$$

$$\gamma = \frac{1}{2}$$

$$k = \frac{w}{L} = \frac{2}{6/12} = 4.$$

Therefore, our equation is:

$$\frac{1}{16}u'' + \frac{1}{2}u' + 4u = 0, \qquad u(0) = \frac{2}{12}, \quad u'(0) = 0.$$

Our characteristic equation is:

$$\frac{1}{16}r^2 + \frac{1}{2}r + 4 = 0$$

$$r^2 + 8r + 64 = 0$$

$$r = -4 \pm 4\sqrt{3}i.$$

Therefore, $u = c_1 e^{-4t}\cos 4\sqrt{3}t + c_2 e^{-4t}\sin 4\sqrt{3}t$. From the initial conditions we can see that $c_1 = \frac{1}{6}$ and:

$$u = \frac{1}{6}e^{-4t}\cos 4\sqrt{3}t + c_2 e^{-4t}\sin 4\sqrt{3}t$$

$$u' = -\frac{2}{3}e^{-4t}\cos 4\sqrt{3}t - \frac{2\sqrt{3}}{3}e^{-4t}\sin 4\sqrt{3}t$$
$$-4c_2 e^{-4t}\sin 4\sqrt{3}t + 4\sqrt{3}c_2 e^{-4t}\cos 4\sqrt{3}t$$

$$0 = -\frac{2}{3} + 4\sqrt{3}c_2$$

$$c_2 = \frac{\sqrt{3}}{18}.$$

So we can see that:

$$u = \frac{1}{6}e^{-4t}\cos 4\sqrt{3}t + \frac{\sqrt{3}}{18}e^{-4t}\sin 4\sqrt{3}t.$$

○○○

A spring whose equation of motion includes exponential functions experiences either decaying or accelerating motion and thus we cannot find frequency, period, or amplitude in the same way. Period is the time it takes for a sinusoidal function to repeat itself. Due to damping, the above equation of motion will not repeat itself as in simple harmonic motion. However, we can calculate the quasi-frequency and quasi-period. The quasi-period is the time between successive minima or successive maxima of the function. These can be calculated as follows.

> For springs with non-harmonic motion, we can calculate:
>
> **Quasi-frequency:** $\mu = \sqrt{1 - \frac{\gamma^2}{4km}}\,\omega$, where $\omega = \sqrt{\frac{k}{m}}$ is the frequency of the undamped motion of the spring.
>
> This can be used to the find:
>
> **Quasi-period:** $T_d = \frac{2\pi}{\mu}$.

Note that in the event our solution takes the form:

$$u(t) = e^{-at}\left(A\cos\mu t + B\sin\mu t\right)$$

the quasi-period is simply the constant μ found in the equation of motion.

○○

Example: For our prior example, with:

$$\frac{1}{16}u'' + \frac{1}{2}u' + 4u = 0, \qquad u(0) = \frac{1}{6}, \quad u'(0) = 0,$$

and which we obtained solution

$$u = \frac{1}{6}e^{-4t}\cos 4\sqrt{3}t + \frac{\sqrt{3}}{18}e^{-4t}\sin 4\sqrt{3}t,$$

we have:

$$\mu = \sqrt{1 - \frac{(1/2)^2}{4(4)(1/16)}}\sqrt{\frac{4}{1/16}} = 4\sqrt{3} \approx 6.928 \text{ rad/sec}$$

and

$$T_d = \frac{2\pi}{4\sqrt{3}} \approx 0.907 \text{ sec.}$$

Note that in the absence of damping we would have $\omega = 8$ rad/sec and period $T_d = \frac{2\pi}{8} = \frac{\pi}{4} \approx 0.7854$ seconds, so the effect of the damping is to elongate the gap between successive maximums of the spring length.

○○

In the case where there is damping, we can also rewrite the solution as before.

The motion of the spring-mass system described by:

$$mu'' + \gamma u' + ku = 0$$

has solution

$$u(t) = e^{-\lambda t} \left(A \cos \mu t + B \sin \mu t\right)$$

when the characteristic equation has complex roots. This can be rewritten as:

$$u(t) = Re^{-\lambda t} \cos(\mu t - \delta),$$

where

- Decay rate, $\lambda = \gamma/2m$

- Quasi-frequency, μ, is defined as $\mu = \sqrt{\omega^2 - \lambda^2}$, where $\omega = \sqrt{\frac{k}{m}}$ is the undamped natural frequency of the spring.

- Damped amplitude is $Re^{-\lambda t}$ where $R = \sqrt{A^2 + B^2}$.

- Phase angle is $\delta = \tan^{-1}\left(\frac{B}{A}\right)$

Note that though the formula given for quasi-frequency here looks different, it is in fact equivalent to the other. This is helpful because the two formulas allow us to view quasi-frequency through different lenses. Specifically,

$$\mu = \sqrt{1 - \frac{\gamma^2}{4km}}\, \omega$$

allows us to think of quasi-frequency as a scaling of the natural frequency ω. This formula also makes it clear that in the event that γ is zero, quasi-frequency reduces to the natural frequency of the spring. Conversely, the formula:

$$\mu = \sqrt{\omega^2 - \lambda^2},$$

in addition to just being a cleaner expression, allows us to view quasi-frequency in terms of the relationship between the natural frequency of the spring, ω, and the decay rate, λ, which will be helpful when considering if a spring is under or over damped. We will explore this concept shortly, and address what happens when the characteristic equation does not give us complex roots. We can use these parameters to gain a better understanding of the behavior of the spring's motion.

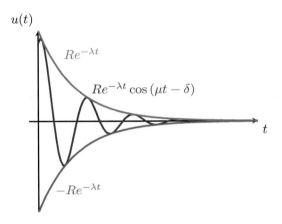

Note that in the graph of $u(t)$, the solution is bounded above and below by $\pm Re^{-\lambda t}$, respectively. This explains why we call $Re^{-\lambda t}$ the damped amplitude. While the local extrema in simple harmonic motion occurs at the same fixed value, damping ensures each subsequent extrema of $u(t)$ is less than the preceding one.

oooooooooooooooooooooooooooooooooooooooo

Example: A 4 kg weight stretches a spring 10 cm and comes to rest. The weight is pulled 5 cm below the equilibrium point and released. There is a damping coefficient of 0.125.

Note that our units must be in kilograms and meters so our displacement of 5cm must be converted to meters first.

Here we have:

$$m = \frac{w}{g} = \frac{4}{9.81} \approx 0.4077$$

$$\gamma = 0.125$$

$$k = \frac{w}{L} = \frac{4}{0.1} = 40.$$

Therefore, our equation is:

$$0.4077u'' + 0.125u' + 40u = 0, \qquad u(0) = 0.05, \quad u'(0) = 0.$$

By using the quadratic formula we see that our characteristic equation has roots:

$$r = -0.1533 \pm 9.9039i$$

after rounding. Therefore, $u = c_1 e^{-0.1533t} \cos 9.9039t + c_2 e^{-0.1533t} \sin 9.9039t$. From the initial conditions we can see that $c_1 = 0.05$ and:

$$u = 0.05e^{-0.1533t} \cos 9.9039t + c_2 e^{-0.1533t} \sin 9.9039t$$

$$u' = -0.0077e^{-0.1533t} \cos 9.9039t - 0.4952e^{-0.1533t} \sin 9.9039t$$
$$-0.1533c_2 e^{-0.1533t} \sin 9.9039t + 9.9039c_2 e^{-0.1533t} \cos 9.9039t$$

$$0 = -0.0077 + 9.9039c_2$$

$$c_2 = 0.0008.$$

So we can see that:

$$u = 0.05e^{-0.1533t} \cos 9.9039t + 0.0008e^{-0.1533t} \sin 9.9039t$$

which has quasi-frequency $\mu = 9.9039$ rad/sec and quasi-period:

$$T_d = \frac{2\pi}{9.9039} = 0.6344 \text{ sec.}$$

We also note that we have:

$$R = \sqrt{0.05^2 + 0.0008^2} = 0.05001$$

and

$$\delta = \tan^{-1}(0.0008/0.05) = 0.0160,$$

thus we can write our equation of motion as:

$$u(t) = 0.05001e^{-0.1533t} \cos(9.9039t - 0.0160).$$

○○

You may note that the characteristic equations for all of our examples so far have had complex roots. These problems are all considered **under damped** springs. A spring is said to be **critically damped** if the characteristic equation of the ODE used to model the spring has one repeated real root and **under damped** if the characteristic equation has two distinct real roots. In summary, we have the following definition.

For a spring-mass system modeled by:

$$mu'' + \gamma u' + ku = 0$$

with roots of the characteristic equation:

$$r_{1,2} = -\frac{\gamma}{2m} \pm \frac{\sqrt{\gamma^2 - 4mk}}{2m}$$

we say that the spring is

- **under damped** if $\gamma^2 - 4mk < 0$, i.e., the characteristic equation has complex roots,

- **critically damped** if $\gamma^2 - 4mk = 0$, i.e., the characteristic equation has repeated roots,

- **over damped** if $\gamma^2 - 4mk > 0$, i.e., the characteristic equation has distinct real roots.

Note that we can also reframe this determination of how damped a spring is in terms of the quasi-frequency. Specifically, recalling that with the natural frequency of the spring, $\omega = \sqrt{k/m}$ and the decay rate $\lambda = \gamma/(2m)$, the quasi-frequency is given by $\mu = \sqrt{\omega^2 - \lambda^2}$. Then we can say that a spring is

- under damped if $\omega^2 - \lambda^2 > 0$;

- critically damped if $\omega^2 - \lambda^2 = 0$;

- over damped if $\omega^2 - \lambda^2 < 0$.

This framing gives helpful intuition for the concept of critical damping. If a spring is under damped, the quasi-frequency is positive; if a spring is critically dampe,d the quasi-frequency is zero and we will see that this corresponds with having repeated roots of the characteristic equation. If a spring is over damped the quasi-frequency would be imaginary, which corresponds with the characteristic equation having distinct real roots. In that case, the general solution will have no sine or cosine terms and no oscillation will occur, thus the notion of quasi-frequency doesn't make sense.

○○

Example: Consider a spring at equilibrium onto which is attached a 20 lb weight which stretches the spring an additional 2 ft before coming to rest. Find the required damping coefficient for this spring to be critically damped.

First we find:

$$m = \frac{w}{g} = \frac{20}{32} = \frac{5}{8}$$

$$k = \frac{w}{L} = \frac{20}{2} = 10.$$

Without even solving the ODE, we can see that we must have:

$$\gamma^2 - 4mk = \gamma^2 - 4\left(\frac{5}{8}\right)(10)$$

$$= \gamma^2 - 25$$

$$= 0.$$

Thus, the spring is critically damped when $\gamma = 5$.

○○

Example: Suppose we have the critically damped spring from the above example. The mass is pulled down 4 inches then pushed down with an initial downward velocity of 2 feet/sec. Find the equation of motion for the mass.

We have $m = \frac{5}{8}$, $\gamma = 5$, and $k = 10$ from above. Thus our IVP is:

$$\frac{5}{8}u'' + 5u' + 10u = 0, \qquad u(0) = 1/3, \quad u'(0) = 2.$$

This has characteristic equation:

$$\frac{5}{8}r^2 + 5r + 10 = 0$$

or

$$r^2 + 8r + 16 = 0$$

which has repeated real root $r = -4$. Thus our general solution is:

$$u(t) = c_1 e^{-4t} + c_2 t e^{-4t}.$$

Taking the derivative we have:

$$u'(t) = -4c_1 e^{-4t} + c_2 e^{-4t} - 4c_2 t e^{-4t},$$

so our initial conditions produce the system of equations:

$$\begin{cases} c_1 = \dfrac{1}{3} \\ -4c_1 + c_2 = 2. \end{cases}$$

This gives the solutions $c_1 = \frac{1}{3}, c_2 = \frac{10}{3}$. Thus, the equation of motion for this spring-mass system is:

$$u(t) = \frac{1}{3}e^{-4t} + \frac{10}{3}te^{-4t}.$$

We can see the graph of this equation of motion below.

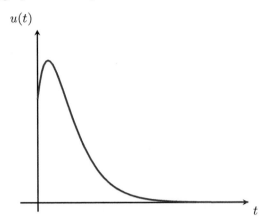

Note that there is no oscillation present. This is due to the fact that we have critically damped the spring, which ensures that it will not oscillate, but simply return to equilibrium. The spring exhausts the initial downward velocity, reaching a peak value before returning to the equilibrium point. At no point does the mass extend above the initial equilibrium point.

4.3.2   Electrical Vibrations: RLC Circuits

Spring mass systems are an example of *mechanical vibrations*. Another similar phenomenon is *electrical vibrations*. These are vibrations in the electrical current across a capacitor in an electrical circuit. The beauty is that all the ideas we have formulated for spring-mass systems translate easily into this setting.

One of the simplest electrical circuits is the RLC circuit, which has a resistor (R), inductor (L), and capacitor (C) as well as an impressed voltage (E).

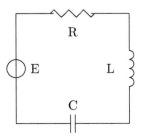

The tool which will allow us to use a second-order ODE to solve this problem is Kirkoff's Second Law, which states that the sum of the voltage drops across all elements of a closed loop is zero. For an RLC circuit these voltage drops are:

- Voltage across the resistor, $V_R = IR$, where I is current in amperes and R is the resistance in ohms.

- Voltage across the inductor, $V_L = LI'$, where L is the inductance in Henries and I' is the derivative of the current.

- Voltage across the capacitor, $V_C = Q/C$, where Q is the charge on the capacitor in coulombs and C is the capacitance in farads.

Thus, we have:

$$V_r + V_L + V_C = 0$$
$$RI + LI' + \frac{Q}{C} = 0.$$

Due to the helpful fact that $I = \dfrac{dQ}{dt}$, we can rewrite this equation in terms of just Q as:

$$RQ' + LQ'' + \frac{1}{C}Q = 0.$$

The charge across a capacitor in an RLC series circuit can be modeled by the second order linear ODE:

$$LQ'' + RQ' + \frac{1}{C}Q = 0,$$

where

- Q is the charge in coulombs (C) across the capacitor
- R is the resistance in ohms (Ω) of the resistor
- L is the inductance in henries (H) of the inductor
- C is the capacitance in farads (F) of the capacitor

Example: Let's find the charge on the capacitor in an RLC circuit with $L = 10$ mH, $R = 5\Omega$, and $C = 400$ μF if the initial charge is $Q(0) = q_0$ $C > 0$ and there is no initial current.

First, we need to convert our units to the standard SI units. Thus we have $L = \frac{10 \text{ milli-henries}}{1000 \text{ milli-henries/henry}} = 0.01H$ and $C = \frac{400 \text{ microfarads}}{10^6 \text{ microfarads/farad}} = 0.0004F.$ Thus, we have the initial value problem:

$$0.01Q'' + 5Q' + \frac{1}{0.0004}Q = 0, \qquad Q(0) = q_0, \quad Q'(0) = 0.$$

This has characteristic equation:

$$0.01r^2 + 5r + 2500 = 0$$

or

$$r^2 + 500r + 250000 = 0,$$

which has complex roots $r = -250 \pm 250i\sqrt{3}$. Thus, the general solution is

$$Q(t) = c_1 e^{-250t} \cos(250\sqrt{3}t) + c_2 e^{-250t} \sin(250\sqrt{3}t).$$

Using the initial conditions we see that:

$$c_1 = q_0$$

and as we have that:

$$
\begin{aligned}
Q'(t) \quad = \quad & -250c_1 e^{-250t} \cos(250\sqrt{3}t) - 250\sqrt{3}c_1 e^{-250t} \sin(250\sqrt{3}t) \\
& -250c_2 e^{-250t} \sin(250\sqrt{3}t) + 250\sqrt{3}c_2 e^{-250t} \cos(250\sqrt{3}t),
\end{aligned}
$$

we must have that $0 = -250c_1 + 250\sqrt{3}c_2$, so $c_2 = \frac{q_0}{\sqrt{3}}$. This gives the solution:

$$Q(t) = q_0 e^{-250t} \cos(250\sqrt{3}t) + \frac{q_0}{\sqrt{3}} e^{-250t} \sin(250\sqrt{3}t),$$

which can be rewritten as:

$$Q(t) = \frac{2q_0}{\sqrt{3}} e^{-250t} \cos\left(250\sqrt{3}t - \frac{\pi}{6}\right).$$

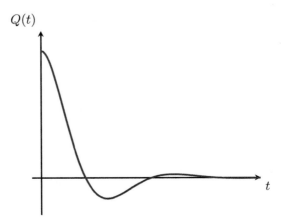

The interpretation here is that the charge on the capacitor dissipates and oscillates between a positive and negative charge until reaching equilibrium at a charge of zero. This dissipation is due to the charge passing through the resistor. In the absence of a resistor (if $R = 0$), then the charge on the capacitor would oscillate forever in simple harmonic motion. However, if we chose a larger resistor such that $R^2 - 4\frac{L}{C} \geq 0$, then the circuit would either be critically or overdamped and the charge on the capacitor would dissipate without oscillation.

○○

In the event there is more than one component of each type in the circuit we can simply sum them and use the same formula. This will not be the case when we consider parallel circuits in Chapter 5.

Example: Consider the circuit diagram below

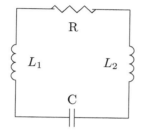

with $L_1 = 15$ mH, $L_2 = 5$ mH, $R = 10\Omega$, and $C = 500$ μF if the initial charge is $Q(0) = 100\mu C$ and there is no initial current. We first convert to standard SI units as before. Thus we have $L_1 = 0.015H$, $L_2 = 0.005H$, $R = 10\Omega$, $C = 0.0005F$, and $Q(0) = 0.0001C$. Thus, we have the initial value problem:

$$(L_1 + L_2)Q'' + RQ' + \frac{1}{C}Q = 0, \qquad Q(0) = q_0, \quad Q'(0) = i_0,$$

which given our values becomes:

$$0.02Q'' + 10Q' + 2000Q = 0, \qquad Q(0) = 0.0001, \quad Q'(0) = 0.$$

This has characteristic equation:

$$r^2 + 500r + 100000 = 0$$

which has complex roots $r = -250 \pm 50i\sqrt{15}$. Thus, the general solution is:

$$Q(t) = c_1 e^{-250t} \cos(50\sqrt{15}t) + c_2 e^{-250t} \sin(50\sqrt{15}t).$$

Using the initial condition $Q(0) = 0.0001$, we see that $c_1 = 0.0001$. Also, as we have that:

$$Q'(t) = \begin{aligned}[t] &-250c_1 e^{-250t} \cos(50\sqrt{15}t) - 50\sqrt{15}c_1 e^{-250t} \sin(50\sqrt{15}t) \\ &-250c_2 e^{-250t} \sin(50\sqrt{15}t) + 50\sqrt{15}c_2 e^{-250t} \cos(50\sqrt{15}t), \end{aligned}$$

we must have that $0 = -250c_1 + 50\sqrt{15}c_2$, so $c_2 = \dfrac{0.0005}{\sqrt{15}}$. This gives the solution:

$$Q(t) = 0.0001e^{-250t}\cos(50\sqrt{15}t) + \frac{0.0005}{\sqrt{15}}e^{-250t}\sin(50\sqrt{15}t).$$

4.3 Exercises

1. A 5 lb weight stretches a spring 3 inches and comes to rest. The weight is pulled 3 inches below the equilibrium point and released. There is no damping.

    (a) Find the equation of motion for this spring.

    (b) Find the period and amplitude of the spring.

    (c) rewrite the equation of motion in $u = R\cos(\omega t - \delta)$ form.

2. If an series circuit has an inductor with $L = 0.25H$ and a capacitor with $C = 0.5mF$ find an equation for the charge on the capacitor if there is an initial charge of $5mC$ and no initial current. (Note there is no resistance).

3. An 8 lb weight stretches a spring 4 inches and comes to rest. The weight is pushed up, compressing the spring, by 3 inches, and released. There is a damping coefficient of $\gamma = 0.25$.

    (a) Find the equation of motion for this spring.

    (b) Find the quasi-period and damped amplitude of the spring.

    (c) rewrite the equation of motion in $u = Re^{-\lambda t}\cos(\mu t - \delta)$ form.

4. If a series circuit has an inductor with $L = 10$ mH, a 10Ω resistor, and a capacitor with $C = 25$ μF find an equation for the charge on the capacitor if there is an initial charge of 5 mC and no initial current.

5. A spring with a 10 lb weight attached has a spring constant of $k = 2$. For a damping coefficient of 0.02 find the equation of motion of the spring if the mass is pushed up by 3 inches.

6. If a series circuit has an inductor with $L = 20$ mH, a capacitor with $C = 10$ μF, and two resistors, one with $R_1 = 10\Omega$ and one with $R_2 = 5\Omega$, find an equation for the charge on the capacitor if there is an initial charge of 5 mC and no initial current.

7. A 20 kg weight stretches a spring 5cm and comes to rest. The weight is pulled down 16 cm then released. There is a damping coefficient of $\gamma = 0.3$.

   (a) Find the equation of motion for this spring.

   (b) Find the quasi-period and damped amplitude of the spring.

   (c) rewrite the equation of motion in $u = Re^{-\lambda t} \cos(\mu t - \delta)$ form.

8. Suppose a spring with spring constant $k = 4$ with a 12 lb weight attached is suspended in a fluid. Find the required damping coefficient for the spring to be critically damped.

9. A 16 lb weight stretches a spring 6 inches and comes to rest. The weight is pushed down with a velocity of 4 feet/second from its equilibrium point. There is a damping coefficient of $\gamma = 0.5$.

   (a) Find the equation of motion for this spring.

   (b) Find the quasi-period and damped amplitude of the spring.

   (c) rewrite the equation of motion in $u = Re^{-\lambda t} \cos(\mu t - \delta)$ form.

10. For a series circuit which has inductor with $L = 1H$ and capacitance $C = 2.5\ \mu F$, find the resistance in ohms necessary for the series to be critically damped.

11. A 5 kg weight stretches a spring 15 cm and comes to rest. The weight is pulled 3 cm below the equilibrium point and released. There is no damping.

    (a) Find the equation of motion for this spring.

    (b) Find the period and amplitude of the spring.

    (c) rewrite the equation of motion in $u = R\cos(\omega t - \delta)$ form.

12. A spring with spring constant $k = 1$ is in a medium with damping coefficient $\gamma = 2.2$. Find the maximum weight allowable for the system to remain either overdamped or critically damped.

13. Suppose two springs both have a 6 kg mass attached and a spring constant of $k = 5$. From equilibrium the first spring is pushed up with a velocity of 2 m/s while simultaneously the second spring is released from having been pulled down by 5 cm.

    (a) Determine the amplitude of the motion of the two springs.

    (b) Do the springs have the same frequency? Why or why not?

    (c) Which mass is moving faster at $t = \pi$?

14. If a series circuit has an inductor with $L_1 = 5$ mH, a second inductor with $L_2 = 20$ mH, a 50Ω resistor, a capacitor with $C_1 = 10$ μF, find an equation for the charge on the capacitor if there is an initial charge of 50 mC and no initial current.

15. A 15 lb weight stretches a spring 6 inches and comes to rest. The weight is pushed down by 3 inches, and released with a downward velocity of 2 ft/sec. There is a damping coefficient of $\gamma = 0.05$.

   (a) Find the equation of motion for this spring.

   (b) Find the quasi-period and damped amplitude of the spring.

   (c) rewrite the equation of motion in $u = Re^{-\lambda t}\cos(\mu t - \delta)$ form.

   (d) Find the time after which the spring does not vary more than 0.5 inches from the equilibrium position.

16. Suppose two springs both have a spring constant of $k = 2$. They are suspended from the same object in a medium with damping coefficient. The first spring has a 2 lb mass attached while the second has an 8 lb mass attached. From equilibrium the first spring is pushed up with a velocity of 3 ft/s while simultaneously the second spring is released from having been pulled down by 5 cm.

   (a) Determine the damped amplitude of the motion of the two springs.

   (b) Do the springs have the same quasi-frequency? Why or why not?

   (c) Which mass is moving faster at $t = \pi$?

   (d) Determine which mass will come to rest first (vary by less than 0.1 inches from the equilibrium for all subsequent time).

17. Suppose a spring with spring constant $k = 7$ with a 10 lb weight attached is suspended in a fluid. The fluid's viscosity can be varied to increase or decrease the damping coefficient for the spring. Suppose that according to engineering specifications, the spring must be damped to such a degree that if the weight is pulled down by 8 inches and released, that it should come to rest (defined as being within one inch of equilibrium) within 20 seconds. Determine the damping coefficient required to meet these specifications.

4.4 Reduction of Order

So far we have restricted ourselves to second-order ODEs with constant coefficients. At times you may want to find the solutions of a second-order ODE with non-constant coefficients. For example, consider the slightly more general ODE:

$$y'' + py' + qy = 0,$$

where $p(t)$, $q(t)$ are continuous functions of t.

While finding the solutions of a differential equation of this general form might be challenging, at times you may be able to find one solution of this equation. If that is the case you will then need a method to find the second. One such method is the method of **reduction of order**. This method is outlined below.

If we know that y_1 is a nonzero solution of:

$$y'' + p(t)y' + q(t)y = 0,$$

where p and q are continuous functions of t, then we can find a second solution of the form $y_2 = v(t)y_1$.

How does this work in practice? We start by knowing that y_1 is a solution of the ODE, and compute:

$$
\begin{aligned}
y_2 &= vy_1 \\
y_2' &= v'y_1 + vy_1' \\
y_2'' &= v''y_1 + 2v'y_1' + vy_1''.
\end{aligned}
$$

We substitute these into the original ODE and find that we have:

$$
\begin{aligned}
y_2'' + py_2' + qy_2 &= v''y_1 + 2v'y_1' + vy_1'' + p[v'y_1 + vy_1'] + qvy_1 \\
&= v''y_1 + 2v'y_1' + pv'y_1 + v[y_1'' + py_1' + qy_1] \\
&= v''y_1 + 2v'y_1' + pv'y_1
\end{aligned}
$$

as $y_1'' + py_1' + qy_1 = 0$ because y_1 is a known solution of the ODE. Thus if we can find a solution of the ODE:

$$v''y_1 + v'(2y_1' + py_1) = 0$$

we will have the second solution. You might notice that this is again a second-order ODE with non-constant coefficients, so did we make this problem easier at all?

In fact we did. At times this differential equation can be solved easily, but no matter what, we can choose to use the substitution $w = v'$ after which this becomes:

$$w'y_1 + w(2y_1' + py_1) = 0$$

which is actually a separable first-order ODE!

As such, we have reduced the order of this problem from solving a second order ODE to solving a first order one (thus reduction of order). Note however that in order to obtain y_2 we then must integrate w. If w is a function without an easy to find anti-derivative this might be challenging. However, if we are willing to leave the second solution in terms of an integral we are always able to find this second solution.

○○

Example: As an example, we can use reduction of order to find a second solution of $y'' - 6y' + 9y = 0$, if we know that $y_1 = e^{3t}$ is one solution. This example is substantially easier than the full process outlined above. You might also notice that this differential equation could be solved using methods you already have. There is a repeated root of the ODE, $r = 3$, so we know the solution should just be $y = c_1 e^{3t} + c_2 t e^{3t}$.

However, note that when we showed that was the general solution, we just made an educated guess that $y_2 = te^{rt}$ and it happened to work.

If instead you started with the solution $y = e^{3t}$ and no idea how to make an educated guess, reduction of order gives you a tool to construct that second solution.

We can set $y_2 = vy_1 = ve^{3t}$ and see that:

$$y_2 = ve^{3t}$$

$$y_2' = v'e^{3t} + 3ve^{3t}$$

$$y_2'' = v''e^{3t} + 6v'e^{3t} + 9ve^{3t}.$$

Using these values for the derivatives of y, we can see that:

$$\begin{aligned} y_2'' - 6y_2' + 9y_2 &= v''e^{3t} + 6v'e^{3t} + 9ve^{3t} - 6(v'e^{3t} + 3ve^{3t}) + 9(ve^{3t}) \\ &= v''e^{3t} + 6v'e^{3t} + 9ve^{3t} - 6v'e^{3t} - 18ve^{3t} + 9ve^{3t} \\ &= v''e^{3t} = 0. \end{aligned}$$

Thus $v'' = 0$, as e^{3t} never takes the value zero. Note that this is a very trivial second order ODE, so there's not even a need to reduce the order of it. We can simply see that $v' = c_1$, $v = c_1 t + c_2$, and:

$$
\begin{aligned}
y_2 &= vy_1 \\
&= (c_1 t + c_2)e^{3t} \\
&= c_1 t e^{3t} + c_2 e^{3t}.
\end{aligned}
$$

However, as $c_2 e^{3t}$ is a scalar multiple of y_1, we discard it, and let $y_2 = te^{3t}$ and have general solution:

$$
y = c_1 t e^{3t} + c_2 e^{3t}.
$$

○○○

The power of reduction of order really lies in its ability to help us find solutions of ODEs we otherwise don't have the tools to solve. For example, ODEs which have non-constant coefficients like the below example.

Example: Using the method of reduction of order and the fact that $y_1 = t^2$, we can find the second solution of the ODE:

$$
t^2 y'' - 4ty' + 6y = 0.
$$

This can be written as:

$$
y'' - \frac{4}{t}y' + \frac{6}{t^2}y = 0,
$$

thus our solutions will hold only for $t \neq 0$. Let $y_2 = vy_1 = vt^2$. Then:

$$
\begin{aligned}
y_2 &= vt^2 \\
y_2' &= v't^2 + 2vt \\
y_2'' &= v''t^2 + 4v't + 2v.
\end{aligned}
$$

We can then use these values and see that:

$$
\begin{aligned}
t^2 y_2'' - 4ty_2' + 6y_2 &= t^2(v''t^2 + 4v't + 2v) - 4t(v't^2 + 2vt) + 6vt^2 \\
&= v''t^4 + 4v't^3 + 2vt^2 - 4v't^3 - 8vt^2 + 6vt^2 \\
&= v''t^4 = 0.
\end{aligned}
$$

As $v''t^4 = 0$, and $t \neq 0$, $v'' = 0$. Thus $v' = c_1$, and $v = c_1 t + c_2$. As $y_2 = vy_1$, we have that the second solution is $y_2 = t^3$ after removing the scalar multiple of y_1, and the general solution is:

$$y = c_1 t^3 + c_2 t^2.$$

ooooooooooooooooooooooooooooooooooooooooooo

In both of these examples we got pretty lucky and ended up with a very easy differential equation to solve in v, namely $v'' = 0$. Note that in this last example, instead of substituting y_2 into the original ODE we could have recalled that we will always end up with the ODE:

$$v'' y_1 + v'(2y_1' + py_1) = 0.$$

If you did, we would have found that:

$$v'' t^2 + v' \left(4t + \left(-\frac{4}{t} \right) t^2 \right) = 0$$

$$v'' t^2 = 0$$

$$v'' = 0$$

as above. The following example will show that unfortunately it is not always the case that this ODE simplifies so easily.

ooooooooooooooooooooooooooooooooooooooooooo

Example: Using the method of reduction of order and the fact that $y_1 = e^t$ for $t > 1$, we will find the second solution of the ODE:

$$(t - 1)y'' - ty' + y = 0.$$

Let $y_2 = vy_1 = ve^t$. Then:

$$y_2 = ve^t$$

$$y_2' = v'e^t + ve^t$$

$$y_2'' = v''e^t + 2v'e^t + ve^t.$$

We can use these values and see that:

$$
\begin{aligned}
(t-1)y_2'' - ty_2' + y_2 &= (t-1)(v''e^t + 2v'e^t + ve^t) - t(v'e^t + ve^t) + ve^t \\
&= v''te^t + 2v'te^t + vte^t - v''e^t - 2v'e^t - ve^t - v'te^t \\
&\quad -vte^t + ve^t \\
&= v''te^t - v''e^t + v'te^t - 2v'e^t \\
&= e^t(t-1)v'' + e^t(t-2)v' = 0.
\end{aligned}
$$

Thus $e^t(t-1)v'' + e^t(t-2)v' = 0$, so that $(t-1)v'' + (t-2)v' = 0$. Let $w = v'$. Then $w' = v''$. We can rewrite our equation as $(t-1)w' + (t-2)w = 0$. This equation is separable as:

$$(t-1)w' + (t-2)w = 0$$

$$(t-1)\frac{dw}{dt} + (t-2)w = 0$$

$$(t-1)\frac{dw}{dt} = -(t-2)w$$

$$\frac{1}{w}\,dw = -\frac{t-2}{t-1}\,dt$$

$$\int \frac{1}{w}\,dw = -\int \frac{t-2}{t-1}\,dt$$

$$\ln|w| = -t + \ln|t-1| + C$$

$$w = C(t-1)e^{-t}.$$

Hence $v' = w = C(t-1)e^{-t}$, so that:

$$v = \int C(t-1)e^{-t}\,dt = c_1 t e^{-t} + c_2$$

and the solution is $y = vy_1 = (c_1 t e^{-t} + c_2)e^t = c_1 t + c_2 e^t$. Because we already know that $y_1 = e^t$ is a solution, this means that $y_2 = t$ is a second solution, so the general solution is:

$$y = c_1 e^t + c_2 t.$$

○○

Example: Find a solution of the IVP:

$$x^2 y'' + 2xy' - 30y = 0, \qquad y(1) = 3, \quad y'(1) = 4$$

given that $y = x^5$ is a solution. Note that it must be the case that we have $x > 0$ in order to include the initial condition $y(1) = 3$ and because there's a discontinuity at $x = 0$.

We let $y_2 = v(x)x^5$ and see that:

$$y_2 = vx^5$$

$$y_2' = v'x^5 + 5vx^4$$

$$y_2'' = v''x^5 + 10v'x^4 + 20vx^3.$$

Substituting these into the ODE we see that:

$$\begin{aligned}
x^2 y_2'' + 2xy_2' - 30y_2 &= x^2(v''x^5 + 10v'x^4 + 20vx^3) + 2x(v'x^5 + 5vx^4) \\
&\quad -30(vx^5) \\
&= v''x^7 + 10v'x^6 + 20vx^5 + 2v'x^6 + 10vx^5 - 30vx^5 \\
&= v''x^7 + 12v'x^6
\end{aligned}$$

Thus we have the ODE:
$$v''x^7 + 12v'x^6 = 0,$$
which, after using the change of variable $w = v'$ and dividing by x^6, yields the separable ODE:
$$w'x + 12w = 0.$$

We proceed by solving this differential equation, seeing that:

$$\begin{aligned}
w'x + 12w &= 0 \\
\frac{dw}{dx}x &= -12w \\
\frac{dw}{w} &= -\frac{12dx}{x} \\
\int \frac{dw}{w} &= -\int \frac{12dx}{x} \\
\ln|w| &= -12\ln|x| + C \\
w &= Cx^{-12}.
\end{aligned}$$

Thus $v' = Cx^{-12}$, and we see that $v = c_1 x^{-11} + c_2$. Therefore we have that $y_2 = (c_1 x^{-11} + c_2)x^5 = c_1 x^{-6} + c_2 x^5$. As $y_1 = x^5$ is already a solution we have that the general solution is:

$$y = c_1 x^5 + c_2 x^{-6}.$$

We find that the derivative is:

$$y' = 5c_1 x^4 - 6c_1 x^{-7}$$

and thus using our initial conditions $y(1) = 3$ and $y'(1) = 4$, we have the system of equations:

$$\begin{cases} c_1 & + & c_2 & = & 3 \\ 5c_1 & - & 6c_1 & = & 4 \end{cases}$$

which we can solve to find $c_1 = 2$ and $c_2 = 1$. Thus, we have found the unique solution:

$$y = 2x^5 + x^{-6}.$$

4.4 Exercises

Use the method of reduction of order to find the general solution of the given differential equation, knowing y_1 is also a solution.

1. $x^2 y'' - 5xy' - 7y = 0, \quad y_1(x) = x^7$

2. $x^2 y'' - 6xy' + 6y = 0, \quad y_1(x) = x^6$

3. $t^2 y'' - 12y = 0, \quad y_1(t) = t^{-3}$

4. $xy'' - y' + 4x^3 y = 0, \quad y_1(x) = \cos x^2$

5. $x^2 y'' - 4xy' - 6y, \quad y_1(x) = x^6$

6. $4x^2 y'' - 2xy' + 2y = 0, \quad y_1(x) = \sqrt{x}$

7. $(t - 2)y'' - ty' + 2y = 0, \quad y_1(t) = e^t$

8. $t^2 y'' + 2ty' - 2y = 0, \quad y_1(t) = t^{-2}$

9. $x^2 y'' + 5xy' + 3y = 0, \quad y_1(x) = x^{-1}$

10. $y'' - 5x^{-1} y' + 5x^{-2} y = 0, \quad y_1(x) = x$

11. $t^2 y'' + 9ty' + 12y = 0, \quad y_1(t) = t^{-2}$

12. $x^2 y'' - 2xy' + 2y = 0, \quad y_1(x) = x^2$

4.5 Linear Nonhomogeneous Second-Order ODEs: Undetermined Coefficients

So far we have largely focused on homogeneous ODEs. However, we can extend the technology we have developed thus far to allow us to deal with the nonhomogeneous case as well. We will see soon that, aside from being interesting in their own right, linear nonhomogeneous differential equations are great for modeling certain forced spring mass systems.

We will explore now the process for finding solutions of these ODEs. This approach is summarized in the following theorem.

The general solution of the nonhomogeneous equation:

$$y'' + py' + qy = g(t)$$

can be written in the form:

$$y = c_1 y_1 + c_2 y_2 + y_p,$$

where y_1 and y_2 form a fundamental set of solutions of the homogeneous equation $y'' + py' + qy = 0$ and y_p is some specific solution of the nonhomogeneous equation.

The theorem above allows us to follow the below steps to find the general solution of a nonhonogeneous ODE:

Step 1: Find the **complementary solution**, y_c, which is the solution of $y'' + py' + qy = 0$.

Step 2: Try to find a **particular solution**, y_p, of $y'' + py' + qy = g(t)$.

Step 3: $y = y_c + y_p$

The difficult step in the above process is, of course, step 2. How do we go about finding such a solution? The **method of undetermined coefficients** will prove helpful when g is relatively "simple" (e.g., polynomials, exponentials, trigonometric functions). This method is, in effect, to make an educated guess about the particular solution. We saw before that making an educated guess can be a worthwhile approach. In fact, we arrived at the characteristic polynomial by "guessing" that $y = e^{rt}$ would be a solution of the homogeneous second-order ODEs we were concerned with before this section.

So the real question is: how do we guess? Our strategy will be to look at the function $g(x)$, and guess that our particular solution y_p is of the same form.

○○○○○○○○○○○○○○○○○○○○○○○○○○○○○○○○○○○○○○

Example: We will use the method of undetermined coefficients to find the general solution of:

$$y'' + 5y' + 6y = t^2.$$

Step 1: To find the complementary solution, y_c, we find the characteristic equation to be $r^2 + 5r + 6 = 0$, so that $r = -3, -2$ and $y_c = c_1 e^{-3t} + c_2 e^{-2t}$.

Step 2: We "guess" that $y_p = At^2 + Bt + C$. This is because $g(t) = t^2$ is a second-degree polynomial, so we naturally guess that y_p is a second-degree polynomial as well. Then our approach is to find the first two derivatives of y_p, substitute them into the ODE, and try to solve for the coefficients A, B, and C. Thus we have:

$$y_p = At^2 + Bt + C$$

$$y_p' = 2At + B$$

$$y_p'' = 2A.$$

Using these values, we have that:

$$y'' + 5y' + 6y = t^2$$

$$2A + 5(2At + B) + 6(At^2 + Bt + C) = t^2$$

$$6At^2 + (10A + 6B)t + (2A + 5B + 6C) = t^2.$$

Comparing coefficients on the left and right side of this equation gives us the system of equations:

$$\begin{cases} 6A = 1 \\ 10A + 6B = 0 \\ 2A + 5B + 6C = 0 \end{cases}$$

which has solutions $A = \frac{1}{6}$, $B = -\frac{5}{18}$, $C = \frac{19}{108}$, so $y_p = \frac{1}{6}t^2 - \frac{5}{18}t + \frac{19}{108}$.

Step 3: Thus, we can conclude that we have $y = y_c + y_p$, or

$$y = c_1 e^{-3t} + c_2 e^{-2t} + \frac{1}{6}t^2 - \frac{5}{18}t + \frac{19}{108}$$

as the general solution of this ODE.

○○

This was a pretty basic example with just a polynomial. What if we have something more complex? What if we have an exponential function or a sine or cosine? What if the guess y_p is already contained in y_c? Let's see an example with this repeated behavior first, then we'll present our general guidelines for making y_p guesses.

Example: Consider the ODE $y'' - 6y' + 9y = e^{3x}$.

Step 1: We find the complementary solution of be $y_c = c_1 e^{3x} + c_2 x e^{3x}$ as the only root to the characteristic equation is $r = 3$ with multiplicity two.

Step 2: At this point you might notice the complication. Until this example our function $g(x)$ was in a real sense independent from the functions in our complementary solution. Here, we see that $g(x) = e^{3x}$ appears in the complementary solution. Essentially, if we naively press forward and make a guess that $y_p = Ae^{3x}$, as that's an exponential function with the same exponent, we would just be guessing that part of the complementary solution is a solution, which we already know to be true.

In the past, and in this example, when we have a repeated root in the characteristic equation, we added a power of x or t to the second solution to obtain a fundamental set of solutions. Let's try that same approach here. Thus, we should just add a power of x to y_p, but then we arrive at $y_p = Ax e^{3x}$ which also appears in the complementary solution.

The way forward is to increase the power of x in y_p and so we choose $y_p = Ax^2 e^{3x}$. Note that we are not using $y_p = (Ax^2 + Bx + C)e^{3x}$ as if y_p were a product of the guesses for $g(x) = x^2$ and $g(x) = e^{3x}$. It should be clear that were you to do that, the B and C terms will just get absorbed by the arbitrary constants in y_c. Therefore, we can use our correct guess, $y_p = Ax^2 e^{3x}$, and taking the first two derivatives, we have:

$$
\begin{aligned}
y_p &= Ax^2 e^{3x} \\
y_p' &= 2Ax e^{3x} + 3Ax^2 e^{3x} \\
y_p'' &= 2Ae^{3x} + 12Ax e^{3x} + 9Ax^2 e^{3x}
\end{aligned}
$$

which, substituted into the original ODE gives us:

$$
\begin{aligned}
y'' - 6y' + 9y &= e^{3x} \\
2Ae^{3x} + 12Ax e^{3x} + 9Ax^2 e^{3x} - 6(2Ax e^{3x} + 3Ax^2 e^{3x}) + 9(Ax^2 e^{3x}) &= e^{3x} \\
2Ae^{3x} &= e^{3x}.
\end{aligned}
$$

This gives us that $2A = 1$. Thus $A = \frac{1}{2}$ and $y_p = \frac{1}{2}x^2 e^{3x}$.

Step 3: Therefore, our general solution is:

$$
y = c_1 e^{3x} + c_2 x e^{3x} + \frac{1}{2}x^2 e^{3x}.
$$

○○

Let's now consolidate our rules for making educated y_p guesses. We have three parent functions we can use: polynomials, exponential functions, and sines and cosines. Additionally, we have a few rules for linear combinations and products of these, as well as the rule we just developed above for repeated terms from y_c.

How to "guess" to find y_p

1. If $g(x) = $ a polynomial of degree n, then $y_p = $ a polynomial of degree n.

   - For example, if $g(x) = x$, then $y_p = Ax + B$;
   - or, if $g(x) = x^2 + 2x$, then $y_p = Ax^2 + Bx + C$.

2. If $g(x) = e^{ax}$, then $y_p = Ae^{ax}$.

3. If $g(x) = \sin ax$ or $g(x) = \cos ax$, then $y_p = A\sin ax + B\cos ax$.

4. If we have any linear combination of #1 – 3, then y_p is a linear combination of their answers.

   - For example, if $g(x) = x + e^x$, then $y_p = Ax + B + Ce^x$.

5. If we have any product of #1 – 3 then y_p is a product of their answers.

   - For example, if $g(x) = xe^x$ then $y_p = (Ax + B)e^x$;
   - or, if $g(x) = xe^x \sin 3x$, then
     $y_p = (Ax + B)e^x \sin 3x + (Cx + D)e^x \cos 3x$.

6. If our guess for y_p already appears in y_c, add powers of x to y_p.

   - For example, if $g(x) = e^x$ but $y_c = c_1 e^x + c_2 e^{-x}$, then $y_p = Axe^x$;
   - or, if $g(x) = e^x$ but $y_c = c_1 e^x + c_2 x e^x$, then $y_p = Ax^2 e^x$;
   - or, if $g(x) = \sin bx$ but $y_c = c_1 \sin bx + c_2 \cos bx$, then $y_p = Ax \sin bx + Bx \cos bx$.

Let's see an example of this in practice.

Example: We will use the method of undetermined coefficients to find the general solution of
$$y'' + 5y' + 6y = \sin 2x + e^{3x}.$$

<u>Step 1</u>: This is identical to a previous example, so we know that we have

$$y_c = c_1 e^{-3x} + c_2 e^{-2x}.$$

Step 2: Our "guess" from $\#2 - 3$ should include $A \sin 2x + B \cos 2x$ and Ce^{3x}. From $\#4$, y_p is a linear combination of the guesses for $\sin 2x$ and e^{3x}, so we let $y_p = A \sin 2x + B \cos 2x + Ce^{3x}$. Then:

$$
\begin{aligned}
y_p &= A \sin 2x + B \cos 2x + Ce^{3x} \\
y'_p &= 2A \cos 2x - 2B \sin 2x + 3Ce^{3x} \\
y''_p &= -4A \sin 2x - 4B \cos 2x + 9Ce^{3x}.
\end{aligned}
$$

Using these values, we have that:

$$
\begin{aligned}
y'' + 5y' + 6y &= -4A \sin 2x - 4B \cos 2x + 9Ce^{3x} \\
&\quad +5(2A \cos 2x - 2B \sin 2x + 3Ce^{3x}) \\
&\quad +6(A \sin 2x + B \cos 2x + Ce^{3x}) \\
&= (-4A - 10B + 6A) \sin 2x + (-4B + 10A + 6B) \cos 2x \\
&\quad +(9C + 15C + 6C)e^{3x} \\
&= (2A - 10B) \sin 2x + (10A + 2B) \cos 2x + 30Ce^{3x}.
\end{aligned}
$$

We also know that $y'' + 5y' + 6y = \sin 2x + e^{3x}$, so we have:

$$(2A - 10B) \sin 2x + (10A + 2B) \cos 2x + 30Ce^{3x} = \sin 2x + e^{3x}.$$

This gives us the system of equations:

$$
\begin{cases}
2A - 10B &= 1 \\
10A + 2B &= 0 \\
30C &= 1
\end{cases}
$$

which has solutions $A = 1/52$, $B = -5/52$, and $C = 1/30$. Thus we have $y_p = \frac{1}{52} \sin 2x - \frac{5}{52} \cos 2x + \frac{1}{30}e^{3x}$.

Step 3: $y = y_c + y_p = c_1 e^{-3x} + c_2 e^{-2x} + \frac{1}{52} \sin 2x - \frac{5}{52} \cos 2x + \frac{1}{30}e^{3x}.$

ooooooooooooooooooooooooooooooooooooooooooo

Example: Consider the IVP:

$$y'' - 4y = \sin t, \qquad y(0) = 1, \quad y'(0) = 2.$$

<u>Step 1</u>: We find the complementary solution to be $y_c = c_1 e^{2t} + c_2 e^{-2t}$ as the roots of the characteristic equation are $r = \pm 2$.

<u>Step 2</u>: As $g(t) = \sin t$, we "guess" $y_p = A \sin t + B \cos t$. Taking the first and second derivative we have:

$$y_p = A \sin t + B \cos t$$

$$y_p' = A \cos t - B \sin t$$

$$y_p'' = -A \sin t - B \cos t.$$

Substituting this into the original differential equation, we have:

$$y'' - 4y = \sin t$$

$$-A \sin t - B \cos t - 4(A \sin t + B \cos t) = \sin t$$

$$-5A \sin t - 5B \cos t = \sin t,$$

from which we obtain the system of equations:

$$\begin{cases} -5A = 1 \\ -5B = 0 \end{cases}$$

which has solutions $A = -\frac{1}{5}$, $B = 0$. Therefore, $y_p = -\frac{1}{5} \sin t$.

<u>Step 3</u>: We now have general solution:

$$y = y_c + y_p = c_1 e^{2t} + c_2 e^{-2t} - \frac{1}{5} \sin t.$$

In the interest of solving for the arbitrary constants, we find the derivative to be:

$$y' = 2c_1 e^{2t} - 2c_2 e^{-2t} - \frac{1}{5} \cos t.$$

Substituting our initial conditions $y(0) = 1$ and $y'(0) = 2$, we obtain the system of equations:

$$\begin{cases} c_1 + c_2 = 1 \\ 2c_1 - 2c_2 - \frac{1}{5} = 2. \end{cases}$$

This has solutions $c_1 = \frac{21}{20}, c_2 = -\frac{1}{20}$.

Therefore, the unique solution of this IVP is:

$$y = \frac{21}{20}e^{2t} - \frac{1}{20}e^{-2t} - \frac{1}{5}\sin t.$$

○○○

Example: Consider the IVP:

$$y'' + 4y = \sin 2t, \qquad y(0) = 0, \quad y'(0) = 0.$$

Step 1: We find the complementary solution to be $y_c = c_1 \cos 2t + c_2 \sin 2t$ as the roots of the characteristic equation are $r = \pm 2i$.

Step 2: We would typically "guess" that $y_p = A \sin 2t + B \cos 2t$ but these terms appear in our complementary solution. As a result, we choose the function $y_p = At \sin 2t + Bt \cos 2t$. Then we have that:

$$
\begin{aligned}
y_p &= At \sin 2t + Bt \cos 2t \\[1em]
y_p' &= A \sin 2t + 2At \cos 2t + B \cos 2t - 2Bt \sin 2t \\[1em]
y_p'' &= 2A \cos 2t + 2A \cos 2t - 4At \sin 2t - 2B \sin 2t - 2B \sin 2t - 4Bt \cos 2t \\
&= (-4B - 4At) \sin 2t + (4A - 4Bt) \cos 2t.
\end{aligned}
$$

Using these, we have that:

$$
\begin{aligned}
y'' + 4y &= \sin 2t \\[1em]
(-4B - 4At) \sin 2t + (4A - 4Bt) \cos 2t + 4(At \sin 2t + Bt \cos 2t) &= \sin 2t \\[1em]
-4B \sin 2t + 4A \cos 2t &= \sin 2t
\end{aligned}
$$

This gives us the system of equations:

$$
\begin{cases}
-4B &= 1 \\[1em]
4A &= 0
\end{cases}
$$

which, of course, has solutions $A = 0$, $B = -\frac{1}{4}$. Therefore, $y_p = -\frac{1}{4}t \cos 2t$.

Step 3: Combining y_c and y_p, our general solution is:

$$y = c_1 \cos 2t + c_2 \sin 2t - \frac{1}{4}t \cos 2t.$$

In order to solve the initial value problem, we find the derivative to be:

$$y' = -2c_1 \sin 2t + 2c_2 \cos 2t - \frac{1}{4} \cos 2t + \frac{1}{2}t \sin 2t.$$

Substituting our initial conditions $y(0) = 0$ and $y'(0) = 0$, we see that we have:

$$\begin{cases} 0 &=& c_1 \\ \\ 0 &=& 2c_2 - \frac{1}{4}. \end{cases}$$

Therefore we have $c_1 = 0$ and $c_2 = \frac{1}{8}$. Thus, the unique solution of the IVP is:

$$y = \frac{1}{8} \sin 2t - \frac{1}{4}t \cos 2t.$$

4.5 Exercises

Find the general solution of the following ODEs.

1. $y'' - 2y' - 8y = t + 7$

2. $y'' - y' - 6y = e^{3t}$

3. $6y'' + 13y' - 5 = 25t^2 + 2$

4. $4y'' - 12y' + 9y = 8te^{3t/2}$

5. $y'' + 9y = \cos 3t + 2$

6. $y'' + 5y' + 6y = te^{-2t}$

7. $y'' - 21y' + 90y = t$

8. $y'' - 5y' + 6y = 3e^{3t} + t + 7 - \sin t$

9. $2y'' - 3y' + y = 2e^t + t$

10. $y'' + 2y' + 5y = e^{-t} \sin 2t$

11. $6y'' + 5y' - 4y = e^{-2t}$

12. $y'' - 5y' + 4y = \cos 2t$

13. $y'' - 2y + 50y = e^{-3t}$

14. $y'' - 2y' + y = \sin 2t$

Find the unique solution of the following IVPs.

15. $y'' - 2y' + y = te^t$, $\quad y(0) = 0$, $\quad y'(0) = 1$

16. $y'' - 24y' + 169y = e^{12t} \cos 5t$, $\quad y(0) = 1$, $\quad y'(0) = 0$

17. $y'' + 10y' + 25y = e^{-5t}$, $\quad y(0) = -2$, $\quad y'(0) = 1$

18. $y'' - 4y' + 4y = t^2 + \sin t$, $\quad y(0) = 5$, $\quad y'(0) = 0$

19. $y'' + 9y = \cos 3t + 2$, $\quad y(0) = 0$, $\quad y'(0) = 2$

20. $y'' - 9y' - 36y = e^{-3t}$, $\quad y(0) = 0$, $\quad y'(0) = 0$

21. Consider the IVP:

$$4y'' + 2y' + 5y = f(t), \qquad y(0) = 0, \quad y'(0) = 0$$

where

$$f(t) = \begin{cases} 50t & 0 \le t < 1 \\ 0 & t \ge 1 \end{cases}$$

Note that there is no undetermined coefficients "guess" for piecewise functions. Your strategy should follow the following steps:

(a) Solve the IVP with $f(t) = 50t$ and initial conditions $y(0) = 0$, $y'(0) = 0$.

(b) Compute $y(1)$ and $y'(1)$

(c) Solve the IVP with $f(t) = 0$ and the initial conditions you found in the above step.

Your solution should be a piecewise function of t.

22. Consider the IVP:

$$y'' + 8y' + 25y = f(t), \qquad y(0) = -2, \quad y'(0) = 1$$

where

$$f(t) = \begin{cases} \cos 3t & 0 \le t < 2\pi/3 \\ 1 & t \ge 2\pi/3 \end{cases}$$

(a) Solve the IVP with $f(t) = \cos 3t$ and initial conditions $y(0) = -2$, $y'(0) = 1$.

(b) Compute $y(2\pi/3)$ and $y'(2\pi/3)$

(c) Solve the IVP with $f(t) = 1$ and the initial conditions you found in the above step.

Your solution should be a piecewise function of t.

4.6 Linear Nonhomogeneous Second-Order ODEs: Variation of Parameters

For some ODEs $y'' + p(t)y' + q(t)y = g(t)$, there are no good "guesses" to be had for y_p. Variation of parameters is a method which allows us to deal with ODEs which are hard to "guess" the y_p. For example, the ODE:

$$y'' + y = \sec t$$

doesn't have a good "guess" available for us. The idea is to assume that y_p is a linear combination of y_1 and y_2, similar to y_c. However, we choose to allow the "parameters" to vary, so instead of $y_c = c_1y_1 + c_2y_2$ we choose $y_p = u_1(t)y_1(t) + u_2(t)y_2(t)$. Essentially, we're presuming that the particular solution will look similar to the complementary solution so long as we allow the arbitrary constants to be replaced by arbitrary functions.

That approach sounds reasonable, but how can we be sure that such an approach will even yield a solution? How could we go about finding those functions u_1, u_2? We proceed in a similar fashion as undetermined coefficients in that we will take the first two derivatives of y_p, substitute them into the ODE, and hope to arrive at a system of equations of some sort that will allow us to find the functions we are after.

In that interest, presume that we have the ODE $y'' + p(t)y' + q(t)y = g(t)$ with complementary solution $y_c = c_1y_1 + c_2y_2$. Note we are choosing to work in full generality, where the linear second order ODE need not have constant coefficients. Let $y_p = u_1y_1 + u_2y_2$. Then we have:

$$y_p = u_1y_1 + u_2y_2$$

$$y_p' = u_1'y_1 + u_1y_1' + u_2'y_2 + u_2y_2'$$

$$y_p'' = (u_1''y_1 + u_1'y_1') + (u_1'y_1' + u_1y_1'') + (u_2''y_2 + u_2'y_2') + (u_2'y_2' + u_2y_2'').$$

Substituting these into our ODE, and grouping terms which contain u_1 or u_2 together, we have:

$$y_p'' + py_p' + qy_p = u_1[y_1'' + py_1' + qy_1] + u_2[y_2'' + py_2' + qy_2]$$

$$+ (u_1''y_1 + u_1'y_1') + (u_2''y_2 + u_2'y_2')$$

$$+ p[u_1'y_1 + u_2'y_2] + u_1'y_1' + u_2'y_2'.$$

Now note that both u_1 and u_2 are multiplied by either $y_1'' + py_1' + qy_1$ or $y_2'' + py_2' + qy_2$, respectively. As y_1, y_2 are solutions of the homogeneous ODE $y'' + py' + qy = 0$, these terms are zero. Also note that both $(u_1''y_1 + u_1'y_1')$ and $(u_2''y_2 + u_2'y_2')$ are the derivatives of the products of $u_1'y_1$ and $u_2'y_2$. Thus, we have that:

$$y_p'' + py_p' + qy_p = \frac{d}{dx}[u_1'y_1 + u_2'y_2] + p[u_1'y_1 + u_2'y_2] + u_1'y_1' + u_2'y_2'$$

$$= g(t).$$

Note that if $u_1'y_1 + u_2'y_2 = 0$, then its derivative is zero as well. Therefore, if we can find u_1 and u_2 such that the system of equations:

$$\begin{cases} u_1' y_1 + u_2' y_2 & = & 0 \\ u_1' y_1' + u_2' y_2' & = & g(t) \end{cases}$$

is satisfied, then $y_p = u_1 y_1 + u_2 y_2$ will be a solution of the ODE. Here, we are treating the system of equations as if u_1', u_2' are the variables to be solved for and y_1, y_2, y_1', y_2' are the coefficients.

We could try using substitution to solve this system of equations, but that would be a bit laborious. Instead, we will employ a tool from linear algebra.

Cramer's Rule: The system of equations:

$$\begin{cases} ax + by & = & f \\ cx + dy & = & g \end{cases}$$

has solutions:

$$x = \frac{W_1}{W}, \quad y = \frac{W_2}{W},$$

where

$$W_1 = \begin{vmatrix} f & b \\ g & d \end{vmatrix}, \quad W_2 = \begin{vmatrix} a & f \\ c & g \end{vmatrix}, \quad W = \begin{vmatrix} a & b \\ c & d \end{vmatrix}.$$

Note that for our system, u_1' and u_2' are playing the role of x and y, respectively. Thus, by Cramer's Rule, the solutions of the system of equations:

$$\begin{cases} y_1 u_1' + y_2 u_2' & = & 0 \\ y_1' u_1' + y_2' u_2' & = & g \end{cases}$$

are

$$u_1' = \frac{\begin{vmatrix} 0 & y_2 \\ g & y_2' \end{vmatrix}}{W}, \quad u_2' = \frac{\begin{vmatrix} y_1 & 0 \\ y_1' & g \end{vmatrix}}{W},$$

where $W = \begin{vmatrix} y_1 & y_2 \\ y_1' & y_2' \end{vmatrix}$ is the Wronskian of y_1 and y_2. Evaluating the determinants in the numerators gives us:

$$u_1' = -\frac{y_2 g}{W} \quad \text{and} \quad u_2' = \frac{y_1 g}{W}.$$

Thus, if we choose $y_p = u_1 y_1 + u_2 y_2$, we can use Cramer's Rule to find the requisite functions, u_1' and u_2' to solve the resulting system of equations. Then, taking the anti-derivatives, we arrive at u_1 and u_2, which gives us y_p and the solution of the ODE. This is summarized below.

Variation of Parameters:
If $y'' + p(t)y' + q(t) = g(t)$ and $y_c = c_1 y_1 + c_2 y_2$, then a particular solution is:
$$y(t) = y_p = u_1 y_1 + u_2 y_2,$$
where
$$u_1' = -\frac{y_2 g}{W} \quad \text{and} \quad u_2' = \frac{y_1 g}{W},$$
where W is the Wronskian of y_1 and y_2.

It is worth noting that Cramer's Rule generalizes to higher dimensional matrices. The whole method of variation of parameters also generalizes. For example, given a third-order nonhomogeneous ODE, $y_c = c_1 y_1 + c_2 y_2 + c_3 y_3$ and $y_p = u_1 y_1 + u_2 y_2 + u_3 y_3$. Following our same steps we would arrive at a system of three variables and apply Cramer's Rule to find the solution. I think you will agree that the bookkeeping for the second order case was work enough.

Example: Consider the ODE:
$$y'' - 9y = te^{-3t}.$$
The complementary solution is $y_c = c_1 e^{-3t} + c_2 e^{3t}$, as the characteristic equation has roots ± 3. Then $y_1 = e^{-3t}$ and $y_2 = e^{3t}$ have Wronskian:
$$W = \begin{vmatrix} e^{-3t} & e^{3t} \\ -3e^{-3t} & 3e^{3t} \end{vmatrix} = 6.$$
Note that $g = te^{-3t}$, the original right-hand side of the ODE. From Cramer's Rule, we have that:
$$u_1' = -\frac{y_2 g}{W} = -\frac{t}{6} \quad \text{and} \quad u_2' = \frac{y_1 g}{W} = \frac{te^{-6t}}{6},$$
so after integration we have that:
$$u_1 = -\frac{1}{12}t^2 \quad \text{and} \quad u_2 = -\frac{1}{36}te^{-6t} - \frac{1}{216}e^{-6t}.$$
Then
$$y_p = u_1 y_1 + u_2 y_2$$
$$= \left(-\tfrac{1}{12}t^2\right)e^{-3t} + \left(-\tfrac{1}{36}te^{-6t} - \tfrac{1}{216}e^{-6t}\right)e^{3t}$$
$$= -\tfrac{1}{12}t^2 e^{-3t} - \tfrac{1}{36}te^{-3t} - \tfrac{1}{216}e^{-3t}.$$

As $\frac{1}{216}e^{-3t}$ is a scalar multiple of y_1, we have that the general solution is:

$$y = y_c + y_p = c_1 e^{-3t} + c_2 e^{3t} - \frac{1}{12} t^2 e^{-3t} - \frac{1}{36} t e^{-3t}.$$

○○○

Example: Consider the ODE:

$$y'' + 6y' + 9y = e^{-3t} \sin t.$$

Note that though we could use undetermined coefficients in this case as well, it would be a bit of a pain with a number of product rules. As such we will treat this using the method of variation of parameters instead.

First we find the complementary solution, $y_c = c_1 e^{-3t} + c_2 t e^{-3t}$, due to the repeated root in the characteristic equation. With $y_1 = e^{-3t}$ and $y_2 = t e^{-3t}$, we have Wronskian:

$$W = \begin{vmatrix} e^{-3t} & t e^{-3t} \\ -3e^{-3t} & e^{-3t} - 3t e^{-3t} \end{vmatrix} = e^{-6t}.$$

Therefore, we have that:

$$u_1' = -\frac{y_2 g}{W} = -\frac{t e^{-3t}(e^{-3t}\sin t)}{e^{-6t}} = -t \sin t$$

and

$$u_2' = \frac{y_1 g}{W} = \frac{e^{-3t}(e^{-3t}\sin t)}{e^{-6t}} = \sin t$$

so after integration we have that:

$$u_1 = -\sin t + t \cos t \quad \text{and} \quad u_2 = -\cos t.$$

Then:

$$\begin{aligned} y_p &= u_1 y_1 + u_2 y_2 \\ &= (-\sin t + t \cos t) e^{-3t} + (-\cos t) t e^{-3t} \\ &= -e^{-3t} \sin t. \end{aligned}$$

Thus the general solution is:

$$y = y_c + y_p = c_1 e^{-3t} + c_2 t e^{-3t} - e^{-3t} \sin t$$

○○○

Example: Consider the ODE:

$$y'' + 2y' + 2y = e^{-t} \sec t.$$

First we find the complementary solution, $y_c = c_1 e^{-t} \cos t + c_2 e^{-t} \sin t$. Note that $y_1 = e^{-t} \cos t$ and $y_2 = e^{-t} \sin t$ have Wronskian:

$$W = \begin{vmatrix} e^{-t} \cos t & e^{-t} \sin t \\ -e^{-t} \cos t - e^{-t} \sin t & -e^{-t} \sin t + e^{-t} \cos t \end{vmatrix} = e^{-2t}.$$

Therefore, we have that:

$$u_1' = -\frac{y_2 g}{W} = -\frac{e^{-t} \sin t (e^{-t} \sec t)}{e^{-2t}} = -\frac{\sin t}{\cos t}$$

and

$$u_2' = \frac{y_1 g}{W} = \frac{e^{-t} \cos t (e^{-t} \sec t)}{e^{-2t}} = 1$$

so after integration we have that:

$$u_1 = -\ln|\cos t| \quad \text{and} \quad u_2 = t.$$

Then:

$$\begin{aligned} y_p &= u_1 y_1 + u_2 y_2 \\ &= -e^{-t} \cos t \ln|\cos t| + t e^{-t} \sin t. \end{aligned}$$

Thus the general solution is:

$$y = y_c + y_p = c_1 e^{-t} \cos t + c_2 e^{-t} \sin t - e^{-t} \cos t \ln|\cos t| + t e^{-t} \sin t.$$

○○○

Example: Consider the ODE:

$$y'' + 2y + y = e^{-t} \ln t.$$

First we find the complementary solution, $y_c = c_1 e^{-t} + c_2 t e^{-t}$. Then $y_1 = e^{-t}$ and $y_2 = t e^{-t}$ have Wronskian:

$$W = \begin{vmatrix} e^{-t} & t e^{-t} \\ -e^{-t} & e^{-t} - t e^{-t} \end{vmatrix} = e^{-2t}.$$

Therefore, we have that:

$$u_1' = -\frac{y_2 g}{W} = -\frac{t e^{-t} (e^{-t} \ln t)}{e^{-2t}} = -t \ln t$$

and

$$u_2' = \frac{y_1 g}{W} = \frac{e^{-t} (e^{-t} \ln t)}{e^{-2t}} = \ln t$$

so after integration we have that:

$$u_1 = \frac{t^2}{4} - \frac{t^2}{2} \ln t \quad \text{and} \quad u_2 = t \ln t - t.$$

Then:

$$y_p = u_1 y_1 + u_2 y_2$$
$$= \frac{t^2}{2} e^{-t} \ln t - \frac{3}{4} t^2 e^{-t}.$$

Thus the general solution is:

$$y = y_c + y_p = c_1 e^{-t} + c_2 t e^{-t} + \frac{t^2}{2} e^{-t} \ln t - \frac{3}{4} t^2 e^{-t}.$$

○○

Example: Consider the IVP:

$$y'' + 4y = \tan 2t, \qquad y(0) = 1, \quad y'(0) = 0.$$

First we find the complementary solution, $y_c = c_1 \cos 2t + c_2 \sin 2t$. Then $y_1 = \cos 2t$ and $y_2 = \sin 2t$ have Wronskian:

$$W = \begin{vmatrix} \cos 2t & \sin 2t \\ -2 \sin 2t & 2 \cos 2t \end{vmatrix} = 2.$$

Therefore, we have that:

$$u_1' = -\frac{y_2 g}{W} = -\frac{\sin 2t (\tan 2t)}{2} = -\frac{\sin^2 2t}{2 \cos 2t} = \frac{\cos^2 2t - 1}{2 \cos 2t} = \frac{1}{2} \cos 2t - \frac{1}{2} \sec 2t$$

and

$$u_2' = \frac{y_1 g}{W} = \frac{\cos 2t (\tan 2t)}{2} = \frac{1}{2} \sin 2t$$

so after integration we have that:

$$u_1 = \frac{1}{4} \sin 2t - \frac{1}{4} \ln |\sec 2t + \tan 2t| \quad \text{and} \quad u_2 = -\frac{1}{4} \cos 2t.$$

Then:

$$y_p = u_1 y_1 + u_2 y_2$$
$$= \left(\frac{1}{4} \sin 2t - \frac{1}{4} \ln |\sec 2t + \tan 2t| \right) \cos 2t + \left(-\frac{1}{4} \cos 2t \right) \sin 2t$$
$$= -\frac{1}{4} \cos 2t \ln |\sec 2t + \tan 2t|.$$

Thus the general solution is:

$$y = y_c + y_p = c_1 \cos 2t + c_2 \sin 2t - \frac{1}{4} \cos 2t \ln |\sec 2t + \tan 2t|.$$

In order to solve for c_1 and c_2, we first take the derivative to find:

$$y = c_1 \cos 2t + c_2 \sin 2t - \frac{1}{4} \cos 2t \ln |\sec 2t + \tan 2t|$$

$$y' = -2c_1 \sin 2t + 2c_2 \cos 2t + \frac{1}{2} \sin 2t \ln |\sec 2t + \tan 2t| - \frac{1}{2}.$$

Then, using our initial conditions $y(0) = 1$ and $y'(0) = 0$, we have that:

$$\begin{cases} y(0) &= 1 = c_1 \\ y'(0) &= 0 = 2c_2 - \frac{1}{2}. \end{cases}$$

Hence we have that $c_1 = 1$ and $c_2 = \frac{1}{4}$. Therefore, we have the unique solution:

$$y(t) = \cos 2t + \frac{1}{4} \sin 2t - \frac{1}{4} \cos 2t \ln |\sec 2t + \tan 2t|$$

of our IVP.

○○○○○○○○○○○○○○○○○○○○○○○○○○○○○○○○○○○○○○○

Example: In a previous example in Section 4.4, we used reduction of order to show that $y_1 = t^2$ and $y_2 = t^3$ are solutions of the homogeneous differential equation:

$$t^2 y'' - 4ty' + 6y = 0.$$

Given that information, we now seek to find the general solution of the differential equation:

$$t^2 y'' - 4ty' + 6y = \ln t.$$

First we find the Wronskian of y_1 and y_2:

$$W = \begin{vmatrix} t^2 & t^3 \\ 2t & 3t^2 \end{vmatrix} = t^4.$$

Therefore, we have that:

$$u_1' = -\frac{y_2 g}{W} = -\frac{t^3 \ln t}{t^4} = -\frac{\ln t}{t}$$

and

$$u_2' = \frac{y_1 g}{W} = \frac{t^2 \ln t}{t^4} = \frac{\ln t}{t^2}$$

so after integration we have that:

$$u_1 = -\frac{(\ln t)^2}{2} \quad \text{and} \quad u_2 = -\frac{\ln t}{t} - \frac{1}{t}.$$

Then:

$$
\begin{aligned}
y_p &= u_1 y_1 + u_2 y_2 \\
&= -\frac{(\ln t)^2}{2} t^2 + \left(-\frac{\ln t}{t} - \frac{1}{t} \right) t^3 \\
&= -\frac{1}{2} t^2 (\ln t)^2 - t^2 \ln t - t^2.
\end{aligned}
$$

Thus the general solution is:

$$
y = y_c + y_p = c_1 t^2 + c_2 t^3 - \frac{1}{2} t^2 (\ln t)^2 - t^2 \ln t.
$$

ooooooooooooooooooooooooooooooooooooooo

So far the examples we have used have been carefully selected so that both $-y_2 g/W$ and $y_1 g/W$ are functions with closed-form anti-derivative. This may not always be the case. In the event that these expressions do not have a closed-form anti-derivative, variation of parameters can still yield a solution. So long, that is, as we are comfortable expressing our solution in terms of an integral function.

Example: Consider the IVP:

$$
y'' - 4y' + 4y = \frac{e^{2t} \sin t}{t}, \qquad y(t_0) = 0, \quad y(t_0) = 1.
$$

We see that $y_c = c_1 e^{2t} + c_2 t e^{2t}$ and find the Wronskian of $y_1 = e^{2t}$ and $y_2 = t e^{2t}$ to be:

$$
W = \begin{vmatrix} e^{2t} & t e^{2t} \\ 2e^{2t} & e^{2t} + 2t e^{2t} \end{vmatrix} = e^{4t}.
$$

Thus we have:

$$
u_1' = -\frac{y_2 g}{W} = -\frac{t e^{2t} \left(\frac{e^{2t} \sin t}{t} \right)}{e^{4t}} = -\sin t
$$

and

$$
u_2' = \frac{y_1 g}{W} = \frac{e^{2t} \left(\frac{e^{2t} \sin t}{t} \right)}{e^{4t}} = \frac{\sin t}{t}.
$$

Thus, we have $u_1 = \cos t$ pretty easily, but notice that $\sin t/t$ does not have an obvious anti-derivative. In fact, this function doesn't have an elementary anti-derivative at all. As such, to proceed we can simply write that:

$$
u_2 = \int_{t_0}^{t} \frac{\sin x}{x} \, dx,
$$

defining u_2 as the function of t resulting from integrating $\sin x/x$ from our initial value, t_0, to some value t. We might not be able to write down this function, but using either a table of values (this is a pretty common function) or a computer algebra system, we can still graph this function and determine the long-term behavior of it. Thus, we have the general solution:

$$y = c_1 e^{2t} + c_2 t e^{2t} + e^{2t} \cos t + t e^{2t} \int_{t_0}^{t} \frac{\sin x}{x} dx.$$

In fact, this particular integral is sometimes called the sine integral and some computer software simply defines a new function,

$$\text{Si}(t) = \int_{0}^{t} \frac{\sin x}{x} dx,$$

so we can write our solution as:

$$y = c_1 e^{2t} + c_2 t e^{2t} + e^{2t} \cos t + t e^{2t} (\text{Si}(t) - \text{Si}(t_0))$$

which, as $\text{Si}(t_0)$ is a constant, can be simplified as:

$$y = c_1 e^{2t} + c_2 t e^{2t} + e^{2t} \cos t + t e^{2t} \text{Si}(t).$$

4.6 Exercises

Using variation of parameters, find the solution of the following ODEs. When the domain of the solution is not the whole real line, state the domain of the solution.

1. $y'' + 5y' + 4y = te^{-2t}$

2. $y'' - 2y' + 9y = e^t \sin 2\sqrt{2}t$

3. $y'' + 9y = \cos 3t$

4. $16y'' - 24y' + 9y = e^{3t/4} \ln t$

5. $y'' + 7y' + 12y = te^{4t}$

6. $y'' + 6y' + 9y = \frac{e^{-3t}}{t}$

7. $y'' + 4y' + 4y = \frac{e^{-2t}}{t^5}$

8. $y'' + 9y = \sin^2 3t$

9. $y'' - 6y' + 9y = e^{3t} \cos 2t$

10. $y'' - 10y' + 25y = e^{5t}\sqrt{9 - t^2}$

Using variation of parameters, find the solution of the following IVPs. When the domain of the solution is not the whole real line, state the domain of the solution.

11. $4y'' + 4y' + 17y = te^{-t/2}$, $\quad y(0) = 0, \quad y'(0) = 0$

12. $y'' + 2y' + y = \dfrac{\sin t}{e^t}$, $\quad y(0) = 0, \quad y'(0) = 1$

13. $y'' - 2y' + y = \dfrac{e^t}{t^2},$ $y(1) = 0,$ $y'(1) = 0$

14. $y'' + 5y' + 6y = e^{-2t} \sin 2t,$ $y(0) = 0,$ $y'(0) = 0$

15. $4y'' + 4y' + y = e^{-t/2} \ln t^2,$ $y(1) = 0,$ $y'(1) = 1$

16. $y'' + y = \csc t,$ $y(\pi/2) = 0,$ $y'(\pi/2) = 0$

17. $y'' + 2y' + y = \dfrac{e^{-t}}{\sin t},$ $y(\pi/2) = 0,$ $y'(\pi/2) = 0$

18. $y'' - 8y' + 16y = e^{4t}\sqrt{t+1},$ $y(0) = 0,$ $y'(0) = 0$

Find the solution of the following ODEs using variation of parameters. Note that you will need to leave the solution in terms of an integral function. Use t_0 as a placeholder initial value.

19. $y'' - 5y' + 6y = t^{-1}$ 20. $y'' - 2y' + 5y = \dfrac{e^t}{t}$

21. $y'' - 6y' + 8y = \sin x^2$ 22. $y'' + 7y' - 8y = \ln x$

4.7   Forced Vibrations

In Section 4.3.1, we saw the spring-mass system. The springs being modeled were free-hanging from a fixed position. Another possibility is that the spring is acted upon by some external force. The idea being that the object the spring is hanging from is in motion, exerting force on the mass. This occurs frequently in engineering, for example when modeling the motion of the tuned mass damper of a building during an earthquake.

In Section 4.3.2, we modeled RLC circuits all of which had no impressed voltage. This voltage can come from a battery or other power source. This is why all of our examples began with a charge on the capacitor. We were essentially modeling how that charge either alternates harmonically, or dissipates due to the resistance from the resistor.

Here we will explore the situation where there is an impressed voltage. This might be modeled with a constant function such as the case where there is a battery or other direct current source, or a sinusoidal function as in the case with alternating current sources. These are not the only options, but they are common ones.

Exploring first the case of forced springs, the motion can be modeled as follows.

Equation of Motion of Forced Vibrations

$$mu'' + \gamma u' + ku = F(t),$$

where $m = \text{mass} = \frac{\text{weight}}{\text{gravity}} = \frac{w}{g}$,

$\gamma = \text{damping coefficient}$,

$k = \text{spring constant} = \frac{F}{L} = \frac{w}{L}$.

$F(t) = \text{an external force applied to the spring-mass system}$.

Problems of this type can often be solved using the method of variation of parameters or undetermined coefficients.

Example: A 16 lb weight stretches a spring 6 inches and comes to rest. The weight is pushed up 2 inches above the equilibrium point and then set in motion with an initial upward velocity of 2 ft/sec. There is no damping, but there is an external force $F(t) = \cos 5t$ acting on the system.

This is the same spring as an example from Section 4.3.1, but with a force being applied to it. As such, we have that:

$$\frac{1}{2}u'' + 32u = F(t) = \cos 5t$$

and our initial value problem is:

$$\frac{1}{2}u'' + 32u = \cos 5t, \qquad u(0) = -\frac{1}{6}, \quad u'(0) = -2.$$

The complementary solution is:

$$u_c = c_1 \cos 8t + c_2 \sin 8t$$

and to utilize the method of undetermined coefficients, we will "guess," based on $F(t) = \cos 5t$, that:

$$u_p = A \cos 5t + B \sin 5t.$$

Then we have that:

$$u_p = A \cos 5t + B \sin 5t$$

$$u_p' = -5A \sin 5t + 5B \cos 5t$$

$$u_p'' = -25A \cos 5t - 25B \sin 5t.$$

Using these expressions, we have that:

$$\tfrac{1}{2}u_p'' + 32u_p \;=\; \cos 5t$$

$$-\tfrac{25}{2}A\cos 5t - \tfrac{25}{2}B\sin 5t + 32A\cos 5t + 32B\sin 5t \;=\; \cos 5t$$

$$\tfrac{39}{2}A\cos 5t + \tfrac{39}{2}B\sin 5t \;=\; \cos 5t$$

so that $A = \tfrac{2}{39}$ and $B = 0$. Therefore, the general solution is:

$$u = c_1 \cos 8t + c_2 \sin 8t + \frac{2}{39}\cos 5t.$$

From the initial conditions $u(0) = -\tfrac{1}{6}$ and $u'(0) = -2$, we have that:

$$-\frac{1}{6} = c_1 + \frac{2}{39},$$

so that $c_1 = -\tfrac{17}{78}$ and

$$u' \;=\; -8c_1\sin 8t + 8c_2\cos 8t - \tfrac{10}{39}\sin 5t$$

$$=\; \tfrac{68}{39}\sin 8t + 8c_2\cos 8t - \tfrac{10}{39}\sin 5t$$

gives:

$$-2 = 8c_2,$$

so that $c_2 = -\tfrac{1}{4}$. Therefore, the solution is:

$$u = -\frac{17}{78}\cos 8t - \frac{1}{4}\sin 8t + \frac{2}{39}\cos 5t.$$

○○

In the event that we have a spring which is at rest until acted on by an external force, we may observe a **beat** pattern in its motion. This is a phenomenon which you have likely observed. If not, grab a spring toy and hold it upside down at rest, then move your hand in an oscillatory pattern and observe. You will likely see that the spring reaches a maximum, then the next few times it drops down it won't quite reach the same maximum, then it will reach it again. That is exactly a beat pattern! Let's see an example where we can explicitly find the equation of motion for such a system.

Example: Consider the IVP:

$$u'' + 4u = 5\cos 3t, \qquad u(0) = 0, \quad u'(0) = 0$$

which is modeling a spring with a 32 lb mass attached to it with a spring constant of $k = 4$ and no damping. The spring is at rest then acted on by an external force $F(t) = 5 \cos 3t$ which is an oscillating force with maximum force of 5 pound-force.

We can solve this by noting that the complementary solution is:

$$u_c = c_1 \cos 2t + c_2 \sin 2t$$

and then using:

$$u_p = A \cos 3t + B \sin 3t$$

as our guess for undetermined coefficients. Note that:

$$u_p'' = -9A \cos 3t - 9B \sin 3t.$$

Using these expressions, we have that:

$$
\begin{aligned}
u'' + 4u &= 5 \cos 3t \\
-9A \cos 3t - 9B \sin 3t + 4(A \cos 3t + B \sin 3t) &= 5 \cos 3t \\
-5A \cos 3t - 5B \sin 3t &= 5 \cos 3t
\end{aligned}
$$

from which we can easily see that $A = -1$ and $B = 0$. Thus our general solution is:

$$u = c_1 \cos 2t + c_2 \sin 2t - \cos 3t.$$

We compute:

$$u' = -2c_1 \sin 2t + 2c_2 \cos 2t + 3 \sin 3t$$

and using our initial conditions we see that we have $c_1 - 1 = 0$ from the first equation and $2c_2 = 0$ from the second. Thus we can conclude that $c_1 = 1$ and $c_2 = 0$ and we have the equation of motion:

$$u = \cos 2t - \cos 3t.$$

This has the following graph

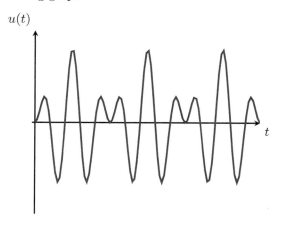

You should likely be able to identify the pattern, but we can make it even more explicit. Recalling the trigonometric identity:

$$\cos \alpha - \cos \beta = -2 \sin \frac{\alpha - \beta}{2} \sin \frac{\alpha + \beta}{2}$$

we can rewrite our solution as:

$$u = 2 \sin \frac{t}{2} \sin \frac{5t}{2}.$$

You can now think about this equation of motion as a sine curve with frequency $\frac{5}{2}$ and a variable amplitude of $2 \sin \frac{t}{2}$. With this viewpoint let's examine the graph of our equation of motion together with the graphs of $\pm 2 \sin \frac{t}{2}$.

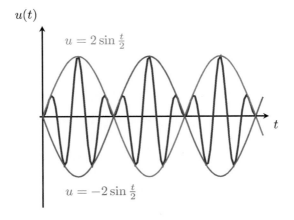

We can see that the graph of our equation of motion is bounded by the graphs of $\pm 2 \sin \frac{t}{2}$! Viewing our equation of motion together with these graphs is a great way to realize this beat frequency behavior.

It is worth noting that in the event that there is an undamped spring such that the frequency in the forcing function does not agree with the natural frequency of the spring, this process will always work. This is detailed below.

Beat Frequencies:

Given an undamped spring mass system at rest with mass m and spring constant k with natural frequency $\omega = \sqrt{\frac{k}{m}}$ which is acted upon by an external force $F(t) = F \cos \nu t$ modeled by the IVP:

$$mu'' + ku = F \cos \nu t, \qquad u(0) = 0, \quad u'(0) = 0,$$

where $\omega \neq \nu$, the equation of motion is given by:

$$u = \frac{F}{m(\omega^2 - \nu^2)} (\cos \omega t - \cos \nu t)$$

which can be rewritten as:

$$u = \frac{2F}{m(\omega^2 - \nu^2)} \sin \frac{\nu - \omega}{2} t \sin \frac{\nu + \omega}{2} t.$$

The derivation of this formula is left to the exercises.

This leaves a natural question: What happens in the event that the frequency of the forcing function is exactly the same as the natural frequency of the spring? In this case, we say that the force is in **resonance** with the spring. Let's look at an example of this in action. In fact, let's consider the exact same spring but with a slightly different forcing function.

○○

Example: Consider the IVP:

$$u'' + 4u = 5 \cos 2t, \qquad u(0) = 0, \quad u'(0) = 0$$

which is modeling a spring with a 32 lb mass attached to it with a spring constant of $k = 4$ and no damping. The spring is at rest then acted on by an external force $f(t) = 5 \cos 2t$ which is an oscillating force with maximum force of 5 pound-force.

Note that these are the exact spring parameters as the previous example. The only difference is the forcing function is now $5 \cos 2t$ rather than $5 \cos 3t$. We see that:

$$u_c = c_1 \cos 2t + c_2 \sin 2t.$$

Thus, the frequency of the forcing function is exactly the same as the natural frequency of the spring.

To find u_p, we choose $u_p = At \sin 2t + Bt \cos 2t$ as the typical guess for u_p would be exactly our complementary solution. Proceeding, we have that:

$$u_p = At \sin 2t + Bt \cos 2t$$

$$u_p' = A \sin 2t + 2At \cos 2t + B \cos 2t - 2Bt \sin 2t$$

$$\begin{aligned} u_p'' &= 2A \cos 2t + 2A \cos 2t - 4At \sin 2t - 2B \sin 2t - 2B \sin 2t - 4Bt \cos 2t \\ &= (-4B - 4At) \sin 2t + (4A - 4Bt) \cos 2t. \end{aligned}$$

Substituting into our ODE, we see that:

$$u_p'' + 4u_p = 5 \cos 2t$$

$$(-4B - 4At) \sin 2t + (4A - 4Bt) \cos 2t + 4(At \sin 2t + Bt \cos 2t) = 5 \cos 2t$$

$$-4B \sin 2t + 4A \cos 2t = 5 \cos 2t.$$

Thus $A = \frac{5}{4}$ and $B = 0$, so our particular solution is $u_p = \frac{5}{4}t \sin 2t$ and we have general solution:

$$u = c_1 \cos 2t + c_2 \sin 2t + \frac{5}{4}t \sin 2t.$$

To find the unique solution we take the derivative, and have:

$$u' = -2c_1 \sin 2t + 2c_2 \cos 2t + \frac{5}{4} \sin 2t + \frac{5}{2}t \cos 2t.$$

Thus, from our initial conditions, we have the equations $c_1 = 0$ and $c_2 = 0$, so we have the unique solution:

$$u = \frac{5}{4}t \sin 2t.$$

Examining the graph of this function may be illuminating.

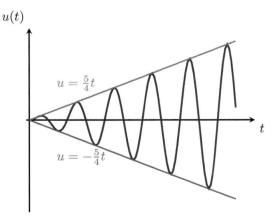

Here we plotted the function $u = \frac{5}{4}t\sin 2t$ together with the functions $\pm\frac{5}{4}t$. You can think of the solution here to be a sine wave with frequency 2 and a variable amplitude of $\frac{5}{4}t$.

○○

Let's examine now an example of an RLC circuit with an impressed voltage. Let's consider a circuit with a constant voltage source.

Example: Find the charge on the capacitor in an RLC circuit if $L = 0.1$ H, $R = 10\ \Omega$, $C = 0.002$ F, there is no initial current or charge on the capacitor, and there is a 9-volt voltage source.

These values give us the IVP:

$$0.1Q'' + 10Q' + \frac{1}{0.002}Q = 9, \qquad Q(0) = 0, \quad Q'(0) = 0$$

which we can rewrite as:

$$Q'' + 100Q' + 5000Q = 90, \qquad Q(0) = 0, \quad Q'(0) = 0.$$

We find that the roots of the characteristic equation are $-50\pm50i$, so we have:

$$y_c = c_1 e^{-50t}\cos 50t + c_2 e^{-50t}\sin 50t.$$

As $g(t) = 9$ is a constant, we have $y_p = A$, $y_p' = 0$, and $y_p'' = 0$. Then we have $y_p'' + 100y_p' + 5000y_p = 5000A = 90$, so $A = 9/500$. Thus we have:

$$y = c_1 e^{-50t}\cos 50t + c_2 e^{-50t}\sin 50t + \frac{9}{500}$$

which has derivative:

$$y' = -50c_1 e^{-50t}\cos 50t - 50c_1 e^{-50t}\sin 50t - 50c_2 e^{-50t}\sin 50t + 50c_2 e^{-50t}\cos 50t.$$

Then, from our initial conditions, we have the system of equations:

$$\begin{cases} c_1 + \frac{9}{500} &= 0 \\[2mm] -50c_1 + 50c_2 &= 0. \end{cases}$$

Solving, we have $c_1 = c_2 = -\frac{9}{500}$. Thus, we have the solution:

$$Q(t) = \frac{9}{500}(1 - e^{-50t}[\cos 50t + \sin 50t]).$$

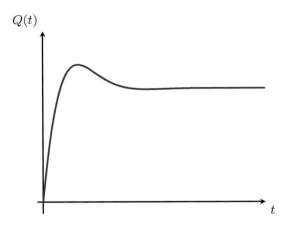

$Q(t)$

t

As you can see, the charge on the capacitor oscillates before stabilizing, with $Q(t) \to \frac{9}{500}$ as $t \to \infty$.

Recall that the voltage drop across the capacitor, $V_c = \frac{Q}{C}$. Thus we can see that:

$$\lim_{t\to\infty} V_c = \frac{\lim_{t\to\infty} Q(t)}{C}$$

$$= \frac{\frac{9}{500}}{0.002}$$

$$= 9.$$

Therefore, we can conclude that with a constant voltage source, the voltage across the capacitor will equal the constant voltage source in the long run.

OOOOOOOOOOOOOOOOOOOOOOOOOOOOOOOOOOOOOOOOO

An RLC circuit model with an alternating current impressed voltage rather than a direct current one can be modeled with a sinusoidal function. In this case, the ODE which models the circuit will be of an identical form to one modeling a spring with a sinusoidal forcing function. Another possible type of impressed voltage function is a sawtooth waveform,

$$F(t) = t - \lfloor t \rfloor$$

so named because the graph resembles a sawtooth. This function, however, cannot be managed with either undetermined coefficients or variation of parameters. The use of Laplace Transforms is sufficient in this case, but is a topic not covered in this text.

4.7 Exercises

1. Suppose a spring with a 2 lb mass attached to it has a spring constant of $k = 4$. Assume the spring is at rest and acted upon by an external force of $F(t) = 2x^2 - x^3$ for time $0 \le t \le 2$.

   (a) Write an IVP to model this spring.

   (b) find the equation of motion for time $0 \le t \le 2$.

   (c) What is the position of the spring at time $t = 2$?

2. Suppose a spring with a 64 lb mass attached has a spring constant of $k = 10$. There is a damping coefficient of $\gamma = 4$. Suppose the spring at rest is acted upon by an external force of $F(t) = 75e^{-t} \cos 2t$.

   (a) Write an IVP to model this spring.

   (b) Is this spring under damped, over damped, or critically damped?

   (c) Using either undetermined coefficients or variation of parameters, solve for the equation of motion.

3. Find the charge on the capacitor in an RLC circuit with $L = 0.005$H, $R = 20\Omega$, and $C = 0.0001$F if there is no current or charge on the capacitor and there is a direct current voltage source $E(t) = 5$.

4. Suppose a spring with an 8 lb mass attached has a spring constant of $k = 9$. There is a damping coefficient of $\gamma = 3$. Suppose the spring at rest is acted upon by an external force of $F(t) = 100e^{-6t} \cos 25t$.

   (a) Write an IVP to model this spring.

   (b) Is this spring under damped, over damped, or critically damped?

   (c) Using either undetermined coefficients or variation of parameters, solve for the equation of motion.

5. Find the charge on the capacitor in an RLC circuit with $L = 0.02$H, $R = 5\Omega$, and $C = 0.001$F if there is no current or charge on the capacitor and there is an alternative current voltage source $E(t) = 75 \cos t$.

6. Write the following sums as products of the form $\sin \alpha \sin \beta$ or $\cos \alpha \cos \beta$.

   (a) $\cos 2t + \cos 3t$

   (b) $\cos 5t - \cos 8t$

   (c) $\cos t + \cos \pi t$

   (d) $\cos 7t - \cos 2\pi t$

7. Suppose a spring with an 8 lb mass attached to it and a spring constant $k = 1$ is at rest. An external force of $F(t) = 3\cos 7t$ is applied to the spring.

   (a) Write an IVP to model this spring-mass system.

   (b) Without solving, will there be a beat or resonant frequency? How can you tell?

   (c) Solve for the equation of motion for the mass.

   (d) Rewrite the equation of motion in the form
   $u(t) = A\sin \alpha t \sin \beta t$, if possible.

8. Suppose a spring with a 4 lb mass attached to it and a spring constant $k = 2$ is at rest. An external force of $F(t) = 2\cos 0.5t$ is applied to the spring.

   (a) Write an IVP to model this spring-mass system.

   (b) Without solving, will there be a beat or resonant frequency? How can you tell?

   (c) Solve for the equation of motion for the mass.

   (d) Rewrite the equation of motion in the form
   $u(t) = A\sin \alpha t \sin \beta t$, if possible.

9. Find the charge on the capacitor in an RLC circuit with $L = 0.05$H, $R = 2\Omega$, and $C = 0.001$F if there is no current or charge on the capacitor and there is a direct current voltage source $E(t) = 25$.

10. Suppose a spring with a 16 lb mass attached to it and a spring constant $k = 4.5$ is at rest. An external force of $F(t) = 2\cos 3t$ is applied to the spring.

    (a) Write an IVP to model this spring-mass system.

    (b) Without solving, will there be a beat or resonant frequency? How can you tell?

    (c) Solve for the equation of motion for the mass.

11. Find the charge on the capacitor in an RLC circuit with $L = 0.01$H, $R = 10\Omega$, and $C = 0.005$F if there is no current or charge on the capacitor and there is an alternative current voltage source $E(t) = 10\sin 2\pi t$.

12. Consider a spring which is stretched by 6 inches by the weight of a 4 lb mass before coming to rest. The mass is pulled down 2 inches then pushed up at a rate of 1.5 ft/sec. Find the equation of motion for this spring and compute its quasi-frequency. How much does this quasi-frequency differ from the natural frequency of the spring? What is the ratio between them?

13. Suppose a spring with spring constant $k = 1$ is in a zero damping environment with a mass of 128 lbs attached is at rest. You can apply a force of $f(t) = F_0 \cos bt$ where $F_0 \le 0.1$ to the spring to induce it to move.

    (a) What is the best strategy for determining the ideal value for b to mover the spring?

    (b) What phenomenon are you observing here? (Beat, resonance, damped amplitude, etc.)

    (c) With your ideal value of b, how long will it take to displace the mass from equilibrium by 10 inches?

14. Derive the formula for the equation of motion of a spring which experiences a beat frequency. That is, given an undamped spring-mass system at rest with a mass m and spring constant k with natural frequency $w = \sqrt{\frac{k}{m}}$, and which is acted upon by an external force $F(t) = F \cos \nu t$, modeled by the IVP:

$$mu'' + ku = F \cos \nu t, \qquad u(0) = 0, \quad u'(0) = 0,$$

where $w \ne \nu$, show that the equation of motion is given by:

$$u = \frac{F}{m(w^2 - \nu^2)} (\cos wt - \cos \nu t).$$

4.7 Project: Variation of Parameters and Power Series Solutions of Forced Vibration

In Section 4.6 we saw that the particular solution, y_p of a nonhomogeneous ODE $ay'' + by' + c = g(t)$ can be written as $y_p = u_1 y_1 + u_2 y_2$ where:

$$u_1 = -\int \frac{y_2(t)g(t)}{W(y_1, y_2)(t)} dt, \quad u_2 = \int \frac{y_1(t)g(t)}{W(y_1, y_2)(t)} dt.$$

The obvious problem is that at times, the above integrals may not have closed-form solutions. However, using the Taylor series representations for some common functions can give us a method for writing solutions of variation of parameters ODEs, not in terms of elementary functions or integrals, but as a power series. In this project we will find power series to represent integrals we otherwise would be unable to find their anti-derivatives. We will then use those power series to express a solution of a forced vibrations problem and determine approximate values for the future behavior of the spring.

Using the Taylor series representations:

$$\sin x = x - \frac{x^3}{3!} + \frac{x^5}{5!} - \cdots = \sum_{n=0}^{\infty} \frac{(-1)^n x^{2n+1}}{(2n+1)!}$$

$$\cos x = 1 - \frac{x^2}{2!} + \frac{x^4}{4!} - \cdots = \sum_{n=0}^{\infty} \frac{(-1)^n x^{2n}}{(2n)!}$$

find power series representations for the following integrals:

1. $\int_0^x \sqrt{t} \sin t \, dt$ 2. $\int_0^x \sqrt{t} \cos t \, dt$

(Hint: Write out the terms of the Taylor series and multiply them by $\sqrt{t}$. Then integrate term by term to find the power series.)

Consider a spring mass system with a 16 lb weight attached to a spring with a spring constant of $k = 5$ in an environment with a damping coefficient of $\gamma = 3$. Suppose the mass is at rest and is acted upon by an external force $F(t) = 10e^{-3t}\sqrt{t}$.

3. Write down an initial value problem to model this scenario.

4. Using variation of parameters, find a general solution of this ODE in terms of the integrals in parts 1 and 2.

5. Write down an approximation of the general solution using the first five terms of the power series representations you found in parts 1 and 2.

6. Solve the initial value problem using these approximations.

7. Use your approximate solution to estimate the position of the spring at time $t = \pi$.

Chapter 5

Modeling with Systems of ODEs

The focus of this chapter is the study of *systems* of differential equations. Up to this point we have only solved single differential equations, though a wide variety of them. Here, we seek to examine a system where we have two dependent variables, typically either x and y or x_1 and x_2, both of which are dependent on an independent variable t.

Of course, we could naturally generalize this to systems of higher-order ODEs. For example, we could consider a system of differential equations in x, y, and z. However, for simplicity, we will focus our attention in this chapter to two dimensional systems.

The most general two-dimensional system of ODEs is the system:

$$\begin{cases} x' &= f(x, y) \\ y' &= g(x, y) \end{cases}$$

where $f(x, y)$ and $g(x, y)$ are arbitrary functions of x and y. Throughout this chapter we will focus on systems where these functions are at least continuous. We will also focus primarily on systems for which f and g are linear in x and y. These linear systems in particular are helpful due to the fact that:

- we can actually find explicit, closed form solutions of them;

- the methods we use to understand the behavior of linear systems can guide us when trying to analyze the behavior of fixed points of nonlinear systems.

In the case where the functions f and g are linear functions of x and y, we can rewrite the above system as:

$$\boldsymbol{x}' = A\boldsymbol{x} + \boldsymbol{b}$$

where

$$\boldsymbol{x}' = \begin{pmatrix} x' \\ y' \end{pmatrix}, \quad \boldsymbol{x} = \begin{pmatrix} x \\ y \end{pmatrix}, \quad \boldsymbol{b} = \begin{pmatrix} b_1 \\ b_2 \end{pmatrix}$$

and A is a 2×2 matrix. In the case where $\boldsymbol{b} = \boldsymbol{0}$, we call the linear system a linear homogeneous system, and we can write the system as:

$$\boldsymbol{x}' = A\boldsymbol{x}.$$

DOI: 10.1201/9781003298663-5

To further understand the notation, consider the two-dimensional system of ODEs:

$$\begin{cases} x' & = & 2x & - & y \\ y' & = & -x & + & 2y \end{cases}.$$

Letting $\boldsymbol{x} = \begin{pmatrix} x \\ y \end{pmatrix}$, and $\boldsymbol{x}' = \begin{pmatrix} x' \\ y' \end{pmatrix}$, we can write this system as:

$$\boldsymbol{x}' = \begin{pmatrix} 2 & -1 \\ -1 & 2 \end{pmatrix} \boldsymbol{x}.$$

Notice that the matrix A is simply the coefficients of the system. In general, for linear systems of the form:

$$\begin{cases} x' & = & ax + by \\ y' & = & cx + dy \end{cases}$$

we can write

$$\boldsymbol{x}' = A\boldsymbol{x}$$

where

$$A = \begin{pmatrix} a & b \\ c & d \end{pmatrix}.$$

We will also explore tools to understand the behaviors of nonlinear systems. These tools will be focused on the use of nullclines and stability analysis of equilibria which will give us a qualitative understanding of the behavior of these systems.

5.1   Systems of ODEs and Their Applications

First, we look at a sample of some applications of systems of differential equations. There is a wide variety of biological applications of systems of ODEs. To begin our study of these system we first consider interacting species models, which are nonlinear models. Then we will return to our analysis of circuit diagrams and consider the problem of modeling parallel circuit diagrams. These circuits serve as the basis for understanding the modeling of neurons, which can be quite complex and even include nonlinearity. Finally, we return to mixing problems in the setting in which we have multiple interacting tanks. These produce linear systems of ODEs in a natural way and have a wide variety of possible configurations, leading to a wealth of problems to consider.

5.1.1 Interacting Species

Our first interesting scenario which can be modeled with a system of differential equations is the predator-prey interaction. Of course, species may also exist harmoniously or compete with one another, and we can model those, too!

In general, for two populations, their basic interactions can be modeled by the system:

$$\begin{cases} x' &= ax + bxy \\ y' &= cy + dxy \end{cases}$$

where x is the population of one species and y is the population of the other. The parameters can be interpreted easily. The term a is the growth or decay rate of the x population in the absence of any y, and c is the growth or decay rate of the y population in the absence of any x. The parameters b and d are interaction terms. Whether they are positive or negative indicate whether the relationship is a predator-prey relationship, a competing species relationship, or a symbiotic relationship.

Example: Consider the system:

$$\begin{cases} x' &= 0.06x + xy \\ y' &= -0.3y + 0.5xy. \end{cases}$$

If $y = 0$, then 0.06 is the exponential growth rate of the population x. Conversely, if $x = 0$, then -0.3 is the exponential decay rate of the population y, so the y population will die out in the absence of the x population. The positive parameters 1 and 0.5 of the xy terms in each equation represent positive interactions between the species. In this case, each species has a positive impact on the growth of the other species, and so the species being described by these differential equations have a mutually beneficial, or symbiotic, relationship.

○○

Example: Consider the system of equations:

$$\begin{cases} x' &= 0.2x - 0.001xy \\ y' &= 0.06y - 0.005xy. \end{cases}$$

We can interpret these equations biologically. As $a = 0.2$, the x population will grow exponentially in the absence of y with growth rate 0.2. The y population will grow exponentially with rate 0.06 ($c = 0.06$) in the absence of x. Finally, because $b = -0.001$, the x population is hindered by the presence of y. As $d = -0.005$, the y population also suffers from the presence of x. These two things together indicate a competitive relationship. This could be due to both species competing for the same resources, food, places to live, etc.

○○○

Sometimes this general framework is not sufficient to model the interactions between two species.

Example: Consider the system of differential equations:

$$\begin{cases} x' &= 0.14x - 0.00035x^2 + 0.0003xy \\ y' &= 0.04y + 0.0006xy. \end{cases}$$

The first thing that should stick out to you is the presence of an x^2 term in the x' equation. Notice that we can rewrite this as:

$$\begin{aligned} x' &= 0.14x - 0.00035x^2 + 0.0003xy \\ &= 0.14x\left(1 - \frac{x}{400}\right) + 0.0003xy. \end{aligned}$$

You might recall that this $1 - \frac{x}{400}$ expression indicates that there is a carrying capacity of 400 for the population of x, just as we saw in Chapter 3. As x approaches the carrying capacity, the growth rate for x diminishes until it becomes a decay rate (if x exceeds 400).

Let's consider what the other terms are telling us. With intrinsic growth rate 0.14 and a carrying capacity of 400, the x population will grow in the absence of y, until it reaches the carrying capacity, provided the population starts at a level lower than 400. The y population will grow exponentially with growth rate 0.04 in the absence of x. To see how these two species interact, we examine the coefficients of the xy terms in both equations. Because we have $0.0003xy$ in the x' equation, the x population benefits from the presence of y. Similarly, we have $0.0006xy$ in the y' equation, so the y population also benefits from the presence of x. Both populations benefit from the presence of the other; thus, we say they have a symbiotic or cooperative relationship.

○○

Example: Consider the system:

$$\begin{cases} x' &= 0.05x - 0.0001x^2 - 0.0002xy \\ y' &= 0.01y - 0.000125y^2 + 0.0001xy \end{cases}.$$

Notice that both of these differential equations can be rewritten in the following way:

$$\begin{cases} x' &= 0.05x\left(1 - \dfrac{x}{500}\right) - 0.0002xy \\ y' &= 0.01y\left(1 - \dfrac{y}{80}\right) + 0.0001xy \end{cases}.$$

As in the last example, the two expressions $1 - \dfrac{x}{500}$ and $1 - \dfrac{y}{80}$ indicate that there is a carrying capacity of 500 for the population of x and a carrying capacity of 80 for the population of y, respectively, in the absence of the other species. In other words, in the absence of y, the x population will grow logistically with a carrying capacity of 500, and in the absence of x, the y population will grow logistically with a carrying capacity of 80.

Let's consider the interaction xy terms. Because the interaction term $-0.0002xy$ in the x' equation is negative, this indicates that the x population is diminished by the presence of y. However, the interaction term $0.0001xy$ in the y' equation is positive, indicating that the y population benefits from the presence of x. This is indicative of a predatory-prey relationship with y being the predator and x being the prey.

5.1.2   Parallel RLC Circuits

We have already seen how we can use second-order ODEs to model the behavior of the charge on a capacitor in a RLC series circuit. Now we want to explore the process for modeling a similar RLC circuit with the components in parallel. Here we will focus only on the setup of a general system and will reserve the setup and solving of specific parallel circuit problems for Section 5.3. Consider the diagram below.

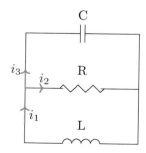

As before, we use Kirkoff's Second Law, which states that the sum of the voltage drops across all elements of a closed loop is zero. For an RLC circuit these voltage drops are:

- Voltage across the resistor $V_R = IR$, where I is current in amperes and R is the resistance in ohms.

- Voltage across the inductor, $V_L = LI'$, where L is the inductance in henries and I' is the derivative of the current.

- Voltage across the capacitor, $V_C = Q/C$, where Q is the charge on the capacitor in coulombs and C is the capacitance in farads.

However, this time we have two loops to consider. We also have multiple currents. Whereas in a series circuit there is only one current to consider, here we have three. As the current into a node must equal the current leaving the node, we see that for our three currents, i_1, i_2, and i_3, we have:

$$i_1 = i_2 + i_3.$$

Across the top loop containing the capacitor and resistor, we sum the voltage drops using Kirkoff's Second Law to get zero. Note that as we sum the voltage drops along the top closed loop, we are considering them in a clockwise orientation. In this case, $V_C = Q/C$ will be positive as the current i_3 also travels clockwise around the top loop, while $V_R = -i_2R$ will be negative as we are traveling against the direction of i_2. Thus, we have the equation:

$$
\begin{aligned}
V_C + V_R &= 0 \\
\frac{Q}{C} - i_2 R &= 0.
\end{aligned}
$$

Recall that $Q' = I$. Then, differentiating this equation with respect to t, and because the charge on the capacitor Q is generated by the current i_3, we have:

$$\frac{i_3}{C} - i_2'R = 0.$$

Note that by using the fact that $i_1 = i_2 + i_3$ and solving for i_2' we have:

$$
\begin{aligned}
\frac{i_3}{C} - i_2'R &= 0 \\
\frac{i_1 - i_2}{C} - i_2'R &= 0 \\
i_2' &= \frac{i_1}{RC} - \frac{i_2}{RC}.
\end{aligned}
$$

Similarly, across the bottom loop we have:

$$V_L + V_R = 0$$

$$Li_1' + i_2R = 0$$

or, rewritten,

$$i_1' = -\frac{i_2 R}{L}.$$

Thus, we can model this parallel circuit with the linear system of equations:

$$\begin{cases} i_1'(t) = & -\dfrac{R}{L}i_2(t) \\ i_2'(t) = \dfrac{1}{RC}i_1(t) & -\dfrac{1}{RC}i_2(t) \end{cases}$$

or

$$x' = \begin{pmatrix} 0 & -\dfrac{R}{L} \\ \dfrac{1}{RC} & -\dfrac{1}{RC} \end{pmatrix} x,$$

where

$$x = \begin{pmatrix} i_1(t) \\ i_2(t) \end{pmatrix}.$$

Here x is a vector which describes the behavior of the current across the inductor and resistor as functions of t. Note that because $i_3 = i_1 - i_2$ we can easily use these to find the current across the capacitor as well.

A similar approach can be used to model more complex circuits, and ones with even more parallel branches and thus more i_k terms, will naturally become higher dimensional systems.

5.1.3 Multiple Tank Mixing Problems

You may recall from Chapter 3 that we can model the amount of dissolved material in solution in a tank using a first-order ODE. At times you may have a situation in which two tanks are present. Furthermore, those tanks might be interacting with each other by each flowing into one another at different rates. An example of this is depicted below.

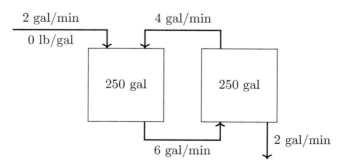

The same methodology from Chapter 3 applies here. For each tank, we want to model the amount of dissolved solids in the tank. Let x_1 be the amount of solids in the first tank and x_2 be the amount of dissolved solids in the second. Then using the principle that:

$$\frac{dx_1}{dt} = R_{in} - R_{out}$$

we can write the system of ODES:

$$\begin{cases} x_1' = (2\,\tfrac{gal}{min})(0\,\tfrac{lb}{gal}) + (4\,\tfrac{gal}{min})\left(\tfrac{x_2}{250}\,\tfrac{lbs}{gal}\right) - (6\,\tfrac{gal}{min})\left(\tfrac{x_1}{250}\,\tfrac{lbs}{gal}\right) \\ x_2' = (6\,\tfrac{gal}{min})\left(\tfrac{x_1}{250}\,\tfrac{lbs}{gal}\right) - (6\,\tfrac{gal}{min})\left(\tfrac{x_2}{250}\,\tfrac{lbs}{gal}\right) \end{cases}$$

or

$$\begin{cases} x_1' = -\tfrac{3}{125}x_1 + \tfrac{2}{125}x_2 \\ x_2' = \tfrac{3}{125}x_1 - \tfrac{3}{125}x_2 \end{cases}$$

which can be represented by the matrix equation:

$$x' = \begin{pmatrix} -\tfrac{3}{125} & \tfrac{2}{125} \\ \tfrac{3}{125} & -\tfrac{3}{125} \end{pmatrix} x.$$

It is worth noting that these multiple tank modeling diagrams are remarkably similar to the compartmental models used to visualize SIR models in Chapter 6. As with parallel RLC circuits, we will reserve the solving of systems of this type for Section 5.3.

5.1 Exercises

1. Consider the system of ODEs:

$$\begin{cases} x' = 0.15x - 0.0012xy \\ y' = -0.05y + 0.001xy \end{cases}.$$

(a) Describe the behavior of the x population in the absence of y.

(b) Describe the behavior of the y population in the absence of x.

(c) Describe what kind of biological relationship exists between the two species.

2. Two separate tanks are used to provide water to different crops. Tank 1 has a capacity of 500 L and tank 2 has a capacity of 1000 L. In order to ensure the crops get an equal mix of nutrients, 5 L/min of water is transferred from tank 1 to tank 2, and 2 L/min is transferred from tank 2 to tank 1. In order to flush the system, pure water is added to tank 1 at a rate of 6 L/min. Effluent is drained from both tank 1 and tank 2 at a rate of 3 L/min.

   (a) Draw a sketch with the tanks and water flow rates labeled to model this situation.

   (b) Construct a linear system of ODEs to describe the amount of dissolved nitrogen in tan 1 (x_1) and tank 2 (x_2) as a function of time.

3. Consider the system of ODEs:

$$\begin{cases} x' &= 0.3x - 0.0008xy \\ y' &= 0.5y - 0.002xy \end{cases}$$

   (a) Describe the behavior of the x population in the absence of y.

   (b) Describe the behavior of the y population in the absence of x.

   (c) Describe what kind of biological relationship exists between the two species.

4. Consider the system of ODEs:

$$\begin{cases} x' &= 0.25x - 0.0004xy \\ y' &= 0.1y - 0.0002y^2 + 0.0002xy \end{cases}$$

   (a) Describe the behavior of the x population in the absence of y.

   (b) Describe the behavior of the y population in the absence of x.

   (c) Describe what kind of biological relationship exists between the two species.

5. Two rain capture tanks are filled with contaminated water. Tank 1 has a capacity of 100 gallons and tank 2 has a capacity of 250 gallons. As part of the normal operation of the tanks, 3 gal/min of water is transferred from tank 1 to tank 2 and 3 gal/min is transferred from tank 2 to tank 1. In order to clean both tanks, pure water is added to tank 1 at a rate of 7 gal/min. Effluent is drained from tank 1 only at a rate of 7 gal/min.

   (a) Draw a sketch with the tanks and water flow rates labeled to model this situation.

   (b) Construct a linear system of ODEs to describe the amount of contaminants in tank 1 (x_1) and tank 2 (x_2) as a function of time.

6. Consider the system of ODEs:

$$\begin{cases} x' &= 0.32x - 0.0064x^2 + 0.005xy \\ y' &= 0.1y - 0.0004y^2 - 0.001xy \end{cases}.$$

   (a) Describe the behavior of the x population in the absence of y.

   (b) Describe the behavior of the y population in the absence of x.

   (c) Describe what kind of biological relationship exists between the two species.

7. Construct a system of ODEs to model the parallel circuit diagram below:

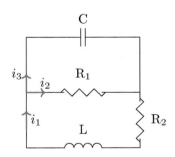

8. Consider the system of ODEs:

$$\begin{cases} x' & = & -0.6x + 0.001xy \\ y' & = & -0.2y + 0.005xy \end{cases}.$$

(a) Describe the behavior of the x population in the absence of y.

(b) Describe the behavior of the y population in the absence of x.

(c) Describe what kind of biological relationship exists between the two species.

9. Construct a system of ODEs to model the parallel circuit diagram below:

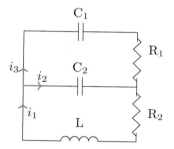

10. Consider the system of ODEs:

$$\begin{cases} x' & = & 0.4x - 0.004x^2 + 0.0008xy \\ y' & = & 0.2y - 0.001xy \end{cases}.$$

(a) Describe the behavior of the x population in the absence of y.

(b) Describe the behavior of the y population in the absence of x.

(c) Describe what kind of biological relationship exists between the two species.

11. Consider the multiple tank model described by the below diagram.

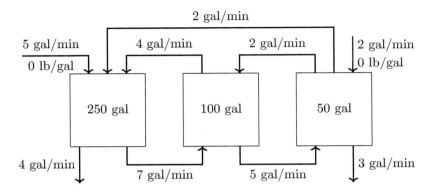

Note that this system has three tanks. Construct a system of ODEs to model this situation. You should have a system of three variables, x_1, x_2, and x_3 which represent the total amount of salt in tanks one, two, and three, respectively.

12. Consider the interaction of two species: species A and species B. Species A has an intrinsic growth rate of 0.5/year while species B has an intrinsic growth rate of 0.3/year. Species A has a carrying capacity of 10,000 while species B has a carrying capacity of 500. Species B benefits from the presence of species A while species A suffers from the presence of species B.

(a) Using positive interaction parameters a, b construct a system of differential equations to model the interaction of these species.

(b) Describe the nature of the interaction between these species (i.e., predator prey, competing species, symbiotic, etc.)

5.2   Stability of Equilibria of Linear Systems Using Eigenvalues

Before finding the general solutions of systems of ODEs, we can use information about their matrix, A, to tell us about the stability of one of their fixed points. Recall that for a first-order ODE, we plotted a phase line where $\frac{dy}{dx} = 0$. In a linear system, a **fixed point** occurs where $\boldsymbol{x}' = \boldsymbol{0}$. This is called a fixed point because in the phase plane, or the graph of x versus y, this occurs at a point $\boldsymbol{x} = (x, y)$. A linear system of ODEs has a fixed point at $\boldsymbol{x}^*$ when

$$\boldsymbol{x}' = A\boldsymbol{x}^* = \boldsymbol{0}.$$

The fixed point can have different behavior. Similar to first order ODEs where phase lines were either stable or unstable, fixed points can be stable or unstable. In systems of two ODEs, the behavior can be classified even further: stable node (also called a sink), stable spiral, center (which is stable), unstable node (also called a source), unstable spiral, and saddle point (which is semistable).

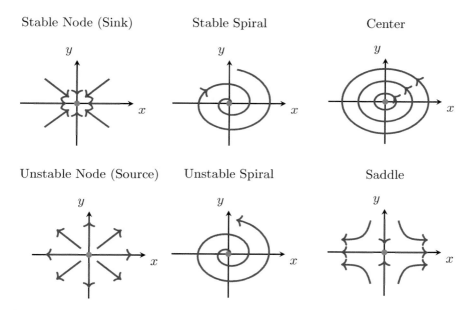

In higher dimensions this classification of fixed points into either node, center, spiral, or saddle is insufficient. However, we will be restricting our focus to two-dimensional systems. Once focus is extended to three dimensions, especially with nonlinearity, we can enter the world of chaos and strange attractors which is beyond the scope of this book.

Note that for a linear system:

$$\boldsymbol{x}' = A\boldsymbol{x},$$

the point $\boldsymbol{x} = \boldsymbol{0}$ at the origin is always a fixed point of the system. This is easy to see, as any matrix multiplied on the right by $\boldsymbol{0}$ is the zero matrix, so $\boldsymbol{x}' = \boldsymbol{0}$.

In order to classify the behavior of the solutions of the linear system near the fixed point at the origin, we will need to introduce some terminology from linear algebra. Associated to every matrix is a set of **eigenvalues** and corresponding **eigenvectors**. These arise naturally. The solution of the equation $Av = \lambda v$ is a vector v such that multiplication by A is the same as multiplication by a scalar λ. This equation is often rewritten as $(A - \lambda I)v = \mathbf{0}$, where I is the identity matrix.

> **Definition:** An **eigenvalue** of a matrix A is a scalar λ such that $Av = \lambda v$ for some non-zero vector v. The vector v which satisfies the above equation with A and λ is the **eigenvector** of A corresponding to λ.

The following theorem from linear algebra is very helpful in finding eigenvalues of a matrix.

> **Theorem:** The equation $Av = \mathbf{0}$ has a non-trivial solution if and only if $\det A = 0$.

For our problem of finding the eigenvalues of a matrix, recall that our equation of interest can be rewritten as $(A - \lambda I)v = \mathbf{0}$. The theorem above then implies that λ is an eigenvalue of A if and only if $\det(A - \lambda I) = 0$.

○○

Example: It will be helpful to return to the system in the introduction to this chapter,

$$x' = \begin{pmatrix} 2 & -1 \\ -1 & 2 \end{pmatrix} x,$$

to illustrate how this process works. In order to find the eigenvalues of A, we need $\det(A - \lambda I) = 0$. Note that:

$$
\begin{aligned}
A - \lambda I &= \begin{pmatrix} 2 & -1 \\ -1 & 2 \end{pmatrix} - \lambda \begin{pmatrix} 1 & 0 \\ 0 & 1 \end{pmatrix} \\
&= \begin{pmatrix} 2 & -1 \\ -1 & 2 \end{pmatrix} - \begin{pmatrix} \lambda & 0 \\ 0 & \lambda \end{pmatrix} \\
&= \begin{pmatrix} 2 - \lambda & -1 \\ -1 & 2 - \lambda \end{pmatrix}.
\end{aligned}
$$

Then we need to solve:

$$
\begin{aligned}
\begin{vmatrix} 2 - \lambda & -1 \\ -1 & 2 - \lambda \end{vmatrix} &= (2 - \lambda)^2 - 1 \\
&= \lambda^2 - 4\lambda + 3 = 0,
\end{aligned}
$$

which has roots $\lambda = 1, 3$. The equation $\lambda^2 - 4\lambda + 3 = 0$ is called the **characteristic equation of the matrix** A. Therefore, the eigenvalues of A are 1 and 3.

○○

A similar process to the one above can be carried out to find the eigenvalues of higher-order matrices. However, due to the difficulty in computing the determinants of larger matrices and of finding roots of higher order characteristic equations, we will restrict our focus to 2×2 matrices.

Example: Consider the matrix:

$$x' = \begin{pmatrix} 5 & -1 \\ 1 & 3 \end{pmatrix} x$$

and find all eigenvalues. We have:

$$\begin{aligned} A - \lambda I &= \begin{pmatrix} 5 & -1 \\ 1 & 3 \end{pmatrix} - \lambda \begin{pmatrix} 1 & 0 \\ 0 & 1 \end{pmatrix} \\ &= \begin{pmatrix} 5 & -1 \\ 1 & 3 \end{pmatrix} - \begin{pmatrix} \lambda & 0 \\ 0 & \lambda \end{pmatrix} \\ &= \begin{pmatrix} 5 - \lambda & -1 \\ 1 & 3 - \lambda \end{pmatrix}. \end{aligned}$$

This has determinant:

$$\begin{aligned} \begin{vmatrix} 5 - \lambda & -1 \\ 1 & 3 - \lambda \end{vmatrix} &= (5 - \lambda)(3 - \lambda) + 1 \\ &= \lambda^2 - 8\lambda + 16. \end{aligned}$$

Hence, the characteristic equation of the matrix A is $\lambda^2 - 8\lambda + 16 = 0$. When simplified to $(\lambda - 4)^2 = 0$, we can see that this equation has a repeated root of $\lambda = 4$. Thus the eigenvalue for the matrix A is $\lambda = 4$.

○○

Now that we are familiar with the eigenvalues of a matrix and how to find them, we can return to the problem at hand: classifying the behavior of a linear system of ODEs at the fixed point at the origin. We will restrict our focus to systems of two ODEs and classify the behavior at the origin in the phase plane.

For a linear homogeneous system of two ODEs, $\boldsymbol{x}' = A\boldsymbol{x}$ with eigenvalues λ_1 and λ_2, the behavior of the equilibrium (or fixed point) at the origin can be classified in the following way:

- **λ_1 and λ_2 are real:**

    - If $\lambda_1 < 0 < \lambda_2$, or in other words if one eigenvalue is positive and one is negative, then the fixed point $(0,0)$ is a saddle.
    - If $\lambda_1 \leq \lambda_2 < 0$, or in other words if both eigenvalues are negative, then the fixed point $(0,0)$ is a stable node.
    - If $\lambda_1 \geq \lambda_2 > 0$, or in other words if both eigenvalues are positive, then the fixed point $(0,0)$ is an unstable node.

- **λ_1 and λ_2 are complex so that $\lambda_1, \lambda_2 = a \pm bi$:**

    - If $a > 0$, then the fixed point $(0,0)$ is an unstable spiral.
    - If $a < 0$, then the fixed point $(0,0)$ is a stable spiral.
    - If $a = 0$, then the fixed point $(0,0)$ is a center.

Example: Continuing with the system $\boldsymbol{x}' = \begin{pmatrix} 2 & -1 \\ -1 & 2 \end{pmatrix} \boldsymbol{x}$, we found the eigenvalues to be $\lambda = 1, 3$. Both λ_1 and λ_2 are real and positive, so we have an unstable node at the origin. Graphing the vector field of the phase plane reflects this finding:

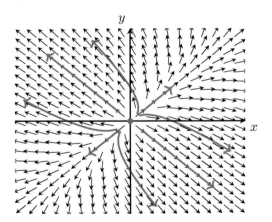

The vector field with a few highlighted solution curves shows that the equilibrium $(0,0)$ is an unstable node because all solutions that start near $(0,0)$ move away from this equilibrium point.

○○○

Let's repeat this process with another linear system.

Example: To classify the fixed point of the system:

$$x' = \begin{pmatrix} 1 & -6 \\ 6 & 1 \end{pmatrix} x,$$

we follow the same procedure as above. Computing $\det(A - \lambda I) = 0$, we find that:

$$\begin{vmatrix} 1-\lambda & -6 \\ 6 & 1-\lambda \end{vmatrix} = (1-\lambda)^2 + 36$$

$$= \lambda^2 - 2\lambda + 37 = 0.$$

Using the quadratic formula, we find that $\lambda = 1 \pm 6i$. As the eigenvalues are complex and the real part is positive ($a = 1$), we find that there is an unstable spiral at the origin.

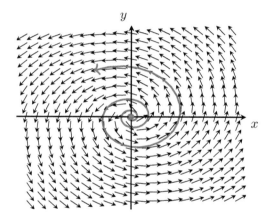

We can also see this unstable spiral by looking at the vector field.

○○○

Classifying the behavior of the fixed point at the origin gives a good qualitative picture of the behavior of the linear system. However, we can use the eigenvalues of the matrix A along with the corresponding eigenvectors to find the solutions of linear systems as well.

5.2 Exercises

For the following questions, determine if the equilibrium $(0,0)$ is a stable/unstable spiral, a stable/unstable node, a saddle, or a center.

1. $\begin{cases} x' = -2x + y \\ y' = 2x - 3y \end{cases}$

2. $\begin{cases} x' = 2x - 13y \\ y' = x - 2y \end{cases}$

3. $x' = \begin{pmatrix} 1 & 4 \\ 0 & 3 \end{pmatrix} x$

4. $x' = \begin{pmatrix} -3 & 5 \\ -1 & 3 \end{pmatrix} x$

5. $\begin{cases} x' = 8x + 3y \\ y' = -3x + 2y \end{cases}$

6. $x' = \begin{pmatrix} 2 & 1 \\ -1 & 2 \end{pmatrix} x$

7. $x' = \begin{pmatrix} -4 & 6 \\ 5 & 3 \end{pmatrix} x$

8. $\begin{cases} x' = -10x - 4y \\ y' = 29x - 2y \end{cases}$

9. $x' = \begin{pmatrix} -1 & -2 \\ 5 & 3 \end{pmatrix} x$

10. $\begin{cases} x' = 2x - 2y \\ y' = 13x \end{cases}$

11. $x' = \begin{pmatrix} 1 & 1 \\ -1 & 3 \end{pmatrix} x$

12. $\begin{cases} x' = 4x - 3y \\ y' = -x + 6y \end{cases}$

13. $\begin{cases} x' = x + 5y \\ y' = 5x + y \end{cases}$

14. $\begin{cases} x' = 8x - 25y \\ y' = 5x - 12y \end{cases}$

15. $x' = \begin{pmatrix} 4 & -2 \\ 1 & 1 \end{pmatrix} x$

16. $\begin{cases} x' = -5x - 4y \\ y' = x - 9y \end{cases}$

17. $\begin{cases} x' = 8x - y \\ y' = 2x + 10y \end{cases}$

18. $x' = \begin{pmatrix} -2 & 7 \\ -4 & 7 \end{pmatrix} x$

5.3   Solutions of Systems of Linear ODEs

In Chapter 3 we discussed methods for finding the solutions of first-order ODEs. One method we could have used for finding the solutions of linear first order ODEs is using the characteristic equation as in Chapter 4. For example, the linear first-order ODE $x' = 2x$ can be written as $x' - 2x = 0$. The characteristic equation for this ODE would be $\lambda - 2 = 0$, so $\lambda = 2$, and $y = Ce^{2t}$ is the general solution of the ODE. This ODE is separable, and were you to have seen it in Chapter 3 you would have arrived at the same general solution by that method. A similar idea will prove fruitful in this section. Rather than finding the roots of the characteristic equation, we will be finding the eigenvalues of the matrix, A, and using those to find the general solution of the linear homogeneous system.

5.3.1   Distinct Eigenvalues

Similar to the approach of second-order ODEs in Chapter 4 where the form of the general solution was dependent on the nature of the roots of the characteristic equation, the general solution of the homogeneous, linear system

$$x' = Ax$$

will be determined by the nature of the eigenvalues of the matrix, A. We are going to restrict our focus to two-dimensional systems, although there is a well developed theory for finding solutions of higher-dimensional linear systems. For linear systems in two dimensions, the eigenvalues can either be real, with differences between distinct and repeated values, or complex. This exactly mirrors the cases for second-order ODEs.

A method for solving linear systems such as $x' = Ax$, is to take a similar characteristic equation approach as in Chapter 4. Here, we will assume that the solution is of the form $x = ve^{\lambda t}$. Then, $x' = \lambda v e^{\lambda t}$, and we have that:

$$x' = Ax$$

$$\lambda v e^{\lambda t} = A v e^{\lambda t}$$

$$-e^{\lambda t}(Av - \lambda v) = 0$$

$$(A - \lambda I)v = 0.$$

Therefore, we need to find a scalar λ such that $(A - \lambda I)v = 0$ has a non-trivial solution. This is exactly the problem of finding the eigenvalues and eigenvectors of A. This leads us to the following result.

> The general solution of a system of linear homogeneous ODEs $\boldsymbol{x}' = A\boldsymbol{x}$ with distinct eigenvalues λ_1 and λ_2 and corresponding eigenvectors $\boldsymbol{u}$ and $\boldsymbol{v}$ is:
>
> $$\boldsymbol{x} = c_1 \boldsymbol{u} e^{\lambda_1 t} + c_2 \boldsymbol{v} e^{\lambda_2 t},$$
>
> where c_1 and c_2 are constants.

Example: In the previous section we found the eigenvalues of the system $\boldsymbol{x}' = \begin{pmatrix} 2 & -1 \\ -1 & 2 \end{pmatrix} \boldsymbol{x}$ to be $\lambda = 1, 3$. Now, let's find the eigenvectors corresponding to these eigenvalues.

The eigenvector corresponding to $\lambda_1 = 1$ is the vector $\boldsymbol{u} = \begin{pmatrix} u_1 \\ u_2 \end{pmatrix}$ such that:

$$
\begin{aligned}
(A - \lambda_1 I)\boldsymbol{u} &= \left(\begin{pmatrix} 2 & -1 \\ -1 & 2 \end{pmatrix} - \begin{pmatrix} 1 & 0 \\ 0 & 1 \end{pmatrix} \right) \boldsymbol{u} \\
&= \begin{pmatrix} 1 & -1 \\ -1 & 1 \end{pmatrix} \boldsymbol{u} \\
&= \begin{pmatrix} u_1 - u_2 \\ -u_1 + u_2 \end{pmatrix} \\
&= \begin{pmatrix} 0 \\ 0 \end{pmatrix}.
\end{aligned}
$$

We may choose $u_1 = 1$, and the first row of the matrix equation then implies that $1 - u_2 = 0$, and thus $u_2 = 1$, so:

$$\boldsymbol{u} = \begin{pmatrix} 1 \\ 1 \end{pmatrix}.$$

Similarly, the eigenvector corresponding to $\lambda_2 = 3$ is the vector $\boldsymbol{v}$ such that:

$$
\begin{aligned}
(A - \lambda_2 I)\boldsymbol{v} &= \left(\begin{pmatrix} 2 & -1 \\ -1 & 2 \end{pmatrix} - 3 \begin{pmatrix} 1 & 0 \\ 0 & 1 \end{pmatrix} \right) \boldsymbol{v} \\
&= \begin{pmatrix} -1 & -1 \\ -1 & -1 \end{pmatrix} \boldsymbol{v} \\
&= \begin{pmatrix} -v_1 - v_2 \\ -v_1 - v_2 \end{pmatrix} \\
&= \begin{pmatrix} 0 \\ 0 \end{pmatrix}.
\end{aligned}
$$

Choosing $v_1 = 1$, the matrix equation implies that $-1 - v_2 = 0$, so $v_2 = -1$, and:

$$\boldsymbol{v} = \begin{pmatrix} 1 \\ -1 \end{pmatrix}.$$

You may wonder why we're able to choose $v_1 = 1$. The reason is twofold.

First, v is the solution of the equation $(A - \lambda I)v = 0$, and you may notice that the vector cv also satisfies this equation for any number c. Intuitively, this is because the vectors v and cv are pointing in the same direction, but have different magnitudes. However, this might make you worry that the choice of v_1 will affect the solution. No matter the choice of v_1, the vector v will only change by a scalar multiple, which will be accounted for in the constant c_2. If we have an initial value problem, we will arrive at the same solution regardless of the choice of v_1 or u_1, as we will see in a following example.

Now, similar to solving second-order ODEs, we have two solutions,

$$x_1 = u e^{\lambda_1 t} = \begin{pmatrix} 1 \\ 1 \end{pmatrix} e^t, \quad x_2 = v e^{\lambda_2 t} = \begin{pmatrix} 1 \\ -1 \end{pmatrix} e^{3t}.$$

To ensure we have a fundamental set of solutions, we can compute the Wronskian,

$$\begin{aligned} W(x_1, x_2) &= \begin{vmatrix} x_1 & x_2 \end{vmatrix} \\ &= \begin{vmatrix} e^t & e^{3t} \\ e^t & -e^{3t} \end{vmatrix} \\ &= -e^{4t} - e^{4t} \\ &= -2e^{4t}, \end{aligned}$$

and because the Wronskian is non-zero, we have a fundamental set of solutions. Thus, the general solution of the system of ODEs is:

$$x = c_1 \begin{pmatrix} 1 \\ 1 \end{pmatrix} e^t + c_2 \begin{pmatrix} 1 \\ -1 \end{pmatrix} e^{3t}.$$

We graphed the vector field for this system in the previous section. We replicate it below with a small change.

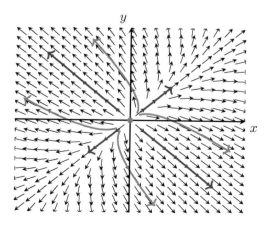

The change is that while the solution plots remain the same, we changed four of them to be blue. You can check, but these correspond with the cases where $c_1 = 0$ and c_2 is either plus or minute one and vice versa. Notice that in these cases, the solution becomes a line pointing exactly in the direction of the eigenvector. Notice that the blue arrows in the first and third quadrants represent the solutions:

$$\boldsymbol{x} = \begin{pmatrix} 1 \\ 1 \end{pmatrix} e^t, \quad \boldsymbol{x} = \begin{pmatrix} -1 \\ -1 \end{pmatrix} e^t$$

while those in the second and fourth quadrants represent the solutions:

$$\boldsymbol{x} = \begin{pmatrix} 1 \\ -1 \end{pmatrix} e^{3t}, \quad \boldsymbol{x} = \begin{pmatrix} -1 \\ 1 \end{pmatrix} e^{3t}.$$

These solutions can be thought of as growth *along the eigenvectors* at a rate proportional to the size of the eigenvalue. Note that the blue lines in the second and fourth quadrants are longer. This is because those solutions are expanding in the direction of those eigenvectors at a greater rate, corresponding to the eigenvector $\lambda_2 = 3$.

All the other possible solution curves are simply linear combinations of these solutions. You can think of them as the building blocks for the solution space as any possible solution of this system of ODEs is a linear combination of them.

○○

You may recall the Wronskian from Chapter 4 having a slightly different definition, where we needed the two solutions and their derivatives to form the fundamental matrix. The definition of Wronskian that we are using here with two vectors is actually equivalent to the Wronskian for solutions of second-order ODEs, and both are used to determine whether or not we have a fundamental set of solutions, even though the components of the matrix used in the Wronskian look different. To understand how these different definitions of Wronskian are related, consider the following example.

Example: In Chapter 4 we found the two solutions of the second order ODE $x'' + 5x' + 6x = 0$ to be $x_1 = e^{-2t}$, and $x_2 = e^{-3t}$, and calculated the Wronskian to be:

$$W(x_1, x_2) = \begin{vmatrix} e^{-2t} & e^{-3t} \\ -2e^{-2t} & -3e^{-3t} \end{vmatrix} = -e^{-5t}.$$

As the Wronskian is non-zero, we concluded that x_1 and x_2 formed a fundamental set of solutions, and that the general solution was $y = c_1 e^{-2t} + c_2 e^{-3t}$.

There is a standard method for converting higher-order ODEs into a system of first-order ODEs. Using our second order ODE example, $x'' + 5x' + 6x = 0$, we will set $y = x'$. Then $y' = x''$, so we have that:

$$x'' + 5x' + 6x = 0$$

$$y' + 5y + 6x = 0$$

$$y' = -6x - 5y.$$

Thus we have the system of ODEs:

$$\begin{cases} x' = y \\ \\ y' = -6x - 5y \end{cases}$$

which can be written as:

$$x' = \begin{pmatrix} 0 & 1 \\ -6 & -5 \end{pmatrix} x.$$

The matrix has characteristic equation $\lambda^2 + 5\lambda + 6 = 0$, and has eigenvalues $\lambda_1 = -2$, and $\lambda_2 = -3$. The corresponding eigenvectors can be computed to be:

$$u = \begin{pmatrix} 1 \\ -2 \end{pmatrix}, \quad v = \begin{pmatrix} 1 \\ -3 \end{pmatrix}.$$

Therefore, we have two solutions,

$$x_1 = \begin{pmatrix} 1 \\ -2 \end{pmatrix} e^{-2t}, \quad \text{and} \quad x_2 = \begin{pmatrix} 1 \\ -3 \end{pmatrix} e^{-3t}.$$

Computing the Wronskian, we find that:

$$W(x_1, x_2) = \begin{vmatrix} e^{-2t} & e^{-3t} \\ -2e^{-2t} & -3e^{-3t} \end{vmatrix} = -e^{-5t}.$$

This Wronskian is identical to the one derived from the second-order ODE. Identically, as the Wronskian is non-zero, the solutions form a fundamental set, and the general solution is:

$$x = c_1 \begin{pmatrix} 1 \\ -2 \end{pmatrix} e^{-2t} + c_2 \begin{pmatrix} 1 \\ -3 \end{pmatrix} e^{-3t}.$$

Furthermore, because $x = \begin{pmatrix} x \\ y \end{pmatrix}$, we can focus our attention on the solution of $x(t)$ and see that $x = c_1 e^{-2t} + c_2 e^{-3t}$, which is the same general solution we obtained in Chapter 4.

As this example demonstrates, the Wronskian as defined in Chapter 4 and the Wronskian as defined in this chapter are, in fact, the same object. Any higher-order ODE, whether second order, third order, etc., can be realized as a system of first order ODEs. The Wronskian in Chapter 4 is simply the Wronskian we obtained here by converting the second order ODE into a system of first order ODEs.

Note that we can reverse the process, and starting with a two dimensional linear system obtain a second order linear ODE.

○○○○○○○○○○○○○○○○○○○○○○○○○○○○○○○○○○○○○○○

Let's examine a system with slightly different behavior.

Example: Consider the system of ODEs:

$$x' = \begin{pmatrix} 1 & 4 \\ 1 & 1 \end{pmatrix} x.$$

We can see that solving $\det(A - \lambda I) = 0$ we have:

$$\begin{vmatrix} 1 - \lambda & 4 \\ 1 & 1 - \lambda \end{vmatrix} = (1 - \lambda)(1 - \lambda) - 4$$

$$= \lambda^2 - 2\lambda - 3$$

$$= (\lambda - 3)(\lambda + 1) = 0.$$

Thus we have eigenvalues $\lambda_{1,2} = 3, -1$. Now, finding the eigenvectors corresponding to these eigenvalues, we see that the eigenvector corresponding to $\lambda_1 = 3$ is the vector $u = \begin{pmatrix} u_1 \\ u_2 \end{pmatrix}$ such that:

$$\begin{aligned} (A - \lambda_1 I)u &= \left(\begin{pmatrix} 1 & 4 \\ 1 & 1 \end{pmatrix} - 3 \begin{pmatrix} 1 & 0 \\ 0 & 1 \end{pmatrix} \right) u \\ &= \begin{pmatrix} -2 & 4 \\ 1 & -2 \end{pmatrix} u \\ &= \begin{pmatrix} -2u_1 + 4u_2 \\ u_1 - 2u_2 \end{pmatrix} \\ &= \begin{pmatrix} 0 \\ 0 \end{pmatrix}. \end{aligned}$$

We may choose $u_1 = 2$, and the second row of the matrix equation then implies that $2 - 2u_2 = 0$, and thus $u_2 = 1$, so:

$$u = \begin{pmatrix} 2 \\ 1 \end{pmatrix}.$$

Note that making our standard choice of $u_1 = 1$ and using the first row of the matrix would have yielded the eigenvector:

$$u = \begin{pmatrix} 1 \\ \frac{1}{2} \end{pmatrix}.$$

This is a constant multiple of the vector we found and also would be an appropriate answer. However, in the interest of having integer eigenvectors when possible, we are choosing $u_1 = 2$ in this case.

Similarly, the eigenvector corresponding to $\lambda_2 = -1$ is the vector v such that:

$$
\begin{aligned}
(A - \lambda_2 I)v &= \left(\begin{pmatrix} 1 & 4 \\ 1 & 1 \end{pmatrix} - (-1) \begin{pmatrix} 1 & 0 \\ 0 & 1 \end{pmatrix} \right) v \\
&= \begin{pmatrix} 2 & 4 \\ 1 & 2 \end{pmatrix} v \\
&= \begin{pmatrix} 2v_1 + 4v_2 \\ v_1 + 2v_2 \end{pmatrix} \\
&= \begin{pmatrix} 0 \\ 0 \end{pmatrix}.
\end{aligned}
$$

Choosing $v_1 = 2$, the matrix equation implies that $2 + 2v_2 = 0$, so $v_2 = -1$, and:

$$v = \begin{pmatrix} 2 \\ -1 \end{pmatrix}$$

is the corresponding eigenvector. Thus, the general solution of this system of ODEs is:

$$x = c_1 \begin{pmatrix} 2 \\ 1 \end{pmatrix} e^{3t} + c_2 \begin{pmatrix} 2 \\ -1 \end{pmatrix} e^{-t}.$$

Let's examine the vector field for this system and observe a few differences. First, note that as we have $\lambda_2 < 0 < \lambda_1$, we should expect to see a saddle-node at the equilibrium point $(0,0)$. How do our eigenvalues and eigenvectors relate to the vector field?

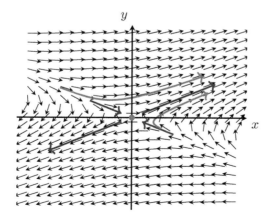

Note that now the blue lines in the second and fourth quadrants which correspond with $\pm v$ have arrows pointing toward the origin. This is due to the fact that the corresponding eigenvalue λ_2 is negative. However, the blue lines in the first and third quadrants which correspond with $\pm u$ have arrows pointing away from the origin as before, due to the fact that λ_1 is positive.

As you can see, the existence of solutions which both converge on the origin and leave the origin results in a vector field with solutions which may at first appear to approach the origin before diverging away. This is the quintessential behavior of saddle nodes.

$\circ\circ\circ\circ\circ\circ\circ\circ\circ\circ\circ\circ\circ\circ\circ\circ\circ\circ\circ\circ\circ\circ\circ\circ\circ\circ\circ\circ\circ\circ\circ\circ\circ\circ\circ\circ\circ\circ\circ\circ$

On occasion you may have a linear system for which you want to find a specific solution, rather than the general one. In these cases you will need initial conditions to set up an initial value problem.

Example: Consider the initial value problem:

$$x' = \begin{pmatrix} -1 & -5 \\ 4 & 8 \end{pmatrix} x, \quad x(0) = \begin{pmatrix} 1 \\ 2 \end{pmatrix}.$$

You'll notice the addition of the initial condition here. This will allow us to find a solution of this system of ODEs, not just the general solution.

Following our steps from before, we can find the characteristic equation for the matrix A by solving the equation $\det(A - \lambda I) = 0$. In this case, we have:

$$\begin{vmatrix} -1 - \lambda & -5 \\ 4 & 8 - \lambda \end{vmatrix} = (-1 - \lambda)(8 - \lambda) + 20$$

$$= \lambda^2 - 7\lambda + 12 = 0.$$

This has roots $\lambda_1 = 4$ and $\lambda_2 = 3$. The eigenvector corresponding to $\lambda_1 = 4$ is the vector u such that:

$$\begin{aligned} (A - \lambda_1 I)u &= \left(\begin{pmatrix} -1 & -5 \\ 4 & 8 \end{pmatrix} - 4 \begin{pmatrix} 1 & 0 \\ 0 & 1 \end{pmatrix} \right) u \\ &= \begin{pmatrix} -5 & -5 \\ 4 & 4 \end{pmatrix} u \\ &= \begin{pmatrix} -5u_1 - 5u_2 \\ 4u_1 + 4u_2 \end{pmatrix} \\ &= \begin{pmatrix} 0 \\ 0 \end{pmatrix}. \end{aligned}$$

Choosing $u_1 = 1$, the matrix equation implies that $-5 - 5u_2 = 0$, so $u_2 = -1$, and:

$$u = \begin{pmatrix} 1 \\ -1 \end{pmatrix}.$$

The eigenvector corresponding to $\lambda_2 = 3$ is the vector $\boldsymbol{v}$ such that:

$$
\begin{aligned}
(A - \lambda_2 I)\boldsymbol{v} &= \left(\begin{pmatrix} -1 & -5 \\ 4 & 8 \end{pmatrix} - 3 \begin{pmatrix} 1 & 0 \\ 0 & 1 \end{pmatrix} \right) \boldsymbol{v} \\
&= \begin{pmatrix} -4 & -5 \\ 4 & 5 \end{pmatrix} \boldsymbol{v} \\
&= \begin{pmatrix} -4v_1 - 5v_2 \\ 4v_1 + 5v_2 \end{pmatrix} \\
&= \begin{pmatrix} 0 \\ 0 \end{pmatrix}.
\end{aligned}
$$

Choosing $v_1 = 1$, the matrix equation implies that $-4 - 5v_2 = 0$, so $v_2 = -\frac{4}{5}$, and:

$$
\boldsymbol{v} = \begin{pmatrix} 1 \\ -\frac{4}{5} \end{pmatrix}.
$$

Therefore, the general solution is:

$$
\boldsymbol{x} = c_1 \begin{pmatrix} 1 \\ -1 \end{pmatrix} e^{4t} + c_2 \begin{pmatrix} 1 \\ -\frac{4}{5} \end{pmatrix} e^{3t}.
$$

Using the initial condition $\boldsymbol{x}(0) = \begin{pmatrix} 1 \\ 2 \end{pmatrix}$, we see that:

$$
\begin{pmatrix} 1 \\ 2 \end{pmatrix} = c_1 \begin{pmatrix} 1 \\ -1 \end{pmatrix} + c_2 \begin{pmatrix} 1 \\ -\frac{4}{5} \end{pmatrix},
$$

so we have the system of equations:

$$
\begin{cases}
1 = c_1 + c_2 \\
2 = -c_1 - \frac{4}{5}c_2.
\end{cases}
$$

Solving this system of equations, we find that $c_1 = -14$ and $c_2 = 15$, so the solution of the initial value problem is:

$$
\begin{aligned}
\boldsymbol{x} &= -14 \begin{pmatrix} 1 \\ -1 \end{pmatrix} e^{4t} + 15 \begin{pmatrix} 1 \\ -\frac{4}{5} \end{pmatrix} e^{3t} \\
&= \begin{pmatrix} -14 \\ 14 \end{pmatrix} e^{4t} + \begin{pmatrix} 15 \\ -12 \end{pmatrix} e^{3t}.
\end{aligned}
$$

Let's return to the comment before about choosing $v_1 = 1$. We could just as easily have chosen $v_1 = 5$, which would give $v_2 = -4$, and general solution:

$$
\boldsymbol{x} = c_1 \begin{pmatrix} 1 \\ -1 \end{pmatrix} e^{4t} + c_2 \begin{pmatrix} 5 \\ -4 \end{pmatrix} e^{3t}.
$$

You'll notice two things. First off, this general solution is different than the one we obtained before, second, this v is just the original vector multiplied by 5. However, if we find the solution of the initial value problem, we will get the same answer. Using the initial condition, we see that:

$$\begin{pmatrix} 1 \\ 2 \end{pmatrix} = c_1 \begin{pmatrix} 1 \\ -1 \end{pmatrix} + c_2 \begin{pmatrix} 5 \\ -4 \end{pmatrix},$$

so we have the system of equations:

$$\begin{cases} 1 = c_1 + 5c_2 \\ \\ 2 = -c_1 - 4c_2. \end{cases}$$

Solving this system of equations, we find that $c_1 = -14$ and $c_2 = 3$, so the solution of the initial value problem is:

$$\begin{aligned} x &= -14 \begin{pmatrix} 1 \\ -1 \end{pmatrix} e^{4t} + 3 \begin{pmatrix} 5 \\ -4 \end{pmatrix} e^{3t} \\ &= \begin{pmatrix} -14 \\ 14 \end{pmatrix} e^{4t} + \begin{pmatrix} 15 \\ -12 \end{pmatrix} e^{3t}. \end{aligned}$$

As you can see, the solution of the initial value problem is independent of our choice of v_1. In general, it is an easier procedure to choose $v_1 = 1$, so that is the process which we will proceed with throughout this section.

5.3.2   Repeated Eigenvalues

Unfortunately, there are some systems for which a fundamental set of solutions cannot be found in this manner. For example, consider the linear system:

$$x' = \begin{pmatrix} 5 & 2 \\ -2 & 1 \end{pmatrix} x.$$

Following the steps from before, we calculate:

$$\begin{aligned} \begin{vmatrix} 5 - \lambda & 2 \\ -2 & 1 - \lambda \end{vmatrix} &= (5 - \lambda)(1 - \lambda) + 4 \\ &= \lambda^2 - 6\lambda + 9 \\ &= (\lambda - 3)^2 = 0. \end{aligned}$$

This has a repeated root $\lambda = 3$. The eigenvector corresponding to $\lambda = 3$ is the vector $\boldsymbol{u}$ such that:

$$
\begin{aligned}
(A - \lambda I)\boldsymbol{u} &= \left(\begin{pmatrix} 5 & 2 \\ -2 & 1 \end{pmatrix} - 3 \begin{pmatrix} 1 & 0 \\ 0 & 1 \end{pmatrix} \right) \boldsymbol{u} \\
&= \begin{pmatrix} 2 & 2 \\ -2 & -2 \end{pmatrix} \boldsymbol{u} \\
&= \begin{pmatrix} 2u_1 + 2u_2 \\ -2u_1 - 2u_2 \end{pmatrix} \\
&= \begin{pmatrix} 0 \\ 0 \end{pmatrix}.
\end{aligned}
$$

Choosing $u_1 = 1$, the matrix equation implies that $2 + 2u_2 = 0$, so $u_2 = -1$, and $\boldsymbol{u} = \begin{pmatrix} 1 \\ -1 \end{pmatrix}$ is the corresponding eigenvector. However, the solution $\boldsymbol{x}_1 = \boldsymbol{u}e^{3t}$ is not itself a fundamental set. We will continue this example in a moment when we have a better idea of how to proceed.

○○

If we have a repeated eigenvalue λ and have already found the corresponding eigenvector $\boldsymbol{u}$, the way forward is to suppose that the second solution is of the form $\boldsymbol{x}_2 = \boldsymbol{u}te^{\lambda t} + \boldsymbol{v}e^{\lambda t}$. However, we must check that this solution will satisfy the equation $\boldsymbol{x}' = A\boldsymbol{x}$. Note that:

$$
\begin{aligned}
\boldsymbol{x}_2 &= \boldsymbol{u}te^{\lambda t} + \boldsymbol{v}e^{\lambda t} \\
\boldsymbol{x}_2' &= \boldsymbol{u}e^{\lambda t} + \lambda \boldsymbol{u}te^{\lambda t} + \lambda \boldsymbol{v}e^{\lambda t}.
\end{aligned}
$$

Then we have:

$$
\begin{aligned}
\boldsymbol{x}_2' &= A\boldsymbol{x}_2 \\
\boldsymbol{u}e^{\lambda t} + \lambda \boldsymbol{u}te^{\lambda t} + \lambda \boldsymbol{v}e^{\lambda t} &= A\left(\boldsymbol{u}te^{\lambda t} + \boldsymbol{v}e^{\lambda t} \right).
\end{aligned}
$$

Grouping together the $te^{\lambda t}$ terms and $e^{\lambda t}$ terms we can see that this equation will be satisfied if:

$$
\lambda \boldsymbol{u}te^{\lambda t} = A\boldsymbol{u}te^{\lambda t} \quad \text{and} \quad \boldsymbol{u}e^{\lambda t} + \lambda \boldsymbol{v}e^{\lambda t} = A\boldsymbol{v}e^{\lambda t}.
$$

Rearranging, we obtain:

$$
A\boldsymbol{u}te^{\lambda t} - \lambda \boldsymbol{u}te^{\lambda t} = 0 \quad \text{and} \quad A\boldsymbol{v}e^{\lambda t} - \lambda \boldsymbol{v}e^{\lambda t} = \boldsymbol{u}e^{\lambda t},
$$

and thus after simplifying, we have:

$$
(A - \lambda I)\boldsymbol{u} = 0 \quad \text{and} \quad (A - \lambda I)\boldsymbol{v} = \boldsymbol{u}.
$$

The first equation we already know to be true for A and u, so our equation will be satisfied if we can find v such that $(A - \lambda I)v = u$. Once we obtain such a v, we can write the general solution as follows.

> The general solution of a system of linear homogeneous ODEs $x' = Ax$ with a repeated eigenvalue λ is:
>
> $$x = c_1 u e^{\lambda t} + c_2 \left[u t e^{\lambda t} + v e^{\lambda t} \right],$$
>
> where u is the eigenvector corresponding to λ and v satisfies:
>
> $$(A - \lambda I)v = u.$$

It is worth noting that such a vector, v, is called a *generalized eigenvalue* of A.

Example: Continuing our previous example, v is the vector such that $(A - \lambda I)v = u$. Note that:

$$
\begin{aligned}
(A - \lambda I)v &= \left(\begin{pmatrix} 5 & 2 \\ -2 & 1 \end{pmatrix} - 3 \begin{pmatrix} 1 & 0 \\ 0 & 1 \end{pmatrix} \right) v \\
&= \begin{pmatrix} 2 & 2 \\ -2 & -2 \end{pmatrix} v
\end{aligned}
$$

and recall that $u = \begin{pmatrix} 1 \\ -1 \end{pmatrix}$. Then v is the vector such that:

$$
\begin{pmatrix} 2 & 2 \\ -2 & -2 \end{pmatrix} v = \begin{pmatrix} 1 \\ -1 \end{pmatrix}.
$$

Choosing $v_1 = 1$, the matrix equation implies that $2 + 2v_2 = 1$, so $v_2 = -\frac{1}{2}$, and:

$$
v = \begin{pmatrix} 1 \\ -\frac{1}{2} \end{pmatrix}.
$$

Then:

$$
\begin{aligned}
x_2 &= u t e^{\lambda t} + v e^{\lambda t} \\
&= \begin{pmatrix} 1 \\ -1 \end{pmatrix} t e^{3t} + \begin{pmatrix} 1 \\ -\frac{1}{2} \end{pmatrix} e^{3t}.
\end{aligned}
$$

It is easy to check that x_1 and x_2 form a fundamental set of solutions, so the general solution of this linear system is:

$$
x = c_1 \begin{pmatrix} 1 \\ -1 \end{pmatrix} e^{3t} + c_2 \left[\begin{pmatrix} 1 \\ -1 \end{pmatrix} t e^{3t} + \begin{pmatrix} 1 \\ -\frac{1}{2} \end{pmatrix} e^{3t} \right],
$$

where c_1 and c_2 are constants.

○○○

This process is more involved than the process for a system with distinct eigenvalues, however it uses the same idea. We simply have two equations to solve, $(A - \lambda I)\boldsymbol{u} = 0$ and $(A - \lambda I)\boldsymbol{v} = \boldsymbol{u}$. Another example may prove helpful.

Example: Consider the initial value problem:

$$\begin{cases} x' &=& 5x &-& y, & x(0) = 0 \\ y' &=& x &+& 3y, & y(0) = 1 \end{cases}.$$

Note that this can be rewritten in matrix form:

$$\boldsymbol{x}' = \begin{pmatrix} 5 & -1 \\ 1 & 3 \end{pmatrix} \boldsymbol{x}, \quad \boldsymbol{x}(0) = \begin{pmatrix} 0 \\ 1 \end{pmatrix}.$$

Any eigenvalues of this matrix, A, will satisfy $(A - \lambda I)\boldsymbol{u} = 0$, so we calculate:

$$\begin{aligned} \begin{vmatrix} 5 - \lambda & -1 \\ 1 & 3 - \lambda \end{vmatrix} &=& (5 - \lambda)(3 - \lambda) + 1 \\ &=& \lambda^2 - 8\lambda + 16 \\ &=& (\lambda - 4)^2 &=& 0. \end{aligned}$$

This quadratic has the repeated root $\lambda = 4$, so we again have a repeated eigenvalue. The eigenvector corresponding to $\lambda = 4$ is the vector $\boldsymbol{u}$ such that:

$$\begin{aligned} (A - \lambda I)\boldsymbol{u} &=& \left(\begin{pmatrix} 5 & -1 \\ 1 & 3 \end{pmatrix} - 4 \begin{pmatrix} 1 & 0 \\ 0 & 1 \end{pmatrix} \right) \boldsymbol{u} \\ &=& \begin{pmatrix} 1 & -1 \\ 1 & -1 \end{pmatrix} \boldsymbol{u} \\ &=& \begin{pmatrix} u_1 - u_2 \\ u_1 - u_2 \end{pmatrix} \\ &=& \begin{pmatrix} 0 \\ 0 \end{pmatrix}, \end{aligned}$$

and we can find that:

$$\boldsymbol{u} = \begin{pmatrix} 1 \\ 1 \end{pmatrix}.$$

Hence, $\boldsymbol{x}_1 = \begin{pmatrix} 1 \\ 1 \end{pmatrix} e^{4t}$ is a solution of the ODE. Now, as we have a repeated eigenvalue, the second solution will be of the form $\boldsymbol{x}_2 = \boldsymbol{u}te^{4t} + \boldsymbol{v}e^{4t}$ where $\boldsymbol{v}$ satisfies the equation $(A - 4I)\boldsymbol{v} = \boldsymbol{u}$, which corresponds to:

$$\begin{pmatrix} 1 & -1 \\ 1 & -1 \end{pmatrix} \boldsymbol{v} = \begin{pmatrix} 1 \\ 1 \end{pmatrix}.$$

Choosing $v_1 = 1$, this implies that $1 - v_2 = 1$, so $v_2 = 0$, and:

$$v = \begin{pmatrix} 1 \\ 0 \end{pmatrix}.$$

Therefore,

$$x_2 = \begin{pmatrix} 1 \\ 1 \end{pmatrix} te^{4t} + \begin{pmatrix} 1 \\ 0 \end{pmatrix} e^{4t}.$$

Therefore, the general solution of this linear system is:

$$\begin{aligned}
x &= c_1 u e^{\lambda t} + c_2 \left[u t e^{\lambda t} + v e^{\lambda t} \right] \\
&= c_1 \begin{pmatrix} 1 \\ 1 \end{pmatrix} e^{4t} + c_2 \left[\begin{pmatrix} 1 \\ 1 \end{pmatrix} te^{4t} + \begin{pmatrix} 1 \\ 0 \end{pmatrix} e^{4t} \right].
\end{aligned}$$

Using the initial value $x(0) = \begin{pmatrix} 0 \\ 1 \end{pmatrix}$, we can see that:

$$\begin{pmatrix} 0 \\ 1 \end{pmatrix} = c_1 \begin{pmatrix} 1 \\ 1 \end{pmatrix} + c_2 \left[\begin{pmatrix} 1 \\ 1 \end{pmatrix} 0 + \begin{pmatrix} 1 \\ 0 \end{pmatrix} \right],$$

or, in other words,

$$\begin{pmatrix} 0 \\ 1 \end{pmatrix} = c_1 \begin{pmatrix} 1 \\ 1 \end{pmatrix} + c_2 \begin{pmatrix} 1 \\ 0 \end{pmatrix}.$$

This results in the system of equations:

$$\begin{cases} 0 &= c_1 + c_2 \\ 1 &= c_1, \end{cases}$$

which has solutions $c_1 = 1$ and $c_2 = -1$. Therefore, the solution of this initial value problem is:

$$\begin{aligned}
x &= 1 \begin{pmatrix} 1 \\ 1 \end{pmatrix} e^{4t} - 1 \left[\begin{pmatrix} 1 \\ 1 \end{pmatrix} te^{4t} + \begin{pmatrix} 1 \\ 0 \end{pmatrix} e^{4t} \right] \\
&= \begin{pmatrix} 1 \\ 1 \end{pmatrix} e^{4t} - \begin{pmatrix} 1 \\ 1 \end{pmatrix} te^{4t} - \begin{pmatrix} 1 \\ 0 \end{pmatrix} e^{4t} \\
&= \begin{pmatrix} 0 \\ 1 \end{pmatrix} e^{4t} - \begin{pmatrix} 1 \\ 1 \end{pmatrix} te^{4t}.
\end{aligned}$$

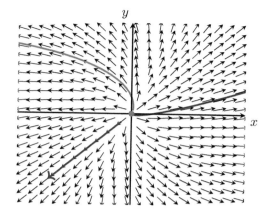

Notice in our graph that the two solutions:

$$\boldsymbol{x}_1 = \begin{pmatrix} 1 \\ 1 \end{pmatrix} e^{4t}, \quad \boldsymbol{x}_2 = \begin{pmatrix} 1 \\ 1 \end{pmatrix} te^{4t} + \begin{pmatrix} 1 \\ 0 \end{pmatrix} e^{4t}$$

are in blue. While $\boldsymbol{x}_2$ is not a line as the eigenvectors in the preceding section, we can still realize any solution of this system of ODEs as a linear combination of these two solutions. In the above IVP we can see that our solution can be realized as $\boldsymbol{x} = \boldsymbol{x}_1 - \boldsymbol{x}_2$.

5.3.3 Complex Eigenvalues

The final case that needs to be addressed is the case when there are complex eigenvalues. For example, consider the linear system:

$$\boldsymbol{x}' = \begin{pmatrix} 2 & 9 \\ -1 & 2 \end{pmatrix} \boldsymbol{x}.$$

Following the same procedure as before, we can find the eigenvalues to be $\lambda = 2 \pm 3i$, with corresponding eigenvectors $\boldsymbol{u} = \begin{pmatrix} -3i \\ 1 \end{pmatrix}$ and $\boldsymbol{v} = \begin{pmatrix} 3i \\ 1 \end{pmatrix}$. You may notice that $\boldsymbol{v} = \overline{\boldsymbol{u}}$ where $\overline{\boldsymbol{u}}$ is the complex conjugate of $\boldsymbol{u}$ (i.e., the complex conjugate of $\boldsymbol{u} = \boldsymbol{p} + \boldsymbol{q}i$ is $\overline{\boldsymbol{u}} = \boldsymbol{p} - \boldsymbol{q}i$). This will hold for any eigenvectors corresponding to complex eigenvalues. We then have two solutions:

$$\boldsymbol{x}_1 = \boldsymbol{u}e^{(2+3i)t}, \quad \boldsymbol{x}_2 = \overline{\boldsymbol{u}}e^{(2-3i)t}.$$

Unfortunately, we have complex solutions! In order to obtain real-valued solutions, we need to borrow a method from Chapter 4, which yields the following result.

The general solution of a system of linear homogeneous ODEs $\boldsymbol{x}' = A\boldsymbol{x}$ with complex eigenvalues $\lambda = a \pm bi$ and corresponding eigenvectors $\boldsymbol{u} = \boldsymbol{p} + \boldsymbol{q}i$ and $\overline{\boldsymbol{u}} = \boldsymbol{p} - \boldsymbol{q}i$ is:

$$\boldsymbol{x} = c_1 \boldsymbol{X}_1 + c_2 \boldsymbol{X}_2,$$

where

$$\boldsymbol{X}_1 = e^{at}\left[\boldsymbol{p}\cos bt - \boldsymbol{q}\sin bt\right], \quad \boldsymbol{X}_2 = e^{at}\left[\boldsymbol{q}\cos bt + \boldsymbol{p}\sin bt\right].$$

To see how this works, let's explore further the method from Chapter 4 that we use here. You may recall that in order to find the general solution of a second-order ODE with complex roots of the characteristic equation, we used Euler's formula and then considered the solutions $y_1 + y_2$ and $y_1 - y_2$. These allowed us to find real-valued solutions, and we were then able to obtain a general solution. The following method is very similar, and will allow us to obtain the real-valued general solution shown above.

First, remember that we are considering the complex solutions $\boldsymbol{x}_1 = \boldsymbol{u}e^{a+bi}$ and $\boldsymbol{x}_2 = \overline{\boldsymbol{u}}e^{a-bi}$. Then the linear combinations $\frac{1}{2}(\boldsymbol{x}_1 + \boldsymbol{x}_2)$ and $-\frac{i}{2}(\boldsymbol{x}_1 - \boldsymbol{x}_2)$ are also solutions of the system, and can be shown to be independent using the Wronskian. Using Euler's formula, we have:

$$\boldsymbol{x}_1 = \boldsymbol{u}e^{at}\left[\cos bt + i\sin bt\right] \quad \text{and} \quad \boldsymbol{x}_2 = \overline{\boldsymbol{u}}e^{at}\left[\cos bt - i\sin bt\right],$$

and writing out these solutions, we have:

$$\frac{1}{2}(\boldsymbol{x}_1 + \boldsymbol{x}_2) = e^{at}\left[\frac{1}{2}(\boldsymbol{u} + \overline{\boldsymbol{u}})\cos bt - \frac{i}{2}(\overline{\boldsymbol{u}} - \boldsymbol{u})\sin bt\right]$$

$$-\frac{i}{2}(\boldsymbol{x}_1 - \boldsymbol{x}_2) = e^{at}\left[\frac{i}{2}(\overline{\boldsymbol{u}} - \boldsymbol{u})\cos bt + \frac{1}{2}(\boldsymbol{u} + \overline{\boldsymbol{u}})\sin bt\right].$$

A fact about complex vectors is that $\frac{1}{2}(\boldsymbol{u} + \overline{\boldsymbol{u}}) = \boldsymbol{p}$ and $\frac{i}{2}(\overline{\boldsymbol{u}} - \boldsymbol{u}) = \boldsymbol{q}$ are both real vectors. In fact, note that $\boldsymbol{p}$ is the real part of $\boldsymbol{u}$ and $\boldsymbol{q}$ is the imaginary part of $\boldsymbol{u}$. Therefore, we can write our solutions as:

$$\boldsymbol{X}_1 = \frac{1}{2}(\boldsymbol{x}_1 + \boldsymbol{x}_2) = e^{at}\left[\boldsymbol{p}\cos bt - \boldsymbol{q}\sin bt\right]$$

$$\boldsymbol{X}_2 = -\frac{i}{2}(\boldsymbol{x}_1 - \boldsymbol{x}_2) = e^{at}\left[\boldsymbol{q}\cos bt + \boldsymbol{p}\sin bt\right].$$

Hence, the general solution is:

$$\boldsymbol{x} = c_1 \boldsymbol{X}_1 + c_2 \boldsymbol{X}_2$$

$$= c_1 e^{at}\left[\boldsymbol{p}\cos bt - \boldsymbol{q}\sin bt\right] + c_2 e^{at}\left[\boldsymbol{q}\cos bt + \boldsymbol{p}\sin bt\right].$$

Example: Returning to our previous example, we had eigenvalues $\lambda = 2 \pm 3i$ and eigenvector $u = \begin{pmatrix} -3i \\ 1 \end{pmatrix} = \begin{pmatrix} 0 - 3i \\ 1 + 0i \end{pmatrix}$, so:

$$p = \begin{pmatrix} 0 \\ 1 \end{pmatrix} \quad \text{and} \quad q = \begin{pmatrix} -3 \\ 0 \end{pmatrix}.$$

Therefore we have:

$$X_1 = e^{2t}\left[\begin{pmatrix} 0 \\ 1 \end{pmatrix}\cos 3t - \begin{pmatrix} -3 \\ 0 \end{pmatrix}\sin 3t\right]$$

and

$$X_2 = e^{2t}\left[\begin{pmatrix} -3 \\ 0 \end{pmatrix}\cos 3t + \begin{pmatrix} 0 \\ 1 \end{pmatrix}\sin 3t\right].$$

Thus, our general solution is:

$$x = c_1 e^{2t}\left[\begin{pmatrix} 0 \\ 1 \end{pmatrix}\cos 3t - \begin{pmatrix} -3 \\ 0 \end{pmatrix}\sin 3t\right]$$
$$+ c_2 e^{2t}\left[\begin{pmatrix} -3 \\ 0 \end{pmatrix}\cos 3t + \begin{pmatrix} 0 \\ 1 \end{pmatrix}\sin 3t\right].$$

Example: Consider the initial value problem:

$$x' = \begin{pmatrix} -1 & 4 \\ -4 & -1 \end{pmatrix}x, \quad x(0) = \begin{pmatrix} 2 \\ 2 \end{pmatrix}.$$

To find the eigenvalues we calculate:

$$\begin{vmatrix} -1-\lambda & 4 \\ -4 & -1-\lambda \end{vmatrix} = (-1-\lambda)(-1-\lambda) + 16$$
$$= \lambda^2 + 2\lambda + 17 = 0,$$

which has roots $\lambda = -1 \pm 4i$. We need to find the eigenvector, u, corresponding to $\lambda = -1 + 4i$. Remember, the second eigenvector will be the complex conjugate of u, so it's alright for us to choose to find the eigenvector corresponding to $-1 + 4i$ rather than $-1 - 4i$. This eigenvector, u, is the vector which satisfies the equation $(A - \lambda I)u = 0$, which corresponds to:

$$\left(\begin{pmatrix} -1 & 4 \\ -4 & -1 \end{pmatrix} - (-1+4i)\begin{pmatrix} 1 & 0 \\ 0 & 1 \end{pmatrix}\right)u = \begin{pmatrix} -4i & 4 \\ -4 & -4i \end{pmatrix}u$$
$$= \begin{pmatrix} -4iu_1 + 4u_2 \\ -4u_1 - 4iu_2 \end{pmatrix}$$
$$= \begin{pmatrix} 0 \\ 0 \end{pmatrix}.$$

Choosing $u_1 = 1$, the matrix equation implies that $-4i + 4u_2 = 0$, so $u_2 = i$. Therefore, we have:

$$\boldsymbol{u} = \begin{pmatrix} 1 \\ i \end{pmatrix} = \begin{pmatrix} 1 \\ 0 \end{pmatrix} + i \begin{pmatrix} 0 \\ 1 \end{pmatrix},$$

so we have that:

$$\boldsymbol{p} = \begin{pmatrix} 1 \\ 0 \end{pmatrix} \quad \text{and} \quad \boldsymbol{q} = \begin{pmatrix} 0 \\ 1 \end{pmatrix}.$$

Therefore, the general solution is:

$$\boldsymbol{x} = c_1 e^{-t} \left[\begin{pmatrix} 1 \\ 0 \end{pmatrix} \cos 4t - \begin{pmatrix} 0 \\ 1 \end{pmatrix} \sin 4t \right] + c_2 e^{-t} \left[\begin{pmatrix} 0 \\ 1 \end{pmatrix} \cos 4t + \begin{pmatrix} 1 \\ 0 \end{pmatrix} \sin 4t \right].$$

Using the initial value, we see that:

$$\begin{pmatrix} 2 \\ 2 \end{pmatrix} = c_1 \begin{pmatrix} 1 \\ 0 \end{pmatrix} + c_2 \begin{pmatrix} 0 \\ 1 \end{pmatrix},$$

so $c_1 = c_2 = 2$. Therefore, the solution of the initial value problem is:

$$\boldsymbol{x} = 2e^{-t} \left[\begin{pmatrix} 1 \\ 0 \end{pmatrix} \cos 4t - \begin{pmatrix} 0 \\ 1 \end{pmatrix} \sin 4t \right] + 2e^{-t} \left[\begin{pmatrix} 0 \\ 1 \end{pmatrix} \cos 4t + \begin{pmatrix} 1 \\ 0 \end{pmatrix} \sin 4t \right],$$

which can be rewritten as:

$$\boldsymbol{x} = \begin{pmatrix} 2 \\ 2 \end{pmatrix} e^{-t} \cos 4t + \begin{pmatrix} 2 \\ -2 \end{pmatrix} e^{-t} \sin 4t.$$

We can see the graph of this solution against the vector field for the system. As the real part of the eigenvalue is negative, the solution converges toward zero. We would categorize the equilibrium point at the origin as a stable spiral in this case.

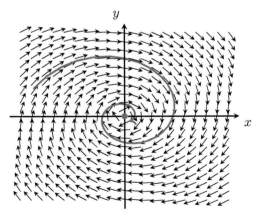

It is worth noting that while the real part of the eigenvalue tells us about the rate at which the solution converges to or diverges from the origin, the imaginary part of the eigenvalue tells us something different. Specifically, it tells us about the rate of oscillation. We could rewrite our solution above as:

$$\boldsymbol{x} = e^{-t} \left(\begin{array}{c} 2\cos 4t + 2\sin 4t \\ 2\cos 4t - 2\sin 4t \end{array} \right).$$

The argument of the cosine and sine functions in our solution is $4t$. That means that the vector:

$$\left(\begin{array}{c} 2\cos 4t + 2\sin 4t \\ 2\cos 4t - 2\sin 4t \end{array} \right)$$

will take the same value for $t = 0, \frac{\pi}{2}, \pi, \frac{3\pi}{2}, \ldots, \frac{k\pi}{2}$ for all integers k. Were the argument to be t rather than $4t$, this would only occur when $t = 2k\pi$, which is a less frequent occurrence.

○○○○○○○○○○○○○○○○○○○○○○○○○○○○○○○○○○○○○○

Given a linear homogeneous system of ODEs, we have seen in this section that the solutions can be determined by finding the eigenvalues and eigenvectors corresponding to the matrix A. These results can be summarized as follows:

The general solution of the linear homogeneous system of two ODEs $\boldsymbol{x}' = A\boldsymbol{x}$ is determined by the nature of the eigenvalues of the matrix A.

- If we have **distinct eigenvalues** λ_1 and λ_2 and corresponding eigenvectors $\boldsymbol{u}$ and $\boldsymbol{v}$, the general solution is:

$$\boldsymbol{x} = c_1 \boldsymbol{u} e^{\lambda_1 t} + c_2 \boldsymbol{v} e^{\lambda_2 t}.$$

- If we have a **repeated eigenvalue** λ, the general solution is:

$$\boldsymbol{x} = c_1 \boldsymbol{u} e^{\lambda t} + c_2 \left[\boldsymbol{u} t e^{\lambda t} + \boldsymbol{v} e^{\lambda t} \right],$$

where $\boldsymbol{u}$ is the eigenvector corresponding to λ and $\boldsymbol{v}$ satisfies $(A - \lambda I)\boldsymbol{v} = \boldsymbol{u}$.

- If we have **complex eigenvalues** $\lambda = a \pm bi$ with corresponding eigenvectors $\boldsymbol{u} = \boldsymbol{p} + \boldsymbol{q}i$ and $\overline{\boldsymbol{u}} = \boldsymbol{p} - \boldsymbol{q}i$, the general solution is:

$$\boldsymbol{x} = c_1 e^{at} \left[\boldsymbol{p} \cos bt - \boldsymbol{q} \sin bt \right] + c_2 e^{at} \left[\boldsymbol{q} \cos bt + \boldsymbol{p} \sin bt \right].$$

Example: Consider the system of ODEs:

$$\boldsymbol{x}' = \left(\begin{array}{cc} 0 & -1 \\ 2 & 0 \end{array} \right) \boldsymbol{x}.$$

This system might seem a bit boring as there's so many zeros in the matrix. However, there's an important behavior that is worth examining here. Following our prior process, we can find that $\lambda = \pm\sqrt{2}i$, so we have purely imaginary eigenvalues. Using $\lambda = \sqrt{2}i$ we have:

$$\left(\begin{pmatrix} 0 & -1 \\ 2 & 0 \end{pmatrix} - (\sqrt{2}i)\begin{pmatrix} 1 & 0 \\ 0 & 1 \end{pmatrix}\right) u = \begin{pmatrix} -\sqrt{2}i & -1 \\ 2 & -\sqrt{2}i \end{pmatrix} u$$

$$= \begin{pmatrix} -\sqrt{2}iu_1 - u_2 \\ 2u_1 - \sqrt{2}iu_2 \end{pmatrix}$$

$$= \begin{pmatrix} 0 \\ 0 \end{pmatrix}.$$

Choosing $u_1 = 1$, this implies that $-\sqrt{2}i - u_2 = 0$, so $u_2 = -\sqrt{2}i$. Thus we have:

$$u = \begin{pmatrix} 1 \\ -\sqrt{2}i \end{pmatrix} = \begin{pmatrix} 1 \\ 0 \end{pmatrix} + i\begin{pmatrix} 0 \\ -\sqrt{2} \end{pmatrix}$$

so we have:

$$p = \begin{pmatrix} 1 \\ 0 \end{pmatrix}, \quad q = \begin{pmatrix} 0 \\ -\sqrt{2} \end{pmatrix}.$$

Thus, our general solution is:

$$x = c_1\left[\begin{pmatrix} 1 \\ 0 \end{pmatrix}\cos\sqrt{2}t - \begin{pmatrix} 0 \\ -\sqrt{2} \end{pmatrix}\sin\sqrt{2}t\right]$$

$$+ c_2\left[\begin{pmatrix} 0 \\ -\sqrt{2} \end{pmatrix}\cos\sqrt{2}t + \begin{pmatrix} 1 \\ 0 \end{pmatrix}\sin\sqrt{2}t\right]$$

or

$$x = \begin{pmatrix} c_1\cos\sqrt{2}t + c_2\sin\sqrt{2}t \\ c_1\sqrt{2}\sin\sqrt{2}t - c_2\sqrt{2}\cos\sqrt{2}t \end{pmatrix}$$

which you might note has no exponential term. As discussed in the previous example, the real part of the eigenvalue tells us about the rate at which the solution converges to or diverges from the origin. As the real part of the eigenvalue here is zero, we should expect that this solution will do neither but just oscillate around the origin.

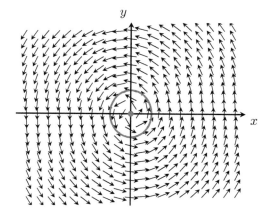

As expected, we see exactly this behavior. Here the origin is a classic example of a center fixed point.

5.3.4 Solutions of Linear Modeling Problems

Let's return to some models we developed in Section 5.1. This will give us a good insight into how finding solutions of linear systems of ODEs can be very helpful. We begin with this model from Section 5.1.3.

Example: Consider the multiple tank model described in the below diagram.

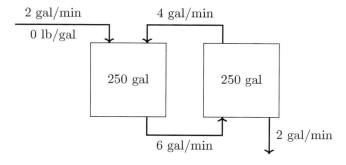

which can be represented by the matrix equation:

$$x' = \begin{pmatrix} -3/125 & 2/125 \\ 3/125 & -3/125 \end{pmatrix} x.$$

as seen in Section 5.1.3. Suppose that the first tank has a concentration of 3 lb/gal of salt while the second tank has a concentration of 10 lb/gal. Then we have the IVP:

$$x' = \begin{pmatrix} -3/125 & 2/125 \\ 3/125 & -3/125 \end{pmatrix} x, \quad x(0) = \begin{pmatrix} 750 \\ 2500 \end{pmatrix}.$$

We find that the matrix A has characteristic equation:

$$\lambda^2 + \frac{6}{125}\lambda + \frac{3}{125^2} = 0$$

which has roots $\lambda = \frac{1}{125}(-3 \pm \sqrt{6})$. The corresponding eigenvectors can be found to be:

$$u = \begin{pmatrix} 1 \\ \frac{\sqrt{6}}{2} \end{pmatrix}, \quad v = \begin{pmatrix} 1 \\ -\frac{\sqrt{6}}{2} \end{pmatrix}$$

respectively. Thus the general solution is:

$$x = c_1 \begin{pmatrix} 1 \\ \frac{\sqrt{6}}{2} \end{pmatrix} e^{(-3+\sqrt{6})t/125} + c_2 \begin{pmatrix} 1 \\ -\frac{\sqrt{6}}{2} \end{pmatrix} e^{(-3-\sqrt{6})t/125}$$

From our initial conditions we have the equations:

$$\begin{pmatrix} 750 \\ 2500 \end{pmatrix} = c_1 \begin{pmatrix} 1 \\ \sqrt{6}/2 \end{pmatrix} + c_2 \begin{pmatrix} 1 \\ -\sqrt{6}/2 \end{pmatrix},$$

or

$$\begin{cases} c_1 + c_2 = 750 \\ \frac{\sqrt{6}}{2}c_1 - \frac{\sqrt{6}}{2}c_2 = 2500 \end{cases}$$

which has solutions $c_1 = 375 + \frac{1250\sqrt{6}}{3}$, $c_2 = 375 - \frac{1250\sqrt{6}}{3}$. Thus we have the solution:

$$\begin{aligned} x = & \left(375 + \frac{1250\sqrt{6}}{3}\right) \begin{pmatrix} 1 \\ \frac{\sqrt{6}}{2} \end{pmatrix} e^{(-3+\sqrt{6})t/125} \\ & + \left(375 - \frac{1250\sqrt{6}}{3}\right) \begin{pmatrix} 1 \\ -\frac{\sqrt{6}}{2} \end{pmatrix} e^{(-3-\sqrt{6})t/125}. \end{aligned}$$

We can observe this solution in the vector field and see that the fixed point is a stable node.

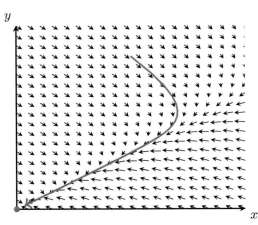

The fact that we have a stable node makes sense in the context of the problem because we are flushing the system with pure water and discharging all the salt in the system in the process.

○○○

Example: Consider the circuit diagram below

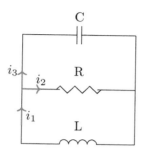

where the resistance is $R = 5\Omega$, the capacitance is $C = 500 \ \mu F$, and the inductance is $L = 10$ mH. Given that the initial currents are $i_1(0) = 2$ amperes and $i_2(0) = 0$ amperes, find a solution for the currents i_1 and i_2.

Note that this is the circuit diagram for which we already found a system of ODEs as a model in Section 5.1.2. As such, we have that:

$$x' = \begin{pmatrix} 0 & -\dfrac{R}{L} \\ \dfrac{1}{RC} & -\dfrac{1}{RC} \end{pmatrix} x, \qquad x(0) = \begin{pmatrix} i_1(0) \\ i_2(0) \end{pmatrix}$$

is a model for this circuit diagram. Substituting in our values, we have the system:

$$x' = \begin{pmatrix} 0 & -500 \\ 400 & -400 \end{pmatrix} x, \qquad x(0) = \begin{pmatrix} 2 \\ 0 \end{pmatrix}$$

which has characteristic equation $\lambda^2 + 400\lambda + 200000 = 0$ with eigenvalues $\lambda = -200 \pm 400i$. The corresponding eigenvector is:

$$u = \begin{pmatrix} 1 + 2i \\ 2 \end{pmatrix}.$$

Thus we have:

$$p = \begin{pmatrix} 1 \\ 2 \end{pmatrix}, \qquad q = \begin{pmatrix} 2 \\ 0 \end{pmatrix}$$

and general solution:

$$x = c_1 e^{-200t} \left[\begin{pmatrix} 1 \\ 2 \end{pmatrix} \cos 400t - \begin{pmatrix} 2 \\ 0 \end{pmatrix} \sin 400t \right]$$
$$+ c_2 e^{-200t} \left[\begin{pmatrix} 2 \\ 0 \end{pmatrix} \cos 400t + \begin{pmatrix} 1 \\ 2 \end{pmatrix} \sin 400t \right].$$

Using our initial conditions we have:

$$\begin{pmatrix} 2 \\ 0 \end{pmatrix} = c_1 \begin{pmatrix} 1 \\ 2 \end{pmatrix} + c_2 \begin{pmatrix} 2 \\ 0 \end{pmatrix}.$$

This gives the system of equations:

$$\begin{cases} c_1 + 2c_2 = 2 \\ 2c_1 = 0 \end{cases}$$

which has solutions $c_1 = 0$, and $c_2 = 1$. Thus, we have the unique solution:

$$x = e^{-200t} \left[\begin{pmatrix} 2 \\ 0 \end{pmatrix} \cos 400t + \begin{pmatrix} 1 \\ 2 \end{pmatrix} \sin 400t \right].$$

Notice that the solution oscillates about the origin with both currents heading to zero in the limit.

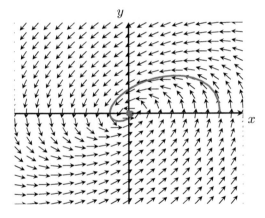

This should match our intuition. We begin with a current on the circuit and as it passes through the resistor over and over it dissipates to zero.

5.3 Exercises

For the following questions, find the general solution of the given system of ODEs.

1. $x' = \begin{pmatrix} -3 & 2 \\ 4 & -1 \end{pmatrix} x$

2. $x' = \begin{pmatrix} 1 & -17 \\ 1 & -7 \end{pmatrix} x$

3. $\begin{cases} x' = -3x + 3y \\ y' = -3x - 9y \end{cases}$

4. $\begin{cases} x' = 5x + 4y \\ y' = 5x + 6y \end{cases}$

5. $x' = \begin{pmatrix} 6 & 2 \\ 3 & 1 \end{pmatrix} x$

6. $x' = \begin{pmatrix} 1 & -2 \\ 3 & -3 \end{pmatrix} x$

7.
$$
\begin{cases}
x' &=& 3x &-& 13y \\
y' &=& 10x &-& 3y
\end{cases}
$$

8.
$$
\begin{cases}
x' &=& 2x &-& 5y \\
y' &=& 17x &-& 20y
\end{cases}
$$

9. $x' = \begin{pmatrix} 1 & -4 \\ 1 & 3 \end{pmatrix} x$

10. $x' = \begin{pmatrix} 2 & 3 \\ 4 & 3 \end{pmatrix} x$

11.
$$
\begin{cases}
x' &=& x &-& 2y \\
y' &=& -2x &+& 4y
\end{cases}
$$

12.
$$
\begin{cases}
x' &=& 4x &+& 2y \\
y' &=& -2x
\end{cases}
$$

For the following questions, find the solution of the given IVP.

13. $x' = \begin{pmatrix} 4 & 1 \\ -2 & 6 \end{pmatrix} x, \quad x(0) = \begin{pmatrix} 2 \\ 1 \end{pmatrix}$

14. $x' = \begin{pmatrix} 10 & -16 \\ 9 & -14 \end{pmatrix} x, \quad x(2) = \begin{pmatrix} 4 \\ 30 \end{pmatrix} e^{-4}$

15.
$$
\begin{cases}
x' &=& 7y, & x(0) = -8 \\
y' &=& 4x + 3y, & y(0) = 14
\end{cases}
$$

16. $x' = \begin{pmatrix} 4 & 2 \\ -2 & 0 \end{pmatrix} x, \quad x(0) = \begin{pmatrix} 4 \\ 7 \end{pmatrix} e^{-4}$

17.
$$
\begin{cases}
x' &=& -20x &-& 8y, & x(\frac{\pi}{2}) = 12e^{-5\pi} \\
y' &=& 13x & & , & y(\frac{\pi}{2}) = -10e^{-5\pi}
\end{cases}
$$

18. Consider the below model for a multi-tank system

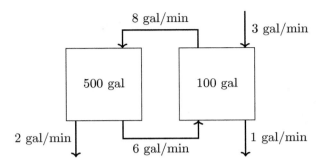

8 gal/min

3 gal/min

500 gal

100 gal

2 gal/min

6 gal/min

1 gal/min

Presuming that the concentration of salt in the first tank is 10 lb/gal and the concentration of salt in the second tank is 5 lb/gal find a system of linear ODEs to model the scenario and find a solution for the amount of salt in each tank as a function of time.

19. Consider the circuit diagram below

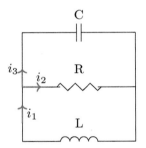

C

i_3 R

i_2

i_1

L

where the resistance is $R = 2\Omega$, the capacitance is $C = 1000 \ \mu F$, and the inductance is $L = 5$ mH. Given that the initial currents are $i_1(0) = 5$ amperes and $i_2(0) = 1$ amperes, find a solution for the currents i_1 and i_2.

20. Consider the below model for a multi-tank system in which a highly concentrated container of chlorinated solution is attached to a pool. It is raining which adds un-chlorinated water at a rate of 10 L/min and the pool is drained at the same rate.

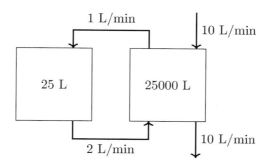

Presuming that the concentration of chlorine in the container is 20 mg/L and the concentration of chlorine in the pool is 0.5 mg/L find a system of linear ODEs to model the scenario and find a solution for the amount of chlorine in each container as a function of time.

21. Consider the circuit diagram below

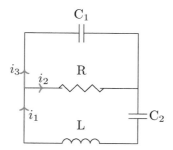

where the resistance is $R = 10\Omega$, the capacitance on C_1 is $C = 500 \ \mu F$, the capacitance on C_2 is $C = 100 \ \mu F$ and the inductance is $L = 10$ mH. Given that the initial currents are $i_1(0) = 10$ amperes and $i_2(0) = 0$ amperes, find a solution for the currents i_1 and i_2.

22.   Consider the below model for a multi-tank system

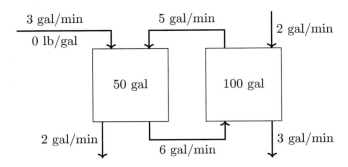

Presuming that the concentration of salt in the first tank is 2 lb/gal and the concentration of salt in the second tank is 30 lb/gal find a system of linear ODEs to model the scenario and find a solution for the amount of salt in each tank as a function of time.

23.   We have seen in this section that any second-order ODE can be rewritten as a system of two linear ODEs. The opposite is also true. For the general linear system of ODES:

$$\boldsymbol{x}' = A\boldsymbol{x}$$

where

$$A = \begin{pmatrix} a & b \\ c & d \end{pmatrix}$$

find a second-order ODE which corresponds with it. (Hint: take the derivative of the equation:

$$x' = ax + by$$

with respect to t then substitute for y' and y to get an equation with solely x terms.)

5.4   Solutions of Linear Nonhomogeneous Systems: Undetermined Coefficients

In the previous section we found a method for determining the solutions of systems of linear homogeneous systems of ODEs. In this section we want to generalize this setting just slightly. We want to find a way to find explicit solutions of some linear systems of the form:

$$\boldsymbol{x}' = A\boldsymbol{x} + \boldsymbol{b}(t)$$

where $b(t)$ is a vector function of t. For example, we may have:

$$b(t) = \begin{pmatrix} 3t + 1 \\ 5 \end{pmatrix}$$

or

$$b(t) = \begin{pmatrix} 5\cos 4t \\ 3\cos 4t \end{pmatrix}.$$

In the case of second order linear ODEs we had two methods for dealing with the nonhomogeneous case: undetermined coefficients and variation of parameters. We will here cover only the generalization of undetermined coefficients to systems of ODEs. This is both because the generalization of variation of parameters requires the concept of inverse matrices, which we have so far not had a need to explore, and because the functions we will encounter in applications can be handled by the method of undetermined coefficients. Notably, for modeling multiple tank mixing problems with an input we will only need a constant function, and for modeling parallel circuits with a voltage source we will again need a constant function or possibly a sinusoidal function to model AC current.

The process mirrors the process for second-order ODEs.

For a linear nonhomogeneous system of ODEs:

$$x' = Ax + b(t),$$

the solution is:

$$x = x_c + x_p,$$

where the complementary solution x_c is the solution of the linear system $x' = Ax$, and the particular solution x_p is a specific solution of the nonhomogeneous equation.

Armed with this fact, we can proceed as we did in Chapter 4 and compile a list of appropriate guesses for x_p.

How to "guess" to find x_p

1. If $b(t)$ = a polynomial of degree n, then y_p = a polynomial of degree n. For example, if

$$b(t) = \begin{pmatrix} 2 \\ 1 \end{pmatrix} x^2 + \begin{pmatrix} 3 \\ 2 \end{pmatrix} x + \begin{pmatrix} -5 \\ 4 \end{pmatrix}$$

then $y_p = Ax^2 + Bx + C$.

2. If $b(t) = u\sin at$ or $b(t) = v\cos at$, then $y_p = A\sin at + B\cos at$.

3. If $b(t) = ue^{at}$ then $y_p = Ae^{at}$

4. If $b(t)$ is a sum or product of these functions then x_p is a sum or product if the guesses for those functions.

5. If our guess for x_p already appears in x_c, add powers of t to x_p. For example, if $b(t) = ae^t$ but $x_c = c_1 ue^t + c_2 ve^{-t}$ then $x_p = Ate^t + Be^t$. Note that if we were truly to generalize the rules from Section 4.5, the Be^t term would not be present. However, for systems of ODEs "adding powers of t" requires us to include the lower powers as well.

Example: Consider the system of ODEs:

$$\begin{cases} x' = 2x + 3y - 7 \\ y' = 4x - 2y + 2. \end{cases}$$

This can be written as:

$$x' = \begin{pmatrix} 2 & 3 \\ 4 & -2 \end{pmatrix} x + \begin{pmatrix} -7 \\ 2 \end{pmatrix}.$$

The homogeneous system:

$$x' = \begin{pmatrix} 2 & 3 \\ 4 & -2 \end{pmatrix} x$$

has eigenvalues $\lambda = 4, -4$ with corresponding eigenvectors $u = (3,2)$ and $v = (1,-2)$. Thus, we have:

$$x_c = c_1 \begin{pmatrix} 3 \\ 2 \end{pmatrix} e^{4t} + c_2 \begin{pmatrix} 1 \\ -2 \end{pmatrix} e^{-4t}.$$

Noticing that $\boldsymbol{b}(t)$ is a constant function we choose:

$$\boldsymbol{x_p} = \begin{pmatrix} a_1 \\ a_2 \end{pmatrix}$$

and have

$$\boldsymbol{x}_p' = \begin{pmatrix} 0 \\ 0 \end{pmatrix}.$$

Substituting into our system we have:

$$\begin{pmatrix} 0 \\ 0 \end{pmatrix} = \begin{pmatrix} 2 & 3 \\ 4 & -2 \end{pmatrix} \begin{pmatrix} a_1 \\ a_2 \end{pmatrix} + \begin{pmatrix} -7 \\ 2 \end{pmatrix}$$

which gives us the system of equations:

$$\begin{cases} 2a_1 + 3a_2 = 7 \\ 4a_1 - 2a_2 = -2 \end{cases}.$$

This has solutions $a_1 = 1/2$ and $a_2 = 2$. Thus we have $\boldsymbol{x_p} = \begin{pmatrix} 1/2 \\ 2 \end{pmatrix}$, and the general solution:

$$\boldsymbol{x} = c_1 \begin{pmatrix} 3 \\ 2 \end{pmatrix} e^{4t} + c_2 \begin{pmatrix} 1 \\ -2 \end{pmatrix} e^{-4t} + \begin{pmatrix} 1/2 \\ 2 \end{pmatrix}.$$

ooooooooooooooooooooooooooooooooooooooooooo

Example: Consider the double tank problem modeled by the below diagram

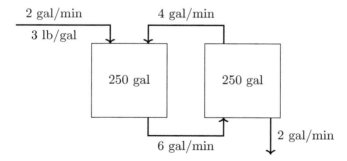

where the first tank has a concentration of 3 lb/gal of salt while the second tank has a concentration of 10 lb/gal.

Note that this is identical to the example in the previous section aside from the fact that the inflow has a nonzero concentration. We can set up the IVP as:

$$\boldsymbol{x}' = \begin{pmatrix} -3/125 & 2/125 \\ 3/125 & -3/125 \end{pmatrix} \boldsymbol{x} + \begin{pmatrix} 6 \\ 0 \end{pmatrix}, \quad \boldsymbol{x}(0) = \begin{pmatrix} 750 \\ 2500 \end{pmatrix}.$$

From the previous section we have the complementary solution:

$$\boldsymbol{x}_c = c_1 \begin{pmatrix} 1 \\ \frac{\sqrt{6}}{2} \end{pmatrix} e^{(-3+\sqrt{6})t/250} + c_2 \begin{pmatrix} 1 \\ -\frac{\sqrt{6}}{2} \end{pmatrix} e^{(-3-\sqrt{6})t/250}.$$

As $\boldsymbol{b}(t)$ is a constant function, we choose:

$$\boldsymbol{x}_p = \begin{pmatrix} a_1 \\ a_2 \end{pmatrix}$$

with

$$\boldsymbol{x}_p' = \begin{pmatrix} 0 \\ 0 \end{pmatrix}.$$

Substituting this into our system we have:

$$\begin{pmatrix} 0 \\ 0 \end{pmatrix} = \begin{pmatrix} -3/125 & 2/125 \\ 3/125 & -3/125 \end{pmatrix} \begin{pmatrix} a_1 \\ a_2 \end{pmatrix} + \begin{pmatrix} 6 \\ 0 \end{pmatrix}$$

which gives us the system of equations:

$$\begin{cases} 3a_1 - 2a_2 &= 750 \\ 3a_1 - 3a_2 &= 0. \end{cases}$$

This has solutions $a_1 = a_2 = 750$, thus giving the solution:

$$\boldsymbol{x} = c_1 \begin{pmatrix} 1 \\ \frac{\sqrt{6}}{2} \end{pmatrix} e^{(-3+\sqrt{6})t/250} + c_2 \begin{pmatrix} 1 \\ -\frac{\sqrt{6}}{2} \end{pmatrix} e^{(-3-\sqrt{6})t/250} + \begin{pmatrix} 750 \\ 750 \end{pmatrix}.$$

Using the initial value we see that:

$$\begin{pmatrix} 750 \\ 2500 \end{pmatrix} = c_1 \begin{pmatrix} 1 \\ \frac{\sqrt{6}}{2} \end{pmatrix} + c_2 \begin{pmatrix} 1 \\ -\frac{\sqrt{6}}{2} \end{pmatrix} + \begin{pmatrix} 750 \\ 750 \end{pmatrix}.$$

This gives us the system of equations:

$$\begin{cases} c_1 + c_2 &= 0 \\ c_1 - c_2 &= \dfrac{3500}{\sqrt{6}} \end{cases}$$

having solutions $c_1 = \dfrac{1750}{\sqrt{6}}$ and $c_2 = -\dfrac{1750}{\sqrt{6}}$. Thus we have unique solution:

$$\boldsymbol{x} = \frac{1750}{\sqrt{6}} \begin{pmatrix} 1 \\ \frac{\sqrt{6}}{2} \end{pmatrix} e^{(-3+\sqrt{6})t/250} - \frac{1750}{\sqrt{6}} \begin{pmatrix} 1 \\ -\frac{\sqrt{6}}{2} \end{pmatrix} e^{(-3-\sqrt{6})t/250} + \begin{pmatrix} 750 \\ 750 \end{pmatrix}$$

or, written a bit more succinctly,

$$x = \begin{pmatrix} 1750/\sqrt{6} \\ 875 \end{pmatrix} e^{(-3+\sqrt{6})t/250} + \begin{pmatrix} -1750/\sqrt{6} \\ 875 \end{pmatrix} e^{(-3-\sqrt{6})t/250} + \begin{pmatrix} 750 \\ 750 \end{pmatrix}.$$

Note that the long term behavior of this system is for both tanks to contain 750 lbs of salt. This converts to a concentration of 3 lb/gal which is the concentration of the inflow. This is as should be expected.

○○○○○○○○○○○○○○○○○○○○○○○○○○○○○○○○○○○○○○

Example: Consider the circuit diagram below where the resistance is $R = 5\Omega$, the capacitance is $C = 500 \ \mu\text{F}$, and the inductance is $L = 10$ mH with an alternating voltage source given by $E(t) = \sin 200tV$ and no initial charges.

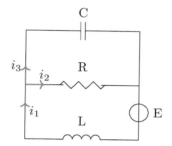

Note that this is the circuit diagram and values for which we already found a system of ODEs as a model in Section 5.3.4. However, we now must include the impressed voltage. For the lower loop, we have that:

$$V_L + V_R = E(t)$$

$$Li_1' + Ri_2 = E(t)$$

$$Li_1' = -Ri_2 + E(t)$$

$$i_1' = -\tfrac{R}{L}i_2 + \tfrac{1}{L}E(t).$$

Substituting in $R = 5\Omega$, $L = 10/1000H$, and $E(t) = \sin 200tV$, we have:

$$i_1' = -500i_2 + 100\sin 200t.$$

The upper loop remains unchanged with:

$$V_C + V_R = 0$$

$$\tfrac{1}{C}i_3 - Ri_2' = 0$$

$$i_2' = \tfrac{1}{CR}i_3.$$

Recall that $i_3 = i_1 - i_2$. Substituting this along with the values $R = 5\Omega$ and $C = 500/10^6 F$, we have that:

$$i_2' = 400i_1 - 400i_2.$$

Hence we have the model for this circuit diagram:

$$x' = \begin{pmatrix} 0 & -500 \\ 400 & -400 \end{pmatrix} x + \begin{pmatrix} 100 \\ 0 \end{pmatrix} \sin 200t, \qquad x(0) = \begin{pmatrix} 0 \\ 0 \end{pmatrix}.$$

From Section 5.3.4 we have that:

$$
\begin{aligned}
x_c &= c_1 e^{-200t} \left[\begin{pmatrix} 1 \\ 2 \end{pmatrix} \cos 400t - \begin{pmatrix} 2 \\ 0 \end{pmatrix} \sin 400t \right] \\
&+ c_2 e^{-200t} \left[\begin{pmatrix} 2 \\ 0 \end{pmatrix} \cos 400t + \begin{pmatrix} 1 \\ 2 \end{pmatrix} \sin 400t \right]
\end{aligned}
$$

is the complementary solution. We choose:

$$x_p = \begin{pmatrix} a_1 \\ a_2 \end{pmatrix} \sin 200t + \begin{pmatrix} b_1 \\ b_2 \end{pmatrix} \cos 200t$$

and have

$$x_p' = \begin{pmatrix} 200a_1 \\ 200a_2 \end{pmatrix} \cos 200t + \begin{pmatrix} -200b_1 \\ -200b_2 \end{pmatrix} \sin 200t.$$

Substituting this into the ODE we have:

$$
\begin{aligned}
&\begin{pmatrix} 200a_1 \\ 200a_2 \end{pmatrix} \cos 200t + \begin{pmatrix} -200b_1 \\ -200b_2 \end{pmatrix} \sin 200t \\
&= \begin{pmatrix} 0 & -500 \\ 400 & -400 \end{pmatrix} \left[\begin{pmatrix} a_1 \\ a_2 \end{pmatrix} \sin 200t + \begin{pmatrix} b_1 \\ b_2 \end{pmatrix} \cos 200t \right] + \begin{pmatrix} 100 \\ 0 \end{pmatrix} \sin 200t.
\end{aligned}
$$

This yields the system of equations:

$$
\begin{cases}
500a_2 & -200b_1 & & = & 100 \\
400a_1 & -400a_2 & +200b_2 & = & 0 \\
200a_1 & & -500b_2 & = & 0 \\
& 200a_2 & -400b_1 & +400b_2 & = & 0
\end{cases}
$$

which can be found to have solutions $a_1 = 1/4$, $a_2 = 3/10$, $b_1 = 1/4$, and $b_2 = 1/10$. Thus, we have general solution:

$$
\begin{aligned}
x &= c_1 e^{-200t} \left[\begin{pmatrix} 1 \\ 2 \end{pmatrix} \cos 400t - \begin{pmatrix} 2 \\ 0 \end{pmatrix} \sin 400t \right] \\
&+ c_2 e^{-200t} \left[\begin{pmatrix} 2 \\ 0 \end{pmatrix} \cos 400t + \begin{pmatrix} 1 \\ 2 \end{pmatrix} \sin 400t \right] \\
&+ \begin{pmatrix} \frac{1}{4} \\ \frac{3}{10} \end{pmatrix} \sin 200t + \begin{pmatrix} \frac{1}{4} \\ \frac{1}{10} \end{pmatrix} \cos 200t
\end{aligned}
$$

Using our initial conditions, we have that:

$$\begin{pmatrix} 0 \\ 0 \end{pmatrix} = c_1 \begin{pmatrix} 1 \\ 2 \end{pmatrix} + c_2 \begin{pmatrix} 2 \\ 0 \end{pmatrix} + \begin{pmatrix} \frac{1}{4} \\ \frac{1}{10} \end{pmatrix}.$$

This yields the system of equations:

$$\begin{cases} c_1 + 2c_2 &= -\frac{1}{4} \\ 2c_1 &= -\frac{1}{10} \end{cases}$$

which has solutions $c_1 = -1/20$ and $c_2 = -1/10$. Thus, we have the unique solution:

$$\begin{aligned} x &= -\tfrac{1}{20}e^{-200t}\left[\begin{pmatrix} 1 \\ 2 \end{pmatrix} \cos 400t - \begin{pmatrix} 2 \\ 0 \end{pmatrix} \sin 400t \right] \\ &\quad -\tfrac{1}{10}e^{-200t}\left[\begin{pmatrix} 2 \\ 0 \end{pmatrix} \cos 400t + \begin{pmatrix} 1 \\ 2 \end{pmatrix} \sin 400t \right] \\ &\quad + \begin{pmatrix} \frac{1}{4} \\ \frac{3}{10} \end{pmatrix} \sin 200t + \begin{pmatrix} \frac{1}{4} \\ \frac{1}{10} \end{pmatrix} \cos 200t \end{aligned}$$

or

$$\begin{aligned} x &= e^{-200t}\left[\begin{pmatrix} -\frac{1}{4} \\ -\frac{1}{10} \end{pmatrix} \cos 400t - \begin{pmatrix} 0 \\ -\frac{1}{5} \end{pmatrix} \sin 400t \right] \\ &\quad + \begin{pmatrix} \frac{1}{4} \\ \frac{3}{10} \end{pmatrix} \sin 200t + \begin{pmatrix} \frac{1}{4} \\ \frac{1}{10} \end{pmatrix} \cos 200t. \end{aligned}$$

5.4 Exercises

For the following questions, find the general solution of the given nonhomogeneous system of ODEs.

1. $x' = \begin{pmatrix} -4 & 3 \\ -3 & 2 \end{pmatrix} x + \begin{pmatrix} 3 \\ 2 \end{pmatrix}$

2. $x' = \begin{pmatrix} -4 & 8 \\ -1 & 2 \end{pmatrix} x + \begin{pmatrix} 1 \\ -3 \end{pmatrix} t$

3. $x' = \begin{pmatrix} 1 & 5 \\ 0 & 2 \end{pmatrix} x + \begin{pmatrix} 5 \\ 6 \end{pmatrix} \sin 2t$

4. $x' = \begin{pmatrix} -2 & 5 \\ -3 & 2 \end{pmatrix} x + \begin{pmatrix} 0 \\ 1 \end{pmatrix} e^{3t}$

5. $x' = \begin{pmatrix} 1 & 4 \\ 5 & 3 \end{pmatrix} x + \begin{pmatrix} 2 \\ -3 \end{pmatrix} e^{2t}$

6. $x' = \begin{pmatrix} 3 & 1 \\ 1 & 1 \end{pmatrix} x + \begin{pmatrix} 0 \\ 1 \end{pmatrix} t^2 + \begin{pmatrix} -3 \\ 2 \end{pmatrix} t$

7. $x' = \begin{pmatrix} 2 & 1 \\ 1 & -1 \end{pmatrix} x + \begin{pmatrix} 1 \\ 0 \end{pmatrix} t^2$

8. $x' = \begin{pmatrix} -1 & 7 \\ 0 & 6 \end{pmatrix} x + \begin{pmatrix} 1 \\ 0 \end{pmatrix} e^{-t}$

9. $x' = \begin{pmatrix} -1 & -5 \\ 1 & 3 \end{pmatrix} x + \begin{pmatrix} 1 \\ 0 \end{pmatrix} e^{2t} + \begin{pmatrix} 2 \\ 1 \end{pmatrix} \sin t$

10. $x' = \begin{pmatrix} -2 & -1 \\ 1 & -4 \end{pmatrix} x + \begin{pmatrix} 2 \\ 1 \end{pmatrix} te^{-3t}$

11. Consider the below model for a multi-tank system

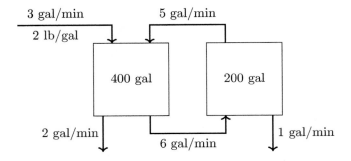

Presuming that the concentration of salt in the first tank is 5 lb/gal and the concentration of salt in the second tank is 0 lb/gal find a system of linear ODEs to model the scenario and find a solution for the amount of salt in each tank as a function of time.

12. Consider the below model for a multi-tank system

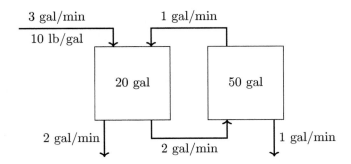

Presuming that the concentration of salt in the first tank is 2 lb/gal and the concentration of salt in the second tank is 10 lb/gal find a system of linear ODEs to model the scenario and find a solution for the amount of salt in each tank as a function of time.

13. Consider the circuit diagram below where the resistance is $R = 10\,\Omega$, the capacitance is $C = 250\ \mu F$, and the inductance is $L = 10$ mH with an alternating voltage source given by $E(t) = 50$ V and no initial charges.

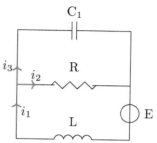

Find a solution of the linear system of ODEs which models this scenario.

14. Consider the circuit diagram below

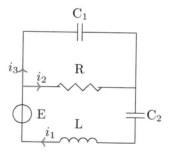

where the resistance is $R = 1\,\Omega$, the capacitance on C_1 is $C = 250\ \mu F$, the capacitance on C_2 is $C = 100\ \mu F$ and the inductance is $L = 10$ mH. Given that the initial currents are $i_1(0) = 0$ amperes and $i_2(0) = 0$ amperes and there is an impressed voltage of $E(t) = 10\sin 20tV$, find a solution for the currents i_1 and i_2.

15.   Consider the below model for a multi-tank system

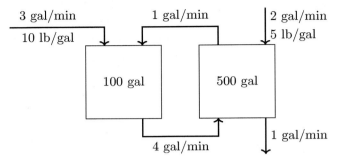

Presuming that the concentration of salt in the first tank is 20 lb/gal and the concentration of salt in the second tank is 1 lb/gal find a system of linear ODEs to model the scenario and find a solution for the amount of salt in each tank as a function of time.

5.5   The Stability Criteria for Linear and Nonlinear ODEs

Recall that during our study of first order ODEs, we found equilibria and determined stability of solutions in two ways: with phase line analysis and by using the Stability Theorem for Autonomous Equations. In the course of this section we will examine the behavior of systems of linear and nonlinear first order ODEs. Similarly as before, we will find equilibria and determine stability of systems of ODEs by using phase plane analysis and by using a stability theorem (here called the Stability Criteria).

In Section 5.2, we focused on determining the stability of the fixed point $(0,0)$. As we were dealing with strictly linear systems of ODEs, this was the only possible fixed point we had to deal with. In this section, we will also consider nonlinear ODEs, and so finding the equilibria will be an important first step. We can no longer guarantee that $(0,0)$ is a fixed point, but rather will identify any fixed points by determining when both differential equations are equal to 0.

ooooooooooooooooooooooooooooooooooooooo

Example: Consider the system of ODEs:

$$\begin{cases} x' &= x(y-1) \\ y' &= (y-2)(x-3) \end{cases}$$

Note that this system of equations is not linear, and so we cannot write it in matrix form. We can find equilibria by identifying where both x' and y' are equal to zero. We know that $x' = x(y - 1) = 0$ when either $x = 0$ or $y = 1$, and $y' = (y - 2)(x - 3) = 0$ when either $x = 3$ or $y = 2$. Therefore, there are equilibrium points at both $(0, 2)$ and $(3, 1)$. Note that these equilibria occur when *both* derivatives are equal to 0, so this is why, for example $(3, 2)$ is *not* an equilibrium.

○○○

Now that we have found equilibrium points, how do we determine stability? We know how to determine stability of the fixed point at zero for a linear system: we use the eigenvalues. However, we don't even have a matrix in the nonlinear context; how can we replicate this approach?

The key idea here is to consider the *local linearization* of the system at the equilibrium points. Our steps will be as follows:

1. Find an equilibrium point of the system.

2. Find the first derivatives of each equation of the system with respect to both x and y.

3. Assemble these derivatives into a matrix called the Jacobian.

4. Find the value of the Jacobian at the equilibrium point.

5. Use this matrix to classify the point.

In effect, we are considering a linear approximation of our system at the equilibrium point in a similar way as considering the linear approximation of a function $f(x)$ by its derivative $f'(x)$ at a point.

Given a system of first order ODEs:

$$\begin{cases} x' & = & f(x, y) \\ y' & = & g(x, y) \end{cases}$$

Stability can be found by computing the **Jacobian**,

$$J(x, y) = \begin{pmatrix} \dfrac{\partial f}{\partial x} & \dfrac{\partial f}{\partial y} \\ \dfrac{\partial g}{\partial x} & \dfrac{\partial g}{\partial y} \end{pmatrix},$$

then evaluating it at the equilibrium points. By checking the trace $\operatorname{tr} J(x, y) = \dfrac{\partial f}{\partial x} + \dfrac{\partial g}{\partial y}$ and determinant $\det J(x, y) = \dfrac{\partial f}{\partial x}\dfrac{\partial g}{\partial y} - \dfrac{\partial g}{\partial x}\dfrac{\partial f}{\partial y}$ of the Jacobian at the equilibrium points, stability can be determined by using the Stability Criteria, which are best visualized by referencing the Stability Diagram.

Stability Diagram

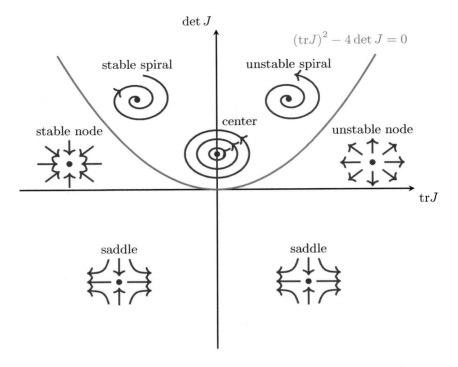

In other words, when det $J < 0$, the equilibrium is a saddle point (unstable). When det $J = 0$, this test is inconclusive. When det $J > 0$, we have several cases which can be seen in the upper half of the Stability Diagram. When tr $J = 0$, the equilibrium is a center; when tr $J > 0$, the equilibrium is unstable; and when tr $J < 0$, the equilibrium is stable. To determine whether the equilibrium is a spiral or a node, we check the criterion $(\text{tr } J)^2 - 4 \det J$. When this quantity is positive, we have a node; when this quantity is negative, we have a spiral.

Example: Let's continue with the system of ODEs:

$$\begin{cases} x' &= x(y-1) \\ y' &= (y-2)(x-3) \end{cases}.$$

We can see that we have Jacobian:

$$J(x,y) = \begin{pmatrix} y-1 & x \\ y-2 & x-3 \end{pmatrix}.$$

For our equilibrium point $(0, 2)$, we have:

$$J(0,2) = \begin{pmatrix} 1 & 0 \\ 0 & -3 \end{pmatrix}.$$

Then $\text{tr} J(0, 2) = 1 - 3 = -2$ and $\det J(0, 2) = 1(-3) - 0(0) = -3$. Therefore, as $\det J(0, 2) < 0$, $(0, 2)$ is a saddle.

Similarly, for the equilibrium at $(3, 1)$, we have:

$$J(3,1) = \begin{pmatrix} 0 & 3 \\ -1 & 0 \end{pmatrix}.$$

Then $\text{tr} J(3, 1) = 0 + 0 = 0$ and $\det J(3, 1) = 0(0) - 3(-1) = 3$. Therefore, because $\det J(3, 1) > 0$ and $\text{tr} J(3, 1) = 0$, $(3, 1)$ is a center.

○○○

Example: For a different system of ODEs, let's say we have equilibrium $(2, 3)$ and that:

$$J(2,3) = \begin{pmatrix} 2 & 3 \\ 1 & 4 \end{pmatrix}.$$

We see that $\det J(2, 3) = 2(4) - 3(1) = 5 > 0$ and $\text{tr } J(2, 3) = 2 + 4 = 6 > 0$. Based on this information, we know that we are in the first quadrant of the Stability Diagram. This means we have either an unstable spiral or an unstable node. To determine which of these cases is correct, we calculate:

$$(\text{tr } J)^2 - 4 \det J = 6^2 - 4(5) = 16 > 0,$$

and thus our equilibrium $(2, 3)$ is an unstable node.

○○○

Example: Consider the system of ODEs:

$$\begin{cases} x' = x - 7y \\ y' = 3x + y \end{cases}.$$

Notice that this system is linear and so if we wanted to find the general solution, we could do so by using the methods in the previous section. However, let's focus our attention on finding the equilibrium and classifying its stability. To find the equilibrium, we set both derivatives equal to 0 and solve. Note that:

$$\begin{cases} x - 7y = 0 \\ 3x + y = 0 \end{cases}$$

only hold for the point $(0,0)$. Hence, $(0,0)$ is our only equilibrium. To determine stability, we must calculate the Jacobian:

$$J(x,y) = \begin{pmatrix} 1 & -7 \\ 3 & 1 \end{pmatrix}.$$

Because the Jacobian has only constant functions, we know $J(0,0)$ and can calculate the determinant and trace. Then $\det J(0,0) = 1(1)-3(-7) = 22 > 0$ and $\operatorname{tr} J(0,0) = 1+1 = 2 > 0$. This tells us that we are in the first quadrant of the Stability Diagram and our equilibrium is unstable. But is it a spiral or node? We must calculate:

$$(\operatorname{tr} J)^2 - 4\det J = 2^2 - 4(22) = -84 < 0,$$

and therefore our equilibrium $(0,0)$ is an unstable spiral.

OOOOOOOOOOOOOOOOOOOOOOOOOOOOOOOOOOOOOOOOO

Example: Let's return to the system of differential equations from Section 5.1:

$$\begin{cases} x' & = & 0.14x - 0.00035x^2 + 0.0003xy \\ \\ y' & = & 0.04y + 0.0006xy. \end{cases}$$

Continuing with this example, let's determine the stability of the equilibrium points for this system. First, notice that:

$$\begin{aligned} y' & = & 0.04y + 0.0006xy. \\ \\ & = & y(0.04 + 0.0006x) \end{aligned}$$

Therefore, stability is obtained when either $y = 0$ or $x = -66.67$. Note that these are the y-nullclines for the system. However, as these equations are modeling the population dynamics of two species, the negative x stability point doesn't make sense and can be discarded.

Therefore, let us consider the x' equation, but with the assumption that $y = 0$, as that's the only possible y value for which stability can be obtained. Then we have:

$$\begin{aligned} x' & = & 0.14x - 0.00035x^2 + 0.0003xy \\ \\ & = & 0.14x - 0.00035x^2 \\ \\ & = & x(0.14 - 0.00035x) \end{aligned}$$

and we can see that stability occurs when either $x = 0$, or $x = 400$, so either the

x population reached its carrying capacity or the population is zero. Therefore, there are two equilibrium points at $(0, 0)$ and $(400, 0)$.

To determine stability, we find that the Jacobian is:

$$J(x, y) = \begin{pmatrix} 0.14 - 0.0007x + 0.0003y & 0.0003x \\ 0.0006y & 0.04 + 0.0006x \end{pmatrix}.$$

Therefore, for $(0, 0)$, we have:

$$J(0, 0) = \begin{pmatrix} 0.14 & 0 \\ 0 & 0.04 \end{pmatrix},$$

which has $\det J(0, 0) = 0.0056$ and $\operatorname{tr} J(0, 0) = 0.18$, both of which are greater than zero, putting us in the first quadrant of the Stability Diagram. Then, we compute:

$$(\operatorname{tr} J)^2 - 4 \det J = (0.18)^2 - 4(0.0056) \doteq 0.01 > 0,$$

therefore $(0, 0)$ is an unstable node. Similarly, for $(400, 0)$ we have:

$$J(400, 0) = \begin{pmatrix} -0.14 & 0.12 \\ 0 & 0.28 \end{pmatrix}$$

which has $\det J(400, 0) = -0.0392 < 0$, meaning we have a saddle point at $(400, 0)$.

5.5 Exercises

For the following ODEs, find all equilibria and classify them as stable/unstable nodes, stable/unstable spirals, centers, or saddles using the Jacobian and Stability Diagram.

1.
$$\begin{cases} x' &= x(y - 6) \\ y' &= x - 3 \end{cases}$$

2.
$$\begin{cases} x' &= y(x - 4) \\ y' &= x - 2 \end{cases}$$

3.
$$\begin{cases} x' &= -x - 3y \\ y' &= 3x + 4y \end{cases}$$

4.
$$\begin{cases} x' &= (x - 1)(y - 5) \\ y' &= (x - 10)(y - 2) \end{cases}$$

5.
$$\begin{cases} x' &= x(x - 4)(y - 8) \\ y' &= (x - 5)(y - 3) \end{cases}$$

6.
$$\begin{cases} x' &= (x - 2)(y - 3) \\ y' &= 2x - y + 3 \end{cases}$$

7.
$$\begin{cases} x' = 2xy \\ y' = (x-3)(y-5) \end{cases}$$

8.
$$\begin{cases} x' = (x-1)(y-1) \\ y' = -x+y+5 \end{cases}$$

9.
$$\begin{cases} x' = x+y-4 \\ y' = x(y-2) \end{cases}$$

10.
$$\begin{cases} x' = y(5-x) \\ y' = x-y \end{cases}$$

11.   Consider the system of ODEs from Exercise 5.1.1:

$$\begin{cases} x' = 0.15x - 0.0012xy \\ y' = -0.05y + 0.001xy \end{cases}$$

Determine the stability and classify any fixed points of this system.

12.   Consider the system of ODEs from Exercise 5.1.3:

$$\begin{cases} x' = 0.3x - 0.0008xy \\ y' = 0.5y - 0.002xy \end{cases}$$

Determine the stability and classify any fixed points of this system.

13.   Consider the system of ODEs from Exercise 5.1.4:

$$\begin{cases} x' = 0.25x - 0.0004xy \\ y' = 0.1y - 0.0002y^2 + 0.0002xy \end{cases}$$

Determine the stability and classify any fixed points of this system.

14.   Consider the system of ODEs from Exercise 5.1.6:

$$\begin{cases} x' = 0.32x - 0.0064x^2 + 0.005xy \\ y' = 0.1y - 0.0004y^2 - 0.001xy \end{cases}$$

Determine the stability and classify any fixed points of this system.

15. Consider the system of ODEs from Exercise 5.1.8:

$$\begin{cases} x' &= -0.6x + 0.001xy \\ y' &= -0.2y + 0.005xy \end{cases}.$$

Determine the stability and classify any fixed points of this system.

16. Consider the system of ODEs from Exercise 5.1.10:

$$\begin{cases} x' &= 0.4x - 0.004x^2 + 0.0008xy \\ y' &= 0.2y - 0.001xy \end{cases}.$$

Determine the stability and classify any fixed points of this system.

5.6 Phase Planes and Nullclines

Beyond simply finding equilibrium points and classifying their stability, you may find yourself wanting a more robust description of the behavior of the solutions of a system of ODEs. In Section 3.2 we used phase line analysis for first order ODEs. Now, we will use phase planes and **nullclines** to obtain qualitative descriptions of the behavior of solutions of systems of ODEs.

> **Definition: Nullclines** are curves where $x' = 0$ (x-nullcline) or $y' = 0$ (y-nullcline).

The best way to visualize possible solutions of a system of ODEs is to generate a full vector field. However, in the absence of suitable technology, nullclines serve as a method to provide a good picture of the possible behavior of solutions.

○○○○○○○○○○○○○○○○○○○○○○○○○○○○○○○○○○○○○○○

Example: Consider the system of ODEs:

$$\begin{cases} x' &= x(y-1) \\ y' &= (y-2)(x-3). \end{cases}$$

We found the equilibrium points to be $(0, 2)$ and $(3, 1)$ in the previous section, but we can further advance our understanding of this system by finding the nullclines.

To find the x-nullclines we set $x' = 0$ and find that $0 = x(y-1)$. Therefore, there are x−nullclines at $x = 0$ and $y = 1$.

To find the y-nullclines we set $y' = 0$ and find that $0 = (y-2)(x-3)$. Therefore, there are y-nullclines at $y = 2$ and $x = 3$.

We can sketch a graph of these nullclines along with the equilibrium values and will then focus on determining stability of the equilibria using information about the nullclines.

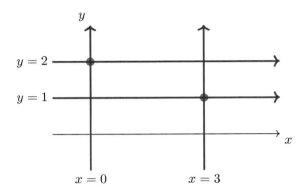

We want to draw arrows (or vectors) on the nullclines indicating which direction a particle on the nullcline would flow. For y-nullclines, we draw horizontal arrows. To determine their direction, we will evaluate x' at the y-nullclines. For example, for the y-nullcline $x = 3$, we have $x' = 3(y-1)$, so $x' > 0$ when $y > 1$ and $x' < 0$ when $y < 1$. In the nullcline graph, this will correspond to right arrows on the line $x = 3$ when $y > 1$ and left arrows on the line $x = 3$ when $y < 1$.

When $y = 2$ (another y-nullcline), $x' = x(2-1) = x$, so $x' > 0$ when $x > 0$ and $x' < 0$ when $x < 0$. In the nullcline graph, this will correspond to right arrows on the line $y = 2$ when $x > 0$ and left arrows on the line $y = 2$ when $x < 0$.

Similarly, we will evaluate y' at the x-nullclines to determine the direction of vertical arrows. For example, we evaluate y' at the x-nullcline $x = 0$ to get $y' = (y-2)(0-3) = -3(y-2)$. So $y' > 0$ when $y < 2$ and $y' < 0$ when $y > 2$. In the nullcline graph, this will correspond to up arrows on the line $x = 0$ when $y < 2$ and down arrows on the line $x = 0$ when $y > 2$.

Finally, for the x-nullcline $y = 1$, we have $y' = (1-2)(x-3) = 3 - x$, so $y' > 0$ when $x < 3$ and $y' < 0$ when $x > 3$. In the nullcline graph, this will correspond to up arrows on the line $y = 1$ when $x < 3$ and down arrows on the line $y = 1$ when $x > 3$.

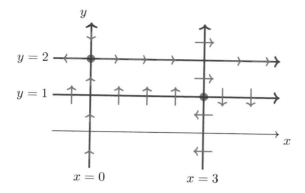

The nullclines and vectors we drew in our graph can give an indication of the stability of the equilibrium. For instance, if we look near the equilibrium $(0, 2)$, we see that some of the vectors are pointing toward the equilibrium, and some are pointing away from the equilibrium. This corresponds to $(0, 2)$ being a saddle point. For the equilibrium $(3, 1)$, we have vectors going in the same direction (clockwise) around the equilibrium. This could mean that $(3, 1)$ is a center, but could also mean that $(3, 1)$ is a spiral. The nullcline graph does not give us any indication of which stability is correct for this equilibrium, so we would have to determine stability using another method, such as the Stability Diagram in the previous section. We could also sketch the vector field and some phase trajectories near the equilibria to get a better idea of the stability in this case.

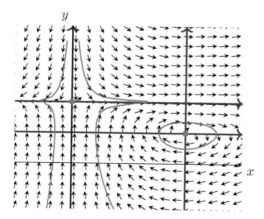

Of course, the vector field gives us a more thorough view of the stability of the equilibria. By looking at the vector field for this example, we can confirm that $(0, 2)$ is a saddle like we originally thought. But more importantly, we can determine the stability of $(3, 1)$, which was inconclusive based on the nullcline graph alone. The vector field shows that $(3, 1)$ is, in fact, a center.

○○

The nullcline graph can be extremely helpful in determining stability quickly, as long as the results are conclusive. Plotting based only off of the nullcline information does not necessarily give you a full picture of what is happening in the phase plane, but is a quick and useful tool that can give an intuitive grasp of the behavior of the system. We can always confirm our results and check stability of equilibria that are inconclusive by other means: the Stability Diagram or the vector field.

Example: To sketch the nullcline graph, and use it to (hopefully) determine the stability of the equilibria for the system of ODEs:

$$\begin{cases} x' & = & (x+2)(x-5) \\ y' & = & xy(y+3) \end{cases},$$

we must first find the equations of the nullclines by setting both the x' and y' equations equal to 0. In this case, the x-nullclines are $x = -2$ and $x = 5$, and the y-nullclines are $x = 0$, $y = 0$, and $y = -3$. Being careful to pair x-nullclines with y-nullclines, we can see that the equilibria of this system are $(-2, 0)$, $(-2, -3)$, $(5, 0)$, and $(5, -3)$. We can then begin to sketch the graph with our nullclines.

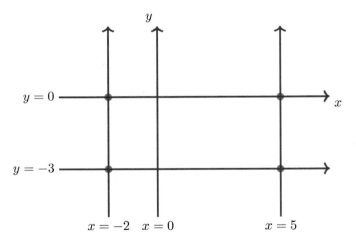

Note that in our nullcline graph, even though there are nullclines intersecting at the points $(0, -3)$ and $(0, 0)$, there are not equilibria in these places because these are intersections of y-nullclines (and not an x-nullcline with a y-nullcline).

We must now determine how to draw the vectors in order to complete our nullcline graph and determine the stability (if possible) of the equilibria. Consider the x-nullclines of $x = -2$ and $x = 5$ and remember that x-nullclines will correspond to vertical vectors. We identify whether these vectors are going up or down based on the y' equation. For example, for the x-nullcline $x = -2$, we have $y' = xy(y+3) = -2y(y+3) > 0$ when $-3 < y < 0$. Similarly, for the x-nullcline $x = 5$, $y' = xy(y+3) = 5y(y+3) > 0$ when $y < -3$ or $y > 0$. Because in these instances we have focused on when $y' > 0$, these inequalities correspond to increasing vectors in our nullcline graph.

We will repeat this process for the y-nullclines, this time substituting into the x' equation and using $x' > 0$ to signal the drawing of vectors pointing to the right. For the y-nullcline $x = 0$, we have $x' = (x+2)(x-5) = -10 < 0$, so we will always draw vectors pointing to the left on the nullcline $x = 0$. For our other two y-nullclines ($y = 0$ and $y = -3$), we have $x' = (x+2)(x-5) > 0$ when $x < -2$ or $x > 5$. Both of these nullclines have the same result because the x' equation is autonomous (a function of x alone). Now we may proceed in sketching these vectors onto our nullcline graph.

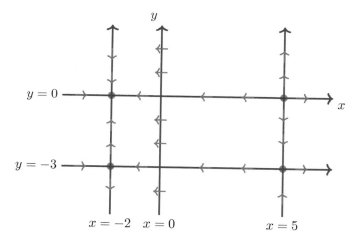

By examining our nullcline graph, we can see that $(-2, 0)$ is a stable node because all nearby vectors are pointing toward that equilibrium point. There is a saddle at $(-2, -3)$ because some vectors are going toward and some vectors are going away from $(-2, -3)$. At $(5, 0)$, we have an unstable node because all nearby vectors are pointing away from that equilibrium. Finally, $(5, -3)$ is a saddle. We have successfully identified the stability of all four equilibria just by using the nullcline graph – no Stability Diagram or vector field needed!

○○

Example: Consider the system of ODEs:

$$\begin{cases} x' &= y^2 - x \\ y' &= (x-2)(y-3) \end{cases}.$$

To find the nullclines for this system, we once again set both equations equal to 0. In this case, y-nullclines include $x = 2$ and $y = 3$. For the x-nullclines, notice that $0 = y^2 - x$, and so the x-nullclines are now curves in the plane $y = \pm\sqrt{x}$. To find the equilibria for this system, we look for intersections between x- and y-nullclines. These intersections occur at the equilibria $(9, 3)$, $(2, \sqrt{2})$, and $(2, -\sqrt{2})$. We can sketch the nullclines and equilibria and will then focus on determining the vectors that will complete the nullcline graph.

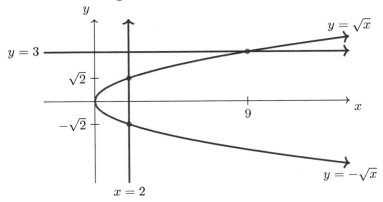

Now we can determine the behavior of solutions near the equilibria by finding the direction vectors. For the y-nullcline $x = 2$, note that $x' = y^2 - 2 > 0$ if $y < -\sqrt{2}$ or $y > \sqrt{2}$. Similarly, for the y-nullcline $y = 3$, we have $x' = 9 - x > 0$ when $x < 9$. Because these are y-nullclines and we've determined when $x' > 0$, these intervals correspond to right arrows in the nullcline graph.

For the x-nullcline $y = \sqrt{x}$, we have $y' = (x - 2)(\sqrt{x} - 3) > 0$ when $0 \le x < 2$ or $x > 9$. Similarly, for the x-nullcline $y = -\sqrt{x}$, we have $y' = (x - 2)(-\sqrt{x} - 3) > 0$ when $0 \le x < 2$. Note that neither x-nullcline is defined when $x < 0$. Because these are x-nullclines and we've determined when $y' > 0$, these intervals correspond to up arrow in the nullcline graph.

Finally, we can sketch the phase plane using the nullclines and direction arrows, which will give us an idea of the stability of the equilibria.

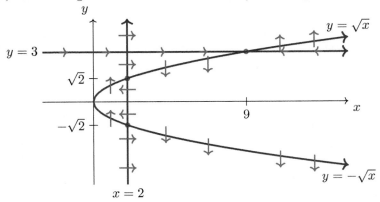

Based on our nullcline graph, we can make the following conclusions. The equilibria $(9, 3)$ and $(2, -\sqrt{2})$ are saddles. Of course, we can confirm this by evaluating the Jacobian and using the Stability Diagram. For this system of ODEs,

$$J(x, y) = \begin{pmatrix} -1 & 2y \\ y - 3 & x - 2 \end{pmatrix}.$$

Then for the equilibrium $(9, 3)$, we have:

$$J(9, 3) = \begin{pmatrix} -1 & 6 \\ 0 & 7 \end{pmatrix},$$

which has $\det J(9, 3) = -7 < 0$. This confirms our nullcline graph that $(9, 3)$ is a saddle point.

Similarly, we have:

$$J(2, -\sqrt{2}) = \begin{pmatrix} -1 & -2\sqrt{2} \\ -\sqrt{2} - 3 & 0 \end{pmatrix},$$

which has $\det J(2, -\sqrt{2}) = -2\sqrt{2}(\sqrt{2}+3) < 0$. Thus, $(2, -\sqrt{2})$ is also a saddle, just as we thought.

Unfortunately, the nullcline graph is a little less helpful when it comes to the equilibrium $(2, \sqrt{2})$. We can see that the direction vectors consistently point in a clockwise direction around the equilibrium point, but this could indicate a center or a stable/unstable spiral. We want to make the correct classification and the nullcline graph does not give us enough information to do this. We must determine stability for this equilibrium a different way, and the Jacobian and Stability Diagram can typically provide us with an answer in this instance. For this equilibrium, we have:

$$J(2, \sqrt{2}) = \begin{pmatrix} -1 & 2\sqrt{2} \\ \sqrt{2} - 3 & 0 \end{pmatrix},$$

which has $\det J(2, \sqrt{2}) = -2\sqrt{2}(\sqrt{2} - 3) > 0$ and $\operatorname{tr} J(2, \sqrt{2}) = -1 < 0$. Thus our equilibrium is in the second quadrant of the Stability Diagram and must be stable. To determine whether it is a node or spiral, we can use the Stability Criteria. Note that we actually already know this based on the nullcline graph, but we will confirm our result using the Stability Criteria for completeness. We have $(\operatorname{tr} J)^2 - 4 \det J = (-1)^2 - 4(-2\sqrt{2}(\sqrt{2} - 3)) < 0$. Hence, the equilibrium $(2, \sqrt{2})$ is a stable spiral.

○○

Example: Let's return once more to the system of differential equations from Sections 5.1 and 5.5:

$$\begin{cases} x' &= 0.14x - 0.00035x^2 + 0.0003xy \\ y' &= 0.04y + 0.0006xy. \end{cases}$$

We previously found that there are two equilibrium points at $(0,0)$ and $(400,0)$. We determined that $(0,0)$ is an unstable node and $(400,0)$ is a saddle point.

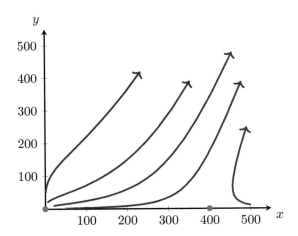

In order to sketch these phase trajectories by hand, without graphing nullclines or using a computer to generate the vector field, we use our information about the stability of the fixed points to give us an idea of the shape of trajectories. To begin, we can focus our attention on the first quadrant as x and y represent population. To draw a trajectory beginning near $(100,10)$, we know that the origin is an unstable node, so the trajectory should head away from the origin, toward the right and slightly up. However, once the trajectory nears the saddle point at $(400,0)$, it is pushed upward and away from the x-axis.

Notice that the trajectories are drawn so that the x value of the trajectory exceeds the carrying capacity of x at 400. This is because the carrying capacity is only for x *in the absence* of y. Because the x population benefits from the presence of y, the interaction term allows for the x population to grow past the carrying capacity.

Now, consider a trajectory beginning at $(600,10)$. This is to the right of the saddle point, and because the trajectories in the x direction are heading toward the saddle (which we know because the origin is a source), the trajectory should head to the left toward the saddle. However, once the trajectory nears the saddle, it will be pushed away from the x-axis and up. As phase trajectories cannot cross, we draw this trajectory to follow roughly parallel to the trajectory starting at $(100,10)$.

For one last trajectory, begin at the point $(10,100)$. As the origin is an unstable node, the trajectory is pushed up and slightly to the right. As there is no other fixed point but the saddle, this trajectory will tend toward the direction of the saddle, but continue upwards.

The biological interpretation for these trajectories is that both the population of x and y will grow without bound. However, if the x population begins too much larger than the carrying capacity without a sufficient amount of y population to sustain it, some of the x population will die off, with the population heading toward the carrying capacity 400, before growing again once the y population increases enough to sustain a larger x population.

○○

Example: Returning to the example from Section 5.1,

$$\begin{cases} x' &= 0.05x - 0.0001x^2 - 0.0002xy \\ y' &= 0.01y - 0.000125y^2 + 0.0001xy, \end{cases}$$

we can determine the stability of the equilibrium points for this system by rewriting the system in the following way:

$$\begin{cases} x' &= x(0.05 - 0.0001x - 0.0002y) \\ y' &= y(0.01 - 0.000125y + 0.0001x) \end{cases}.$$

By writing the system in this way, it's clear to see that $x' = 0$ if $x = 0$ or $0.5 - 0.0001x - 0.0002y = 0$ (the x-nullclines), and $y' = 0$ if $y = 0$ or $0.01 - 0.000125y + 0.0001x = 0$ (the y-nullclines).

If both $x = 0$ and $y = 0$, we have fixed point $(0,0)$. If both $x = 0$ and $0.01 - 0.000125y + 0.0001x = 0$, then $y = 80$, so we have the fixed point $(0, 80)$. Similarly, if $y = 0$ and $0.05 - 0.0001x - 0.0002y = 0$, then $x = 500$, so we have the fixed point $(500, 0)$. The final possibility is that both of the equations $0.05 - 0.0001x - 0.0002y = 0$ and $0.01 - 0.000125y + 0.0001x = 0$. This corresponds to finding the intersection point of the two lines $y = \frac{0.05 - 0.0001x}{0.0002}$ and $y = \frac{0.01 + 0.0001x}{0.000125}$, which occurs at the point $(\frac{1700}{13}, \frac{2400}{13})$.

These fixed points correspond to both populations dying out at $(0,0)$, x dying out and y reaching its carrying capacity at $(0, 80)$, y dying out and x reaching its carrying capacity at $(500, 0)$, and an equilibrium point at $(\frac{1700}{13}, \frac{2400}{13}) \approx (130.765, 184.615)$ at which the populations coexist. For this last equilibrium, we will continue to use the fractional form so that we do not obtain any rounding error in our stability calculations. What remains to be seen is the type (saddle, node, spiral, center) of each fixed point and whether each is stable or unstable.

To determine stability, we find that the Jacobian is:

$$J(x, y) = \begin{pmatrix} 0.05 - 0.0002x - 0.0002y & -0.0002x \\ 0.0001y & 0.01 - 0.00025y + 0.0001x \end{pmatrix}.$$

Therefore, for $(0,0)$, we have:

$$J(0,0) = \begin{pmatrix} 0.05 & 0 \\ 0 & 0.01 \end{pmatrix},$$

which has $\det J(0,0) = 0.0005$ and $\operatorname{tr} J(0,0) = 0.06$. As the determinant is positive, we compute $(\operatorname{tr} J)^2 - 4\det J = 0.0016 > 0$ and see that $(0,0)$ is an unstable node.

Next, for $(0,80)$, we have:

$$J(0,80) = \begin{pmatrix} 0.034 & 0 \\ 0.008 & -0.01 \end{pmatrix},$$

which has $\det J(0,80) = -0.00034 < 0$, so $(0,80)$ is a saddle point.

Similarly, for $(500,0)$, we have:

$$J(500,0) = \begin{pmatrix} -0.05 & -0.1 \\ 0 & 0.06 \end{pmatrix},$$

which has $\det J(500,0) = -0.003 < 0$, so $(500,0)$ is also a saddle point.

Finally, for $(\frac{1700}{13}, \frac{2400}{13})$, we have:

$$J(\tfrac{1700}{13}, \tfrac{2400}{13}) = \begin{pmatrix} -\frac{17}{1300} & -\frac{17}{650} \\ \frac{6}{325} & -\frac{3}{130} \end{pmatrix},$$

which has $\det J(\frac{1700}{13}, \frac{2400}{13}) = \frac{51}{65000} > 0$ and $\operatorname{tr} J(\frac{1700}{13}, \frac{2400}{13}) = -\frac{47}{1300} < 0$. Because the determinant is positive, we compute:

$$(\operatorname{tr} J)^2 - 4\det J = -\frac{619}{338000} < 0$$

and see that $(\frac{1700}{13}, \frac{2400}{13}) \approx (130.765, 184.615)$ is a stable spiral.

We can interpret these results biologically. As the only stable equilibrium is the spiral at approximately $(130.765, 184.615)$, no matter the initial populations of x and y, the species will eventually stabilize with the population of x being around 130.765 and the population of y being around 184.615, and these populations will coexist at these levels. We can also sketch some phase trajectories to illustrate this behavior:

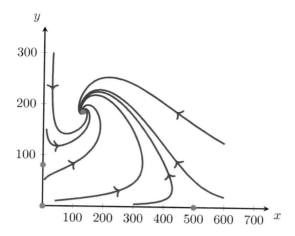

In order to sketch, by hand, the phase trajectories which appear in the graph, we can use the information we have about the stability of the fixed points and work from there.

To begin, we know that $(0,0)$ is a source (unstable node), and $(130.765, 184.615)$ is a stable spiral, so we can start by drawing phase trajectories which begin near the origin and end up at the stable spiral. For example, you could begin drawing a phase trajectory near the point $(10, 50)$. Because the origin is a source, you know you must move away from the origin, and start heading up and to the right. However, there is a saddle at $(0, 80)$, so your trajectory moves away from the y-axis to avoid the saddle point. As the stable fixed point is a spiral, not a node, your trajectory first passes below the fixed point, then loops around to arrive at $(130.765, 184.615)$.

Again, because we know that the origin is a source, we know that if you start above the saddle point at $(0, 80)$, say the initial conditions are $(10, 120)$, you must start by moving down toward the saddle point before then being pushed away from the y-axis toward the stable spiral.

Similarly, beginning at the point $(200, 25)$, your trajectory begins moving away from the source at the origin and toward the saddle point at $(0, 500)$ before being repulsed away from it and up toward the stable spiral. Because the trajectories in the x direction are heading toward the saddle from the left, the same will hold true from the right. So starting your trajectory at the point $(600, 25)$, for example, the trajectory will tend towards the saddle and then be pushed upwards toward the stable spiral.

These are the thought processes we would go through to arrive at a sketch of phase trajectories starting at various initial conditions nearby the equilibria. In doing so, we can obtain such a graph of the phase trajectories without nullclines or computer-generated vector fields, both of which we have discussed in previous sections.

5.6 Exercises

For the following ODEs, sketch the nullcline graph and use it to determine the stability of any equilibria. Confirm results using the Jacobian and Stability Diagram.

1.
$$\begin{cases} x' = (x-5)(y-3) \\ y' = 4x \end{cases}$$

2.
$$\begin{cases} x' = (x-1)(y-7) \\ y' = (x-3)(y-4) \end{cases}$$

3.
$$\begin{cases} x' = x(y-6) \\ y' = x+y-2 \end{cases}$$

4.
$$\begin{cases} x' = (x-5)(y-2) \\ y' = (x-8)(y-6) \end{cases}$$

5.
$$\begin{cases} x' = 3x-y \\ y' = (x-1)(y-6) \end{cases}$$

6.
$$\begin{cases} x' = y-x-1 \\ y' = (x-4)(y-6) \end{cases}$$

7. Consider the system of ODEs from Exercises 5.1.1 and 5.5.11:

$$\begin{cases} x' = 0.15x - 0.0012xy \\ y' = -0.05y + 0.001xy \end{cases}$$

(a) Sketch a phase portrait of this system using nullclines and your classification of fixed points.

(b) For an initial population of $(x_0, y_0) = (50, 75)$, draw a phase trajectory on your phase portrait. Describe the behavior of this trajectory in words.

8. Consider the system of ODEs from Exercises 5.1.3 and 5.5.12:

$$\begin{cases} x' = 0.3x - 0.0008xy \\ y' = 0.5y - 0.002xy \end{cases}$$

(a) Sketch a phase portrait of this system using nullclines and your classification of fixed points.

(b) Draw phase trajectories on your phase portrait for the following initial populations: $(x_0, y_0) = (250, 100)$, $(50, 375)$, $(250, 500)$, and $(400, 375)$. Describe the behavior of one of these trajectories in words.

9. Consider the system of ODEs from Exercises 5.1.4 and 5.5.13:

$$\begin{cases} x' & = & 0.25x - 0.0004xy \\ y' & = & 0.1y - 0.0002y^2 + 0.0002xy \end{cases}$$

(a) Sketch a phase portrait of this system using nullclines and your classification of fixed points.

(b) Describe the long-term behavior of these species.

10. Consider the system of ODEs from Exercises 5.1.6 and 5.5.14:

$$\begin{cases} x' & = & 0.32x - 0.0064x^2 + 0.005xy \\ y' & = & 0.1y - 0.0004y^2 - 0.001xy \end{cases}$$

(a) Sketch a phase portrait of this system using nullclines and your classification of fixed points.

(b) Describe the long-term behavior of these species.

11. Consider the system of ODEs from Exercises 5.1.8 and 5.5.15:

$$\begin{cases} x' & = & -0.6x + 0.001xy \\ y' & = & -0.2y + 0.005xy \end{cases}$$

(a) Sketch a phase portrait of this system using nullclines and your classification of fixed points.

(b) Describe the two possible long-term behaviors of these two populations.

12. Consider the system of ODEs from Exercises 5.1.10 and 5.5.16:

$$\begin{cases} x' & = & 0.4x - 0.004x^2 + 0.0008xy \\ y' & = & 0.2y - 0.001xy \end{cases}$$

(a) Sketch a phase portrait of this system using nullclines and your classification of fixed points.

(b) Describe the long-term behavior of these species.

5.6 Project: One ODE or Two?

1. Solve $y'' + 4y = 4$ using the methods from Chapter 4 (i.e., by using the characteristic equation).

2. Here's a new approach! Introduce a new variable, called x, and assume that $x = y'$. Construct a system of ODEs that is equivalent to the differential equation $y'' + 4y = 4$. Hint: $y'' = x'$

3. Find the nullclines and equilibria for this new system of equations.

4. Sketch the nullcline graph and use this to determine the stability of the equilibria. Draw a few phase trajectories near the equilibria. If the nullcline graph does not determine stability, state the possibilities for the type of equilibria you could have and then confirm stability using the Jacobian and Stability Diagram. Don't forget to sketch a few phase trajectories near the equilibria once you have determined the type.

5. Sketch a graph of y versus t assuming $y(0) = 2$. Compare this graph to the graph of your answer in Question 1 assuming $c_1 = c_2 = 1$. Explain how these graphs are related and why this makes sense.

Chapter 6

SIR-Type Models

In this chapter we are going to shift focus to a specific application of systems of nonlinear differential equations. This application uses systems of ODEs to model disease spread using Susceptible-Infected-Recovered (SIR)-type models. These models are extremely flexible and can become very complicated to model the behavior of complex diseases. However, the basic SIR model and a few other, more general models, are well within our scope. We will explore how we can apply some of the previous concepts in this text, along with learning a few more tools, in an effort to understand how differential equations can be used to study problems in epidemiology. After reading this chapter, you should be able to read articles with simple models from mathematical biology and epidemiology literature, and we recommend that you do this to further understand these types of models.

One notational aside needs to be made. In this field, and a few other fields, it is common to use dot notation rather than prime notation to denote derivatives with respect to time. For example, in Chapter 5, we would write a system of ODEs as:

$$\begin{cases} x' &= 3x - 2y \\ y' &= 4x + 6y \end{cases},$$

while in the dot notation used in this chapter, the exact same system would instead be written as:

$$\begin{cases} \dot{x} &= 3x - 2y \\ \dot{y} &= 4x + 6y \end{cases}.$$

With this aside out of the way, let's move forward and explore the first, most basic SIR model.

DOI: 10.1201/9781003298663-6

6.1 The Basic SIR Model with Birth and Death

The SIR (Susceptible, Infected, Recovered) model is a system of differential equations that can be used to study the transmission of various diseases.

Let's use SIR to describe the population infected with a disease. Assume that the only thing we know about this disease is that it is passed from person to person, infection spreads pretty quickly so that we are really only interested in seeing what happens over a short period of time, and that recovery is possible. We might draw a **compartmental diagram** like this:

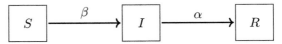

Each compartment or box in the diagram corresponds to a category of individuals. These categories are called the **classes** or **variables** of the model. In this particular example, the classes are S (susceptible individuals), I (infected individuals), and R (recovered individuals). We will also define N to be the total population of our model so that $N = S + I + R$. The arrows in the diagram illustrate how individuals move from one category to the next. The labels on these arrows are the **parameters** of the model. These include the transmission (or infection) rate for the disease β and the recovery rate α. It is possible to construct models that include a variety of other factors as well, such as different treatment and control measures, hospitalization, quarantine, death from the disease, etc., and some models include a birth and natural death rate due to the time-frame under consideration and a slower-paced disease progression, so these models are completely customizable. We will analyze the basic SIR model with birth and natural death a little bit later in this section and will consider a few other cases in the following sections.

The model shown in the compartmental diagram above forms a nonlinear system of ODEs. To construct it, we use our compartmental diagram, with arrows going into a class producing positive terms and arrows coming out of a class producing negative terms. For example, to create the $\dot{S}$ equation, we only have one arrow, labeled β, which is leaving the S class. The β term, or transmission term, represents healthy individuals becoming infected with the disease. One of the few things we know about this disease is that it passes from person to person. In other words, in order to get sick, a healthy person must come into contact with a sick person. In the $\dot{S}$ equation, this will correspond to $-\beta S \frac{I}{N}$. Essentially, we multiply the infection rate by the number of people who are healthy and can become sick times the fraction of people who are sick and can pass the disease on in the first place, and the expression is negative because these individuals are leaving the S class. Generally, this transmission term is the only term which accounts for contact between individuals. This is similar to the interaction term that is used in models describing interacting species, as discussed in Section 5.1.

The remaining terms are simply a progression of the disease. For the $\dot{I}$ equation, we know that we will have a positive $\beta S \frac{I}{N}$ term because the β arrow is pointing in toward I. However, the I population can also change by individuals recovering from the disease. So the $\dot{I}$ equation will have a negative term as well which accounts for the removal of individuals from this I class. Because α indicates the recovery of infected individuals, the corresponding term in the equation is αI. The number of individuals who recover according to the rate α depends on how many infected individuals I exist in our model. Notice that we multiply by the class in which the arrow starts because these are the individuals leaving the class. Using this idea to construct one equation for each class, we obtain:

$$\begin{cases} \dot{S} &= -\beta S \dfrac{I}{N} \\ \dot{I} &= \beta S \dfrac{I}{N} - \alpha I \\ \dot{R} &= \alpha I \end{cases} \cdot$$

We can now use our model to see how the disease affects the population over time. For example, we can determine if the population size changes over time. Remember that we defined $N = S + I + R$ to be the total population for our model. Then to see what happens to this population size over time, we can calculate $\dot{N}$:

$$\begin{aligned} \dot{N} &= \dot{S} + \dot{I} + \dot{R} \\ &= \underbrace{-\beta S \frac{I}{N}}_{\dot{S}} + \underbrace{\beta S \frac{I}{N} - \alpha I}_{\dot{I}} + \underbrace{\alpha I}_{\dot{R}} \\ &= 0 \end{aligned}$$

As $\dot{N} = 0$, we know that N must be a constant. Hence our population size remains constant over time. This is often the case with models that exclude birth and death, and we'll see in a moment how this affects the population dynamics.

Notice that the SIR model is a nonlinear system of ODEs. You may be tempted to apply some of the tools from Chapter 5 to determine stability, or use nullclines to sketch a phase plane. However, there are a few barriers to this method. First, once there are more than two variables, the classification of fixed points into nodes, saddles, spirals, and centers is no longer sufficient. Second, because we now have three variables, S, I, and R, we would want to plot a vector field in three dimensional space, which can be done using computer software but would be challenging, to say the least, by hand. Finally, most of the tools you would want to use will be severely complicated.

For example, if you wanted to try to use nullclines to understand the behavior of the SIR model, you would notice that the two I nullclines are, in fact, planes! Notice that because R does not appear in either the $\dot{S}$ or $\dot{I}$ equations, you could, in fact, graph the phase plane for S and I (ignoring R altogether) using tools from Chapter 5. You would discover S-nullclines at $S = 0$ and $I = 0$, I-nullclines at $I = 0$ and $S = \frac{\alpha N}{\beta}$, and a fixed point at the origin. However, the vector field will tell a slightly more complicated story. At $I = 0$, both $\dot{S}$ and $\dot{I}$ are equal to zero, regardless of the value of S ($I = 0$ is both an S- and I-nullcline). Therefore, the whole S axis is a "fixed point," something our discussion in Chapter 5 didn't address at all. This line in the SI plane is an *attractor*, a topic which was beyond the scope of Chapter 5. Therefore, in order to understand these systems we will focus on solution plots of the individual variables over time. These plots can be obtained using numerical methods and mathematical software.

Let's consider some hypothetical parameter values to see what the behavior of a typical SIR model. Using our equations with transmission rate $\beta = 0.12$ and recovery rate $\alpha = \frac{1}{14}$, which corresponds to a 14 day infectious period where $\alpha = 1/$infectious period, we obtain the following solutions.

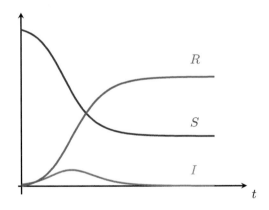

In this figure, on which we've plotted the graphs of S, I, and R with respect to time, you can see the long term behavior of the system. As more people become infected, the number of people in the I class increases while the number of people in the S class decreases. Also, as people recover from the infection, the number of people in the I class decreases and the number of people in the R class increases. Eventually, people recover and move out of the I class faster than new people are infected, so the number of people in the I class goes to zero. Once there are no infectious individuals, the disease cannot infect any of the susceptible people, so the disease dies out without having infected everyone in the population.

○○○

Similarly, let's consider the SIR model that includes birth and natural death to see how these demographic parameters change our study of disease. Here's a typical SIR compartmental diagram that includes birth and death:

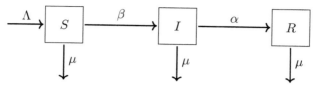

In this case, Λ is the birth rate and μ is the natural death rate. When constructing the model equations, different assumptions about the birth rate can produce different variations of the $\dot{S}$ equation. We'll examine models with different kinds of birth rates in the exercises throughout this chapter. In this particular model, Λ is a constant influx and will be left as a constant in the $\dot{S}$ equation (i.e., we will not multiply Λ by any variable when we include this term in our equation). Then using the same idea as before, we have new model equations:

$$
\begin{cases}
\dot{S} &= \Lambda - \beta S \dfrac{I}{N} - \mu S \\[2mm]
\dot{I} &= \beta S \dfrac{I}{N} - \alpha I - \mu I \\[2mm]
\dot{R} &= \alpha I - \mu R.
\end{cases}
$$

To see what happens to this new population size over time, we can once again calculate $\dot{N}$:

$$
\begin{aligned}
\dot{N} &= \dot{S} + \dot{I} + \dot{R} \\[2mm]
&= \underbrace{\Lambda - \beta S \frac{I}{N} - \mu S}_{\dot{S}} + \underbrace{\beta S \frac{I}{N} - \alpha I - \mu I}_{\dot{I}} + \underbrace{\alpha I - \mu R}_{\dot{R}} \\[2mm]
&= \Lambda - \mu S - \mu I - \mu R \\[2mm]
&= \Lambda - \mu(S + I + R) \\[2mm]
&= \Lambda - \mu N
\end{aligned}
$$

If the population size is not changing (i.e., a constant), we would expect $\dot{N} = 0$. However, we find that $\dot{N} = \Lambda - \mu N$, so we know that the population size is changing unless $\Lambda = \mu N$. In fact, the population size for this model, provided $\dot{N} \neq 0$, will approach $\frac{\Lambda}{\mu}$ as $t \to \infty$.

Let's produce numerical simulations to examine the behavior of our model. Note that we've used hypothetical parameter values to generate this plot in order to illustrate the possibility that an infectious population will remain.

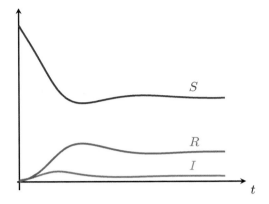

Due to the impact of birth and death, in this model, the infectious population never dies out. There is a constant influx of susceptible individuals through birth, so the susceptible population decreases after the initial outbreak, then increases before leveling off. This increase in susceptible individuals allows for the infectious population to replenish. This is indicative of an *endemic* model, in which a disease may never die out (i.e., it becomes endemic to the population, such as a disease like varicella), as opposed to an *epidemic* model in which the disease eventually does dies out, as in our first example without birth and death and which could describe a disease like the flu.

6.1.1   Equilibria and Stability

Just as we could determine the equilibria and stability of our models in previous chapters, we can do so for SIR models as well.

Example: Let's find the equilibria of the model:

$$
\begin{cases}
\dot{S} &= \Lambda - \beta S \dfrac{I}{N} - \mu S \\[2mm]
\dot{I} &= \beta S \dfrac{I}{N} - \alpha I - \mu I \\[2mm]
\dot{R} &= \alpha I - \mu R
\end{cases}
$$

Remember that to find the equilibria we must set all of our model equations equal to 0. Typically, investigating the $\dot{I}$ equation first will provide some useful information. In this case, we have:

$$
0 = \beta S \frac{I}{N} - \alpha I - \mu I = I\left(\beta \frac{S}{N} - \alpha - \mu\right).
$$

Thus either $I = 0$ or $\beta\dfrac{S}{N} - \alpha - \mu = 0$, which gives $S = (\alpha + \mu)\dfrac{N}{\beta}$. These will ultimately lead us to two different equilibria.

Let's first consider $I = 0$. Because I represents the infected population, this equilibrium will be called the **disease-free equilibrium**, or **DFE**. We can substitute $I = 0$ into the $\dot{S}$ and $\dot{R}$ equations to solve for S and R. For instance, the $\dot{R}$ equation gives $0 = \alpha I - \mu R = -\mu R$. Hence $R = 0$. Using the $\dot{S}$ equation, we obtain $0 = \Lambda - \beta\dfrac{SI}{N} - \mu S = \Lambda - \mu S$. Thus $S = \dfrac{\Lambda}{\mu}$. Then our DFE is $\left(\dfrac{\Lambda}{\mu}, 0, 0\right)$.

○○○

A common assumption in these types of models is that the population size is constant. We have seen that this is not the case for our model as is because the population tends to $\dfrac{\Lambda}{\mu}$ as $t \to \infty$. In order to assume the population size is constant, we must assume $\Lambda = \mu N$. Then, because $N = S + I + R$, we can also see that at the DFE, $N = S + 0 + 0$ so that $N = S$ and we can define our DFE as $(N, 0, 0)$ instead. Hence the DFE is the equilibrium in which all individuals are healthy, and in which no one has the disease – it is literally free of disease.

○○○

Continuing our example, we may consider the other equilibrium where $S = (\alpha + \mu)\dfrac{N}{\beta}$ and solve for I and R in a similar way, we will obtain the **endemic equilibrium**, or **EE**, which is the equilibrium with disease still present in the population. In this case, if we substitute our expression into the $\dot{S}$ equation, we arrive at:

$$
\begin{aligned}
0 &= \Lambda - \beta(\alpha + \mu)\frac{N}{\beta}\frac{I}{N} - \mu(\alpha + \mu)\frac{N}{\beta} \\
&= \Lambda - (\alpha + \mu)I - \mu(\alpha + \mu)\frac{N}{\beta},
\end{aligned}
$$

which gives $I = \dfrac{\Lambda}{\alpha + \mu} - \mu\dfrac{N}{\beta}$. Finally, with the $\dot{R}$ equation, we obtain $0 = \alpha I - \mu R$, so that $R = \dfrac{\alpha I}{\mu} = \dfrac{\Lambda\alpha}{(\alpha + \mu)\mu} - \alpha\dfrac{N}{\beta}$. Hence the EE is $\left((\alpha + \mu)\dfrac{N}{\beta}, \dfrac{\Lambda}{\alpha + \mu} - \mu\dfrac{N}{\beta}, \dfrac{\Lambda\alpha}{(\alpha + \mu)\mu} - \alpha\dfrac{N}{\beta}\right).$

For an SIR model, we typically have two kinds of equilibria:

- The **disease-free equilibrium** (or **DFE**) is the equilibrium that consists of only healthy (or recovered) individuals and no infected individuals.

- The **endemic equilibrium** (or **EE**) is the equilibrium in which the disease is endemic, or still present, in the population.

In reality, the disease-free equilibrium is quite easy to find. For instance, in our derivation of the equilibrium in the previous example, we could have just as easily assumed $I = 0$ at the very beginning, without worrying about setting the $\dot{I}$ equation equal to 0. However, finding the endemic equilibrium is often much more complicated and we will usually resort to the methods used a moment ago.

○○

To determine the stability of the equilibria we just found, we find the Jacobian of our original system:

$$
J = \begin{pmatrix} \dfrac{\partial \dot{S}}{\partial S} & \dfrac{\partial \dot{S}}{\partial I} & \dfrac{\partial \dot{S}}{\partial R} \\[2mm] \dfrac{\partial \dot{I}}{\partial S} & \dfrac{\partial \dot{I}}{\partial I} & \dfrac{\partial \dot{I}}{\partial R} \\[2mm] \dfrac{\partial \dot{R}}{\partial S} & \dfrac{\partial \dot{R}}{\partial I} & \dfrac{\partial \dot{R}}{\partial R} \end{pmatrix} = \begin{pmatrix} -\beta \dfrac{I}{N} - \mu & -\beta \dfrac{S}{N} & 0 \\[2mm] \beta \dfrac{I}{N} & \beta \dfrac{S}{N} - \alpha - \mu & 0 \\[2mm] 0 & \alpha & -\mu \end{pmatrix}.
$$

It is, of course, much easier to determine the stability when we have real parameter values. Similar to the process in Chapter 5, in order to determine stability we evaluate the Jacobian at the equilibrium point and then find the eigenvalues. If all eigenvalues have negative real part, then the equilibrium will be stable. Because we have not chosen numerical parameter values for this example, we will only look at the case of the disease-free equilibrium. However, with numerical values and a computer, we can usually determine the stability of any equilibrium pretty quickly.

Evaluating at the disease-free equilibrium for this system, $(N, 0, 0)$, our Jacobian becomes:

$$
J = \begin{pmatrix} -\mu & -\beta & 0 \\ 0 & \beta - \alpha - \mu & 0 \\ 0 & \alpha & -\mu \end{pmatrix}.
$$

The eigenvalues for this matrix are $-\mu$, $-\mu$, and $\beta - \alpha - \mu$. Clearly the eigenvalues $-\mu < 0$ because $\mu > 0$ represents the death rate for the population, so if we can show that $\beta - \alpha - \mu < 0$, then the disease-free equilibrium is stable. In other words, if $\beta - \alpha - \mu < 0$, then the disease will die out in the population. As a preview to what we will ultimately discover in the next section, let's manipulate our equation:

$$\begin{aligned} \beta - \alpha - \mu &< 0 \\ \beta &< \alpha + \mu \\ \beta \cdot \frac{1}{\alpha + \mu} &< 1. \end{aligned}$$

Hence, if $\beta \cdot \dfrac{1}{\alpha + \mu} < 1$, then the DFE is stable and the disease will die out in the population. This will be important in the next section.

6.1.2 The Basic Reproduction Number, R_0

One thing that we often want to know when studying a disease is how quickly it can spread. An important threshold value that determines if a disease will spread or die out is the **basic reproduction number**.

> **Definition: R_0** is the **basic reproduction number**, which is the average number of secondary cases caused by a typical infectious individual over his/her entire infectious lifetime in a completely susceptible population.

For example, if $R_0 = 2.5$, this means that if only one person has the disease, then this infected person will infect, on average, 2.5 other people. In this case, the disease will spread because even if the original infected individual recovers, he has infected 2.5 other people who will then go on to infect others.

On the other hand, if $R_0 = 0.7$, for example, this means that, on average, one infected person will infect 0.7 more people, so eventually the disease will die out.

So how do we know if a disease will spread or die out?

> If $R_0 > 1$, then the disease will continue to spread, with each infected person, on average, infecting more than one other person. However, if $R_0 < 1$, then each infected person will infect less than one other person (on average) and the disease will die out.

In this sense, the goal is to calculate R_0. If we can interpret the expression biologically, then we may find easier ways in which we can reduce R_0, for example with treatment options, and ultimately lower R_0 to below 1 so that the disease dies out.

To calculate R_0, we will use the Next Generation Matrix Approach[a]**:**

① Write down the DFE. You will need this later.

② Write down only the infected population equation(s) from the model.

③ Split the infected equation(s) into 2 parts so that you have $\mathcal{F} - \mathcal{V}$, where:

$\mathcal{F}$ = terms that represent all new infections

$\mathcal{V}$ = terms that represent all other transitions to other classes.

④ Find the Jacobian by taking partial derivatives with respect to the infected population(s).

⑤ Evaluate at the DFE.

⑥ Calculate $R_0 = \rho\left(FV^{-1}\right)$ and then interpret R_0 biologically. In this case, ρ is the spectral radius (meaning dominant eigenvalue) of the matrix, so R_0 is the dominant (or largest in absolute value) eigenvalue of the matrix FV^{-1}.

---

[a]Van den Driessche, P., & Watmough, J. (2002). Reproduction numbers and sub-threshold endemic equilibria for compartmental models of disease transmission. *Mathematical Biosciences*, 180(1):29–48.

Let's try to calculate R_0 for our model:

$$\begin{cases} \dot{S} &= \Lambda - \beta S \dfrac{I}{N} - \mu S \\[2mm] \dot{I} &= \beta S \dfrac{I}{N} - \alpha I - \mu I \\[2mm] \dot{R} &= \alpha I - \mu R \end{cases} \cdot$$

We found the DFE for this model already to be $(N, 0, 0)$, assuming the population size is constant, or $\Lambda = \mu N$. Because we only have one infected class, I, this is the only equation we will consider: $\dot{I} = \beta S \dfrac{I}{N} - \alpha I - \mu I$. Now we must split the infected equation ($\dot{I}$) into 2 parts so that we have $\dot{I} = \mathcal{F} - \mathcal{V}$, where:

$\mathcal{F}$ = terms that represent all new infections
$\mathcal{V}$ = terms that represent all other transitions to other classes.

Sometimes splitting the equation into $\mathcal{F}$ and $\mathcal{V}$ is easier to do based off of the compartmental diagram. For $\mathcal{F}$, we are looking for any terms that represent individuals getting sick for the first time. For this model, $\mathcal{F} = \beta S \dfrac{I}{N}$. To find $\mathcal{V}$, we simply need the remaining terms in the $\dot{I}$ equation, but negated so that $\dot{I} = \mathcal{F} - \mathcal{V}$. In this case, $\mathcal{V} = \alpha I + \mu I$. So we have:

$$\mathcal{F} = \beta S \frac{I}{N}, \qquad \mathcal{V} = \alpha I + \mu I.$$

The next step in the Next Generation Matrix Approach is to take the Jacobian of $\mathcal{F}$ and $\mathcal{V}$ with respect to the infected classes. However, note that in this example these expressions are simply one function, because there was only one infected class I, rather than a vector, which would occur when we have multiple infected classes, as we will see in the following sections. In this case, the Jacobians of $\mathcal{F}$ and $\mathcal{V}$ are simply the partial derivatives of $\mathcal{F}$ and $\mathcal{V}$ with respect to I. Then we have:

$$F = \frac{\partial \mathcal{F}}{\partial I} = \beta \frac{S}{N}, \qquad V = \frac{\partial \mathcal{V}}{\partial I} = \alpha + \mu.$$

Recall that from Step 1, the DFE is $(N, 0, 0)$. Then evaluating F and V at this equilibrium, we have:

$$F = \beta, \qquad V = \alpha + \mu.$$

Finally, we know that $R_0 = \rho\left(FV^{-1}\right)$. Depending on the model, the F and V that result from the last step may be matrices and so calculating FV^{-1} can take a little bit of effort, and then we still need to find the eigenvalues. However, our F and V are constants, not matrices, so we do not need to worry about finding eigenvalues for this example. We will look at more complex models in the following sections. For our model, $R_0 = FV^{-1}$ is just a simple calculation:

$$R_0 = FV^{-1} = \beta \cdot \frac{1}{\alpha + \mu}.$$

If we calculate R_0 correctly, then there should be a nice way to explain the expressions biologically because R_0 should represent the average number of secondary cases from a single infectious person. We obtained $R_0 = \beta \cdot \dfrac{1}{\alpha + \mu}$, where β is the transmission rate for our model and $\dfrac{1}{\alpha + \mu}$ is the average length of time a person is infectious as a part of the I class. Notice that α and μ are the labels on the arrows in the compartmental diagram coming out of the I class. This is not a coincidence! To find the time spent in the I class, we can consider the fraction 1 divided by the removal from that class. Hence, our R_0 depends on how quickly an infectious person can infect others and how many other people that person comes into contact with (β), as well as how long that infectious person is sick and capable of infecting others $\left(\dfrac{1}{\alpha + \mu}\right)$.

You may remember that the expression we just calculated for R_0 is the same expression that we found in Section 6.1.1 when we determined the stability of the disease-free equilibrium. We had previously said that the DFE was stable (and hence the disease would die out) if $\beta \cdot \dfrac{1}{\alpha + \mu} < 1$. It is no coincidence that $R_0 = \beta \cdot \dfrac{1}{\alpha + \mu}$, and, in fact, we reach the same conclusion here. Recall that R_0 is a threshold parameter that determines when a disease spreads or dies out, and $R_0 < 1$ is exactly the condition that we need for a disease to be eradicated from a population.

6.1 Exercises

1. Let S, I, and R represent the number of individuals who are susceptible, infectious, and recovered, respectively, from a certain disease. Assume that a recovered individual can eventually become susceptible again. For the system of ODEs:

$$\begin{cases} \dot{S} &= -\beta S \frac{I}{N} + \varphi R \\[2mm] \dot{I} &= \beta S \frac{I}{N} - \delta I \\[2mm] \dot{R} &= \delta I - \varphi R \end{cases},$$

    where $N = S + I + R$, answer the following questions:

    (a) Draw the compartmental diagram associated with this system of equations.

    (b) Interpret the meaning of β, δ, and φ.

    (c) Prove that the total population size (N) does not change over time.

    (d) Is there a disease-free equilibrium? If so, find it.

    (e) Is there an endemic equilibrium? If so, do not find it, but explain how you know. If not, explain why not.

    (f) Determine the stability of the equilibrium you found in part (d). You may assume $\varphi = \dfrac{1}{15}$, $\delta = \dfrac{1}{7}$, and $\beta = 0.0001$. Use computer software to help you.

2. Write down a possible description for each parameter in the compartmental model below, where S is the susceptible population, I is the infected population, and R is the population of recovered or vaccinated individuals. Then, write the system of equations being represented in the diagram.

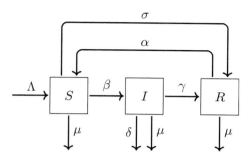

Be aware that we have only shown the parameters in the diagram. Each parameter should be multiplied by the appropriate variables when being used in the equations.

3. Consider the compartmental diagram A below:

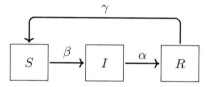

(a) Write down the model equations. Assume that the transmission parameter is divided by N (i.e., $\beta S \frac{I}{N}$).

(b) Find the basic reproduction number R_0 and interpret each component biologically. Use $S = N$ in your calculation of R_0.

4. Consider the compartmental diagram B below:

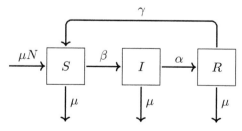

(a) Write down the model equations. Assume that the transmission parameter is divided by N (i.e., $\beta S \frac{I}{N}$).

(b) Find the basic reproduction number R_0 and interpret each component biologically. Use $S = N$ in your calculation of R_0.

5. How does R_0 for model A and B, from the previous two exercises, compare? Explain how the differences in the compartmental models affect the calculation and interpretation of R_0.

6. Consider the system of ODEs that represents susceptible, infectious, and recovered individuals, with a constant birth rate Λ:

$$\begin{cases} \dot{S} &=& \Lambda - \beta S \frac{I}{N} - \mu S \\[2mm] \dot{I} &=& \beta S \frac{I}{N} - (\mu + \rho) I \\[2mm] \dot{R} &=& \rho I - \mu R \end{cases},$$

(a) Assuming that $N = S + I + R$, find the equation $\dot{N}$.

(b) Is the total population N constant? If yes, how do you know? If not, what happens to the population over time?

(c) Solve for N. (Hint: use the method of integrating factors!)

(d) Show that $\lim\limits_{t \to \infty} N = \dfrac{\Lambda}{\mu}$.

6.2 The SEIR Model

Let's look at a more complex model: SEIR. The E class is for the exposed individuals who have come into contact with individuals with a disease and have been infected with this disease, but who are not yet showing symptoms and are not infectious. Even though these individuals are sick, they cannot infect others. We can also describe this class as the latent population. The compartmental diagram could look like this:

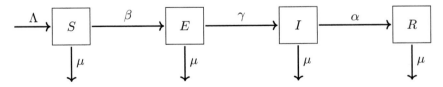

In this example, the classes are S (susceptible individuals), E (exposed individuals), I (infected individuals), and R (recovered individuals). We define N to be the total population of our model: $N = S + E + I + R$. The parameters include β which is the transmission rate, γ which is the rate at which individuals begin to show symptoms and are now infectious, α which is the recovery rate, Λ which is the birth rate, and μ which is the natural death rate. Just as before, we will assume that $\Lambda = \mu N$, that way our population size remains constant.

We can construct the model equations in the same way as the SIR model in the previous section. In the case of SEIR we obtain:

$$\begin{cases} \dot{S} &= \mu N - \beta S \frac{I}{N} - \mu S \\ \dot{E} &= \beta S \frac{I}{N} - \gamma E - \mu E \\ \dot{I} &= \gamma E - \alpha I - \mu I \\ \dot{R} &= \alpha I - \mu R \end{cases}.$$

Notice that the transmission term is still $\beta S \frac{I}{N}$. This is because we still want to consider the contact of healthy individuals with infected, and more importantly, infectious, individuals. Even though both the E and I class are infected with the disease, we only see I appear in the transmission term because only the I individuals are able to infect others.

In order to calculate the basic reproduction number for this model, we first need to find the disease-free equilibrium. By definition, disease-free means that there is no infection, so $E = I = 0$. Then because $\dot{R} = \alpha I - \mu R$, we obtain $R = 0$. Finally, as $N = S + E + I + R$, we know that $N = S$. So our DFE is $(N, 0, 0, 0)$.

We will consider our infected equations:

$$\dot{E} = \beta S \frac{I}{N} - \gamma E - \mu E$$

$$\dot{I} = \gamma E - \alpha I - \mu I$$

and we must split these infected equations into two vectors so that we have $\begin{pmatrix} \dot{E} \\ \dot{I} \end{pmatrix} = \mathcal{F} - \mathcal{V}$, where:

$\mathcal{F} =$ terms that represent all new infections
$\mathcal{V} =$ terms that represent all other transitions to other classes.

For $\mathcal{F}$, we want only the terms that represent individuals getting sick for the first time. For this model, $\mathcal{F} = \begin{pmatrix} \beta S \frac{I}{N} \\ 0 \end{pmatrix}$. This means that

$$\mathcal{V} = \begin{pmatrix} \gamma E + \mu E \\ -\gamma E + \alpha I + \mu I \end{pmatrix}.$$

Next we will take the Jacobian by finding partial derivatives, first with respect to E, and then with respect to I. Because we have more than one infected population, we are calculating the Jacobian matrix for both $\mathcal{F}$ and $\mathcal{V}$. Then we have:

$$F = \begin{pmatrix} \dfrac{\partial \mathcal{F}_1}{\partial E} & \dfrac{\partial \mathcal{F}_1}{\partial I} \\[2mm] \dfrac{\partial \mathcal{F}_2}{\partial E} & \dfrac{\partial \mathcal{F}_2}{\partial I} \end{pmatrix} = \begin{pmatrix} 0 & \beta \dfrac{S}{N} \\[2mm] 0 & 0 \end{pmatrix},$$

$$V = \begin{pmatrix} \dfrac{\partial \mathcal{V}_1}{\partial E} & \dfrac{\partial \mathcal{V}_1}{\partial I} \\[2mm] \dfrac{\partial \mathcal{V}_2}{\partial E} & \dfrac{\partial \mathcal{V}_2}{\partial I} \end{pmatrix} = \begin{pmatrix} \gamma + \mu & 0 \\[2mm] -\gamma & \alpha + \mu \end{pmatrix}.$$

Recall that the DFE is $(N, 0, 0, 0)$. Then evaluating F and V at this equilibrium, we have:

$$F = \begin{pmatrix} 0 & \beta \\ 0 & 0 \end{pmatrix}, \quad V = \begin{pmatrix} \gamma + \mu & 0 \\ -\gamma & \alpha + \mu \end{pmatrix}.$$

In order to obtain an expression for R_0, we must first find the matrix FV^{-1}. Recall that the inverse of a 2×2 matrix $M = \begin{pmatrix} a & b \\ c & d \end{pmatrix}$ is:

$$M^{-1} = \frac{1}{\det M} \begin{pmatrix} d & -b \\ -c & a \end{pmatrix},$$

where $\det M$ is the determinant of the matrix M. Then for our matrix V, we have:

$$V^{-1} = \frac{1}{(\gamma + \mu)(\alpha + \mu)} \begin{pmatrix} \alpha + \mu & 0 \\ \gamma & \gamma + \mu \end{pmatrix}$$

$$= \begin{pmatrix} \dfrac{1}{\gamma + \mu} & 0 \\[2mm] \dfrac{\gamma}{\gamma + \mu} \cdot \dfrac{1}{\alpha + \mu} & \dfrac{1}{\alpha + \mu} \end{pmatrix}.$$

Calculating FV^{-1}, we find:

$$FV^{-1} = \begin{pmatrix} 0 & \beta \\ 0 & 0 \end{pmatrix} \begin{pmatrix} \dfrac{1}{\gamma + \mu} & 0 \\[2mm] \dfrac{\gamma}{\gamma + \mu} \cdot \dfrac{1}{\alpha + \mu} & \dfrac{1}{\alpha + \mu} \end{pmatrix}$$

$$= \begin{pmatrix} \beta \cdot \dfrac{\gamma}{\gamma + \mu} \cdot \dfrac{1}{\alpha + \mu} & \beta \cdot \dfrac{1}{\alpha + \mu} \\[2mm] 0 & 0 \end{pmatrix}.$$

Finally, R_0 is the dominant eigenvalue of the matrix FV^{-1}. This matrix has eigenvalues 0 and $\beta \cdot \dfrac{\gamma}{\gamma + \mu} \cdot \dfrac{1}{\alpha + \mu}$. Therefore,

$$R_0 = \beta \cdot \frac{\gamma}{\gamma + \mu} \cdot \frac{1}{\alpha + \mu},$$

where β is the transmission rate for our model, $\dfrac{\gamma}{\gamma + \mu}$ is the probability that a person becomes infectious (moves from the E class to the I class before natural death), and $\dfrac{1}{\alpha + \mu}$ is the average length of time a person is infectious as a part of the I class. This interpretation is extremely similar to that of the SIR model and we can often tell what the expression for R_0 should be based on the compartmental diagram and a biological explanation.

6.2 Exercises

1. Consider compartmental diagram A below:

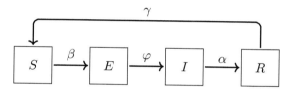

    (a) Write down the model equations. Assume that the transmission parameter is divided by N (i.e., $\beta S \frac{I}{N}$).

    (b) Find the basic reproduction number R_0 and interpret each component biologically. Use $S = N$ in your calculations of R_0.

2. Consider compartmental diagram B below:

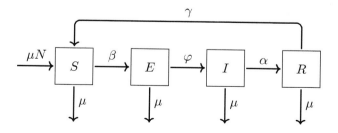

    (a) Write down the model equations. Assume that the transmission parameter is divided by N (i.e., $\beta S \frac{I}{N}$).

    (b) Find the basic reproduction number R_0 and interpret each component biologically. Use $S = N$ in your calculations of R_0.

3.   How does R_0 for model A and B, from the previous two exercises, compare? Explain how the differences in the compartmental models affect the calculation and interpretation of R_0.

4.   For the following compartmental model:

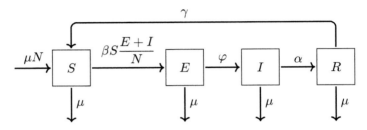

    (a) Interpret the meaning of the transmission term.

    (b) It can be shown that $R_0 = \dfrac{\beta}{\varphi + \mu} + \dfrac{\beta\varphi}{(\varphi + \mu)(\alpha + \mu)}$. Separate R_0 into easily understood components and interpret each one biologically.

6.3   A Model with Vaccination

    We can make the compartmental model as simple or as complicated as we desire. The simpler the model, the easier it is to analyze mathematically, but we can always use numerical and graphical methods to analyze models that are more complicated. In this section, we will investigate a model that includes vaccination. However, we ultimately could include many more control measures to lower the severity of the disease, such as hospitalization and quarantine, or we could include other features of the disease, such as the ability for a person to get sick for a second time or a death rate due to the disease.

○○

For now, let's continue with the SEIR model from the previous section, but this time add in a vaccinated class, V.

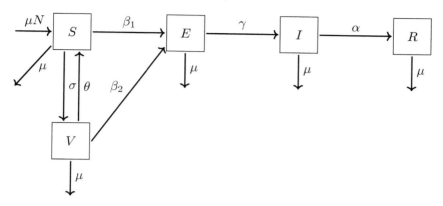

In this model, we have two transmission rates: β_1 is the transmission rate for susceptible individuals and β_2 is the transmission rate for vaccinated individuals. The other new parameters are the vaccination rate σ and the rate of waning vaccination and the return to the susceptible class θ. Notice that in this model the vaccinated class can still become infected with the disease. These individuals have received a vaccine, but the vaccine has not necessarily provided full immunity. In this case, the vaccinated class is really a compartment of partially immune individuals. They are less susceptible than individuals in the S class, but can still be infected at the rate β_2. For this model, we obtain the following system of differential equations:

$$
\left\{
\begin{aligned}
\dot{S} &= \mu N - \beta_1 S \frac{I}{N} - \sigma S + \theta V - \mu S \\
\dot{E} &= \beta_1 S \frac{I}{N} + \beta_2 V \frac{I}{N} - \gamma E - \mu E \\
\dot{I} &= \gamma E - \alpha I - \mu I \\
\dot{R} &= \alpha I - \mu R \\
\dot{V} &= \sigma S - \beta_2 V \frac{I}{N} - \theta V - \mu V
\end{aligned}
\right. ,
$$

where $N = S + E + I + R + V$.

Again, we need to find the disease-free equilibrium, or where we have $E = I = 0$. Then because $\dot{R} = \alpha I - \mu R$, we obtain $R = 0$. We can also solve $\dot{S} = 0$ and $\dot{V} = 0$ as a system of equations to obtain the DFE of $(S^*, 0, 0, 0, V^*) = (\frac{\theta + \mu}{\sigma + \theta + \mu} N, 0, 0, 0, \frac{\sigma}{\sigma + \theta + \mu} N)$. For simplicity in calculating the basic reproduction number, we will just refer to this as $(S^*, 0, 0, 0, V^*)$.

We will consider our infected equations:

$$\dot{E} = \beta_1 S \frac{I}{N} + \beta_2 V \frac{I}{N} - \gamma E - \mu E$$

$$\dot{I} = \gamma E - \alpha I - \mu I$$

and we split these infected equations into the corresponding vectors:

$$\mathcal{F} = \begin{pmatrix} \beta_1 S \dfrac{I}{N} + \beta_2 V \dfrac{I}{N} \\ 0 \end{pmatrix}, \qquad \mathcal{V} = \begin{pmatrix} \gamma E + \mu E \\ -\gamma E + \alpha I + \mu I \end{pmatrix}.$$

Next we will find the Jacobian by taking the partial derivatives, first with respect to E, and then with respect to I, which gives:

$$F = \begin{pmatrix} 0 & \beta_1 \dfrac{S}{N} + \beta_2 \dfrac{V}{N} \\ 0 & 0 \end{pmatrix}, \qquad V = \begin{pmatrix} \gamma + \mu & 0 \\ -\gamma & \alpha + \mu \end{pmatrix}.$$

Evaluating F and V at the DFE, we have:

$$F = \begin{pmatrix} 0 & \beta_1 \dfrac{S^*}{N} + \beta_2 \dfrac{V^*}{N} \\ 0 & 0 \end{pmatrix}, \qquad V = \begin{pmatrix} \gamma + \mu & 0 \\ -\gamma & \alpha + \mu \end{pmatrix}.$$

Finally, we can calculate FV^{-1} so that we can find the expression for the basic reproduction number:

$$FV^{-1} = \begin{pmatrix} 0 & \beta_1 \dfrac{S^*}{N} + \beta_2 \dfrac{V^*}{N} \\ 0 & 0 \end{pmatrix} \begin{pmatrix} \dfrac{1}{\gamma+\mu} & 0 \\ \dfrac{\gamma}{\gamma+\mu} \cdot \dfrac{1}{\alpha+\mu} & \dfrac{1}{\alpha+\mu} \end{pmatrix}$$

$$= \begin{pmatrix} \beta_1 \dfrac{S^*}{N} \cdot \dfrac{\gamma}{\gamma+\mu} \cdot \dfrac{1}{\alpha+\mu} + \beta_2 \dfrac{V^*}{N} \cdot \dfrac{\gamma}{\gamma+\mu} \cdot \dfrac{1}{\alpha+\mu} & \beta_1 \dfrac{S^*}{N} \cdot \dfrac{1}{\alpha+\mu} + \beta_2 \dfrac{V^*}{N} \cdot \dfrac{1}{\alpha+\mu} \\ 0 & 0 \end{pmatrix}.$$

R_0 is the dominant eigenvalue of the matrix FV^{-1} and thus:

$$R_0 = \beta_1 \frac{S^*}{N} \cdot \frac{\gamma}{\gamma+\mu} \cdot \frac{1}{\alpha+\mu} + \beta_2 \frac{V^*}{N} \cdot \frac{\gamma}{\gamma+\mu} \cdot \frac{1}{\alpha+\mu},$$

where β_1 is the transmission rate for the susceptible population, $\dfrac{S^*}{N}$ is the fraction of the population that is susceptible and able to get infected, β_2 is the transmission rate for the vaccinated population, $\dfrac{V^*}{N}$ is the fraction of the population that has been vaccinated but still able to get infected, $\dfrac{\gamma}{\gamma+\mu}$ is the probability that a person becomes infectious (moves from the E class to the I class before natural death), and $\dfrac{1}{\alpha + \mu}$ is the average length of time a person is infectious as a part of the I class.

○○○○○○○○○○○○○○○○○○○○○○○○○○○○○○○○○○○○○○

In an endemic model (one that includes birth and natural death, like we have here), we can also calculate **herd immunity**.

Definition: Herd immunity is the community immunity that occurs when the vaccination of a certain population (or herd!) is at a level that provides protection to the individuals that have not developed immunity.

In order to achieve herd immunity, a population needs two things: a high vaccine efficacy (effectiveness) and a high vaccination rate. In other words, if a certain percentage of the population is vaccinated and the infection rate for vaccinated people is low, then it could be possible to achieve herd immunity. But what percentage of the population needs to be vaccinated for this to work? The short answer is that this completely depends on the population and on the quality of the vaccine, but we can calculate the herd immunity level as follows.

The herd immunity level (or the **critical vaccination level**) is:

$$v_c = \frac{1 - \frac{1}{R_0}}{E},$$

where E represents the vaccine efficacy.

Example: Let's use an example to see how we can compute the herd immunity. The basic reproduction number for measles in England and Wales between 1944 and 1979 was between 13.7 and 18.0.[1] For this example, let's use the lower of these two estimates. If the measles vaccine is 95% effective, what percentage of the population must be immunized in order to lower R_0 to below 1 and thereby reach herd immunity?

We have the requisite information. Here, $R_0 = 13.7$ and $E = 0.95$, so we

---

[1]Delamater, P. L., Street, E. J., Leslie, T. F., Yang, Y. T., & Jacobsen, K. H. (2019). Complexity of the basic reproduction number (R_0). *Emerging Infectious Diseases*, 25(1):1–4.

have:

$$v_c = \frac{1 - \frac{1}{R_0}}{E}$$

$$= \frac{1 - \frac{1}{13.7}}{0.95}$$

$$\approx 0.976.$$

Therefore, in order to reach herd immunity, 97.6% of the population must be vaccinated.

○○○

The herd immunity is extremely sensitive to the reproduction number. As you will see in one of the exercises, even an increase in R_0 by 0.5 can dramatically increase v_c, especially with a low efficacy rate.

6.3 Exercises

1. Let S, I, and V be the number of susceptible, infectious, and vaccinated (or recovered) individuals, respectively, from a certain disease. Assume $N = S + I + V$ is the total population size. Consider the model:

$$\begin{cases} \dot{S} = \mu N - \beta S \frac{I}{N} - \mu S \\ \dot{I} = \beta S \frac{I}{N} - (\mu + \rho) I \\ \dot{V} = \rho I - \mu V \end{cases}$$

   (a) Draw the compartmental model associated with this system of equations.
   (b) Interpret the meaning of μ, β, and ρ.
   (c) Find the basic reproduction number R_0.
   (d) Interpret your answer from part (c) biologically.
   (e) Explain any limitations of this model (i.e., how are vaccinated individuals treated and what other considerations could be made)?

2. Let S, I, and V be the number of susceptible, infectious, and vaccinated (or recovered) individuals, respectively, from a certain disease. Assume $N = S + I + V$ is the total population size. Consider the model:

$$\begin{cases} \dot{S} = \mu N - \beta S \frac{I}{N} - (\mu + \gamma) S \\ \dot{I} = \beta S \frac{I}{N} - (\mu + \rho) I \\ \dot{V} = \rho I + \gamma S - \mu V \end{cases}$$

(a) Draw the compartmental model associated with this system of equations.

(b) Interpret the meaning of μ, β, ρ, and γ.

(c) Find the basic reproduction number R_0.

(d) Interpret your answer from part (c) biologically.

(e) Explain any limitations of this model (i.e., how are vaccinated individuals treated and what other considerations could be made)?

(f) Using the following parameter values, find the critical vaccination level for the model if the vaccine efficacy is 40%: $\beta = 0.4$, $\rho = 0.05$, $\gamma = 0.0002$, and $\mu = 0.00004$.

3. Let S, I, R, and V be the number of susceptible, infectious, recovered, and vaccinated individuals, respectively, from a certain disease. Assume $N = S + I + R + V$ is the total population size. Consider the model:

$$\begin{cases} \dot{S} &=& \mu N - \beta_1 S\frac{I}{N} + \varphi R - (\delta + \mu)S \\ \dot{I} &=& \beta_1 S\frac{I}{N} + \beta_2 V\frac{I}{N} - (\alpha + \mu)I \\ \dot{R} &=& \alpha I - (\varphi + \mu)R \\ \dot{V} &=& \delta S - \beta_2 V\frac{I}{N} - \mu V \end{cases}$$

(a) Draw the compartmental model associated with this system of equations.

(b) Interpret the meaning of μ, β_1, β_2, α, φ, and δ.

(c) Find the basic reproduction number R_0.

(d) Interpret your answer from part (c) biologically.

(e) Explain any limitations of this model (i.e., how are vaccinated individuals treated and what other considerations could be made)?

(f) Using the following parameter values, find the critical vaccination level for the model if the vaccine efficacy is 68%: $\beta_1 = 0.4$, $\beta_2 = 0.3$, $\alpha = 0.15$, $\varphi = 0.00007$, $\delta = 0.00005$ and $\mu = 0.00003$.

4. Let S, E, I, and P be the number of susceptible, exposed, infectious, and partially immune individuals, respectively, from a certain disease. Note that partially immune individuals may still be infected with the disease. Assume $N = S+E+I+P$ is the total population size. Consider the compartmental diagram:

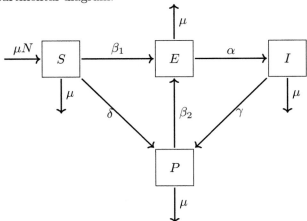

   (a) Write the model equations associated with the compartmental diagram.

   (b) Explain how an individual can become partially immune in this model.

   (c) Interpret the meaning of μ, β_1, β_2, α, γ, and δ.

   (d) Find the basic reproduction number R_0.

   (e) Interpret your answer from part (d) biologically.

5. In this problem, you will find the threshold efficacy for a vaccine. In this scenario, the disease has R_0 between 2 and 4. If the vaccine is developed and you can only ensure that 75% of the population will receive the vaccine, what efficacy must the vaccine have in order to ensure herd immunity is reached with only 75% vaccination? Calculate the threshold efficacy for $R_0 = 2, 2.5, 3, 3.5$, and 4. Based on your answers, what conclusions can you draw about how much a change in R_0 effects the required efficacy for a fixed vaccination level.

6. In this problem, you will find the threshold vaccination rate for a specific vaccine. In this scenario, the disease has R_0 between 2 and 5. If the vaccine has an efficacy of 85%, how much of the population must be vaccinated in order to reach herd immunity? Calculate the critical vaccination level for $R_0 = 2, 2.5, 3, 3.5, 4, 4.5$, and 5. Based on your answers, what conclusions can you draw about how much a change in R_0 effects the critical vaccination level for a fixed vaccine efficacy.

6.4   Sensitivity Analysis

A **sensitivity analysis** is an analysis that shows how sensitive an outcome is to changes in parameter values (because we might not be 100% sure that our parameter values are correct!). For example, we may want to know how a change in the recovery rate for a disease will affect the R_0. A sensitivity analysis can also shed light on the importance of data collection and estimation for certain parameters, whereas other parameters may have very little influence on R_0. There are several ways to perform a sensitivity analysis, but probably the most basic method is outlined below.

A basic method for conducting a sensitivity analysis for R_0 computes the relative change in R_0 divided by the relative change in each parameter λ.

$$\text{Sensitivity Index} = \text{SI}_\lambda = \frac{\frac{\Delta R_0}{R_0}}{\frac{\Delta \lambda}{\lambda}} \approx \frac{\partial R_0}{\partial \lambda} \cdot \frac{\lambda}{R_0}$$

How do we interpret these results?

- Positive SI means a positive relationship (i.e., when the parameter increases, R_0 will increase as well).

- Negative SI means a negative relationship (i.e., when the parameter increases, R_0 will decrease).

- The higher the absolute value of the SI, the stronger the relationship and R_0 is more sensitive to changes in the parameter value.

  - All values will be between -1 and 1, so values closest to -1 and 1 indicate more sensitive parameters, while SI near 0 indicate very little sensitivity.

  - In general, if our SI in absolute value is bigger than 0.3, we will say that our outcome is sensitive to changes in that parameter. Some researchers may have different opinions as to what this threshold should be.

For example, let's consider the basic SIR model that we discussed in the very beginning of this chapter:

$$
\begin{cases}
\dot{S} &= \mu N - \beta S \dfrac{I}{N} - \mu S \\[2mm]
\dot{I} &= \beta S \dfrac{I}{N} - \alpha I - \mu I \\[2mm]
\dot{R} &= \alpha I - \mu R
\end{cases}
\;\cdot
$$

For this model, we previously found $R_0 = \beta \cdot \dfrac{1}{\alpha + \mu}$. Let's say we believe that $\beta = 0.0901$, $\alpha = 0.045$, and $\mu = 0.00005$. Then:

$$
R_0 = 0.0901 \cdot \frac{1}{0.045 + 0.00005} = 2 > 1,
$$

so the disease will spread because on average, one infected individual will infect 2 others. We may want to conduct a sensitivity analysis to determine which parameter values have the strongest effect on the basic reproduction number because this will tell us parameters that we can target when implementing control measures, but also parameters that we should focus our attention on while gathering data and estimating the values used in our model.

Let's start by computing the sensitivity index for R_0 with respect to the transmission rate β:

$$
\begin{aligned}
\mathrm{SI}_\beta &= \frac{\partial R_0}{\partial \beta} \cdot \frac{\beta}{R_0} \\[2mm]
&= \frac{\partial}{\partial \beta}\left(\beta \cdot \frac{1}{\alpha + \mu}\right) \cdot \frac{\beta}{R_0} \\[2mm]
&= \frac{1}{\alpha + \mu} \cdot \frac{\beta}{R_0} \\[2mm]
&= \frac{1}{0.045 + 0.00005} \cdot \frac{0.0901}{2} \\[2mm]
&= 1
\end{aligned}
$$

Because $\mathrm{SI}_\beta = 1 > 0$, we know that an increase in β will increase R_0. This makes sense because β is the transmission rate. A higher transmission rate means that it is more likely for a healthy person to get sick via contact with an infected person, and hence the disease should spread faster, which is in line with an increase in R_0. Furthermore, the sensitivity index with respect to β is 1 which is the highest value that we can obtain. This shows that R_0 is extremely sensitive to any changes in β.

We can similarly find the sensitivity indices for the remaining parameter values:

$$
\begin{aligned}
\text{SI}_\alpha &= \frac{\partial R_0}{\partial \alpha} \cdot \frac{\alpha}{R_0} \\[2mm]
&= \frac{\partial}{\partial \alpha} \left(\beta \cdot \frac{1}{\alpha + \mu} \right) \cdot \frac{\alpha}{R_0} \\[2mm]
&= \beta \cdot \frac{-1}{(\alpha + \mu)^2} \cdot \frac{\alpha}{R_0} \\[2mm]
&= 0.0901 \cdot \frac{-1}{(0.045 + 0.00005)^2} \cdot \frac{0.045}{2} \\[2mm]
&\approx -0.999
\end{aligned}
$$

and

$$
\begin{aligned}
\text{SI}_\mu &= \frac{\partial R_0}{\partial \mu} \cdot \frac{\mu}{R_0} \\[2mm]
&= \frac{\partial}{\partial \mu} \left(\beta \cdot \frac{1}{\alpha + \mu} \right) \cdot \frac{\mu}{R_0} \\[2mm]
&= \beta \cdot \frac{-1}{(\alpha + \mu)^2} \cdot \frac{\mu}{R_0} \\[2mm]
&= 0.0901 \cdot \frac{-1}{(0.045 + 0.00005)^2} \cdot \frac{0.00005}{2} \\[2mm]
&\approx -0.0011
\end{aligned}
$$

For α, our sensitivity index was approximately -0.999. Because this index is, in absolute value, close to 1, we know that R_0 is very sensitive to changes in α. However, unlike β, the sensitivity index with respect to α is negative, so an increase in α will cause R_0 to decrease. If done right, this should make sense biologically: α is the recovery rate, so an increase in α means a faster recovery and thus a shorter time spent infectious. When individuals are infectious for less time, we would expect them to get fewer other people sick, and hence the basic reproduction number R_0 should decrease.

We can similarly interpret the sensitivity index with respect to μ. This index is close to 0, so we would say that R_0 is not sensitive to changes in μ. A small change in the parameter value used for μ will not change the model results drastically, unlike a change in β or α. Because $\text{SI}_\mu < 0$, we also know that an increase in μ will result in a decrease in R_0. This makes sense because μ is the death rate, and an increase in the death rate means that individuals are living for a shorter amount of time. They then have less time to get sick and/or infect others because they have a shorter lifespan, and thus we expect each person to infect fewer others, resulting in a lower R_0.

We can also illustrate these sensitivity analysis results using a bar graph to make it easier to find the parameters with the most influence on our variable, in this case, R_0.

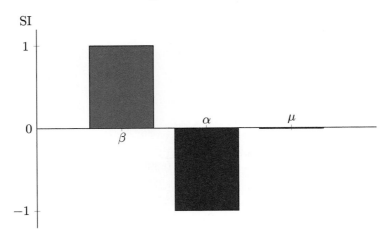

We clearly see in the graph that β and α have a strong impact on the R_0 value. Even a small change in these parameters can result in a large change in the R_0. This tells us that we need to make sure we are estimating the values for these parameters as accurately as possible. Sometimes this may mean that our focus should be in the medical studies related to finding these values. If we are certain that our parameter values are correct, then this high sensitivity tells us that these are parameters we should focus on targeting with any control measures for the disease. For example, in order to use β to lower R_0, we could focus on preventative control measures such as hand washing, quarantining, or even providing vaccines that would make it less likely for an infectious person to infect others. Or, we could focus our attention on increasing α as a way to reduce the R_0 value. This could be in the form of new treatment protocols that help an individual to recover faster from infection and reduce recovery time. We also notice from the graph that a change in the μ value has almost no impact on the change in R_0.

6.4 Exercises

1. For the following compartmental model, conduct a sensitivity analysis for R_0 with respect to each of the parameters (μ, β, φ, α). Use the following values in your calculations and interpret your results (i.e., is R_0 sensitive to the parameters? Do the parameters have a positive/negative effect on R_0? State these results for each parameter.) $\mu = 0.017$, $\beta = 0.3$, $\varphi = 0.25$, $\alpha = 0.08$

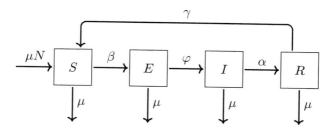

Hint: You may have found the R_0 value if you completed Exercise 2 from Section 6.2.

2. In Exercise 4 of Section 6.2, we interpreted the R_0 expression for the following compartmental model:

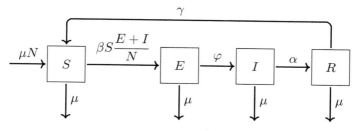

We had previously stated that $R_0 = \dfrac{\beta}{\varphi + \mu} + \dfrac{\beta\varphi}{(\varphi + \mu)(\alpha + \mu)}$, but that it could be separated into easily understood components. Now, conduct a sensitivity analysis for R_0 with respect to each of the parameters (β, φ, μ, α). Use the following values in your calculations and interpret your results (i.e., Is R_0 sensitive to the parameters? Do the parameters have a positive/negative effect on R_0? State these results for each parameter.) $\mu = 0.0014$, $\beta = 0.24$, $\phi = 0.1$, $\alpha = 0.05$.

6.4 Project: Shadowpox

In this project, we will model the spread of a fictitious disease, shadowpox, through a city. When an individual is infected with shadowpox, there are no signs of the disease for about half a year (so symptoms generally appear at the rate **1/180** per day) after which large, infectious, grey pox appear across the body and eyes become sullen. There are doctors all across the world with secret knowledge of how to treat shadowpox. There is no cure, but if treated continually by a doctor, the disease can be halted so that the infected person will no longer spread the disease. Assume the treatment rate is **0.0005** per day. Those who are not treated by a doctor are sometimes exiled from their homes (at a rate of **0.008** per day), but otherwise live a normal life.

At the beginning of the year, there are no infected people in the city, which has a population of **20,000**. On the first day of the year, a ship arrives with **100** passengers who are afflicted with an advanced form of shadowpox. They will spread the disease at a rate of **0.012** per day. You may assume that the lifespan of an average individual in this city is **50** years.

1. Draw a compartmental model for this disease. Use appropriate Greek letters to indicate your parameters: (Λ = birth rate, μ = death rate, β = infection rate, α = progression rate to infectiousness, γ = treatment rate, δ = quarantine rate (rate of exile)).

2. Write equations for your disease that follow from your compartmental model. Confirm that your population size remains constant (or figure out what must be done to make it so).

3. Find the disease-free equilibrium for your model.

4. What is the basic reproduction number? Interpret it biologically. (Hint: your F and V matrices should be 4×4. Use computer software to help compute R_0 if needed.)

5. Using the numbers (in the statement of this project, in bold) determine values for your parameters. Use computer software to graph the solutions of your model and interpret the behavior of the solutions. (Hint: μ is 1/lifespan, but remember to convert to days!)

6. Conduct a sensitivity analysis for R_0 with respect to each parameter and create a bar graph to illustrate results. Which parameters have a positive/negative effect on R_0? To which parameters is R_0 the most sensitive?

Index